Praise for this book

"Professor Ekins gives us the information that we need and does so in a clear, detailed, and authoritative way. It is the handbook that even the most informed will find invaluable but which, at the same time, is wholly accessible to the widest of audiences. Nobody, whether they are already seriously involved or just want to understand the issues for the first time, should be without *Stopping Climate Change*. I will certainly have it close at hand as the battle to save the planet continues with increasing urgency."

From the Foreword by the Rt. Hon. The Lord Deben, *Chairman of the UK Climate Change Committee, 2012–2023*

"With academic rigour and razor-sharp clarity, this heroic effort of covering the entire climate change challenge, from risks, impacts to policy and pathways to a safe landing for humanity, is not only commendable. It is a must read for all in search of realistic light in the rapidly darkening climate tunnel."

Johan Rockström, *Professor of Earth System Science and Director of the Potsdam Institute for Climate Impact Research*

"Paul Ekins has been an astute, incisive and outspoken commentator of climate change policy for many years. In this book, he puts forward a comprehensive and convincing blueprint for tackling the climate crisis. An essential read for anybody interested in climate change policy."

Sam Fankhauser, *Professor of Climate Change Economics and Policy, University of Oxford*

"In battling climate change, it is crucial to understand the economic implications of different policies and courses of action. Paul Ekins provides a key intervention for those engaged in these issues."

Karen Turner, *Professor and Director of the Centre for Energy Policy, University of Strathclyde*

"A comprehensive survey of the state of the climate debate is very welcome and much needed. Professor Ekins spells out what is happening and what is not, and explains what could and should be done – to deliver real net zero and genuine climate security. This book should be essential reading for everyone concerned to find a practical solution to the climate challenge."

Nick Butler, *Visiting Professor and founder of the Policy Institute, King's College London*

"This book is comprehensible to non-experts and has a strong focus on the key issue – what needs to be done to stop climate change. Paul Ekins has used his broad knowledge to set out a clear route map, whilst remaining realistic about the likely rate of political action."

Nick Eyre, *Professor of Energy and Climate Policy, University of Oxford*

"Building on decades of experience, Paul Ekins sets out concisely in this book the science, the politics and the possibilities for zero greenhouse gas emissions. For anyone wanting to get to grips with the climate action challenge in a single read, this book meets your needs in a characteristically clear fashion, spiced with the author's opinions and insights."

Jim Skea, *Professor of Sustainable Energy, Imperial College London*

"If you only read one book about the climate crisis read this one. Informed by many years of direct involvement with climate policy, the book covers all aspects of the technology, policy and market developments needed to solve the climate problem. It deals in detail with the economic and societal requirements for change. The overriding messages are that we have the tools available, that change is both possible and can bring many benefits. I sincerely hope that decision makers around the world take notice."

Rob Gross, *Professor and Director of the UK Energy Research Centre*

Stopping Climate Change

Written by one of the leading experts in the field, Paul Ekins, *Stopping Climate Change* provides a comprehensive overview of what is required to achieve 'real zero' carbon dioxide emissions by 2050, and negative emissions thereafter, which is the only way to stop human-induced climate change.

This will require innovation in socio-technical systems, and in human behaviour, on an unprecedented scale. *Stopping Climate Change* describes the changes required to meet this goal: in technologies, social institutions and individual activities. Paul Ekins examines in detail issues around the supply and demand of energy and materials, and the efficiency of their use. It also analyses greenhouse gas removal technologies, offsetting and geoengineering, and plots the reduction of the non-CO_2 greenhouse gas-emitting activities. Having set out the changes required, Ekins considers the economic implications, in terms of both the innovation and investments that are necessary to bring them about, and the effects that these are likely to have on national economies. The evidence presented points clearly to the economic impacts of decarbonisation being positive for the majority of countries, and for the world as a whole, even before considering the benefits of avoided climate change. When the health benefits of stopping the burning of fossil fuels are factored in, the global net benefits of decarbonisation are unequivocal.

Drawing on examples from the UK and Europe, but with wider relevance at a global scale, *Stopping Climate Change* clearly shows how determined policy action at different levels could stop climate change. It will be of great interest to students, scholars and policymakers researching and working in the field of climate change and energy policy.

Paul Ekins is Professor of Resources and Environmental Policy at the Institute for Sustainable Resources, University College London (UCL). For 14 years he was a Co-Director of the UK Energy Research Centre. He was also the special adviser to the joint Houses of Parliament on the UK Climate Change Bill. At the EU level he was a member of the High-Level Panel of the European Commission's European Decarbonisation Pathways Initiative and was Vice Chair in 2012–13 of the then Environment Commissioner's Expert Economists' Group on Resource Efficiency. He is a member of the International Resource Panel of the United Nations Environment Programme (UNEP), and was a Co-Chair of UNEP's flagship publication, the sixth *Global Environment Outlook*, published in 2019. In 1994 he received UNEP's Global 500 Award 'for outstanding environmental achievement'. In 2015 he received an OBE from the UK Government 'for services to environmental policy'.

Routledge Studies in Environmental Policy Series

The European Union and Global Environmental Protection
Transforming Influence into Action
Edited by Mar Campins Eritja

Environmental Policy and Air Pollution in China
Governance and Strategy
Yuan Xu

Climate Change Law and Policy in the Middle East and North Africa Region
Edited by Damilola S. Olawuyi

European Foreign Policy in a Decarbonising world
Challenges and Opportunities
Sebastian Oberthür, Dennis Tänzler, and Emily Wright, with Gauri Khandekar

How to Successfully Encourage Sustainable Development Policy
Lessons from Germany
Günther Bachmann

Deliberative Governance for Sustainable Development
An Innovative Solution for Environment, Economy and Society
Franz Lehner

Stopping Climate Change
Policies for Real Zero
Paul Ekins

Deploying the European Green Deal
Protecting the Environment Beyond the EU Borders
Edited by Mar Campins Eritja and Xavier Fernández-Pons

For more information about this series, please visit: www.routledge.com/Routledge-Studies-in-Environmental-Policy/book-series/RSEP

Stopping Climate Change
Policies for Real Zero

Paul Ekins

 Routledge
Taylor & Francis Group

LONDON AND NEW YORK

 earthscan
from Routledge

Designed cover image: © Getty images

First published 2024
by Routledge
4 Park Square, Milton Park, Abingdon, Oxon OX14 4RN

and by Routledge
605 Third Avenue, New York, NY 10158

Routledge is an imprint of the Taylor & Francis Group, an informa business

British Library Cataloguing-in-Publication Data
A catalogue record for this book is available from the British Library

ISBN: 978-1-032-57140-9 (hbk)
ISBN: 978-1-032-57141-6 (pbk)
ISBN: 978-1-003-43800-7 (ebk)

DOI: 10.4324/9781003438007

Typeset in Times New Roman
by Newgen Publishing UK

Contents

Figures

Tables

Foreword

The Rt Hon. the Lord Deben

As even the most purblind recognise the reality of climate change, it is becoming a factor in every decision we make. No other concern so engages politicians, campaigners, and polemicists – not to mention the Pope and the King as well as contrarians and conspiracy theorists galore. From being a scientific theory concerning only a few of us in the 1980s it is now a reality that cannot be ignored. The happenstance that the UK elected the first scientist in its history to be Prime Minister has meant that, since Margaret Thatcher gave international political recognition to the Rio Summit and persuaded George Bush to attend, the UK has been at the heart of the change.

Now, when withering temperatures, drought and wildfires, floods and torrential rain engage the attention of so much more of the world, and almost every country is at least signed up to action, the significance of our Climate Change Act continues to resonate. The key has been all-party support for independent advice translated into statute to which Government can be held accountable by the public and the courts. After 14 years, the Climate Change Committee continues to set the carbon budgets, inform Parliament, and keep the Government's feet to the fire that threatens the planet.

So much has changed since Rio not only now with Biden's Inflation Reduction Act, the European Union's Green Deal, and the substantive action in China, but also the achievements of the Kyoto, Paris, and Glasgow COPs; the continuing insights of the IPCC; and significant actions by industry and the financial world. There is so much more to be done if the battle is to be won but recognising those achievements is essential if we are to have the courage to press on and the world is to accept that it can indeed save itself.

The problem for all of us who are in the thick of it – a problem we share with the newest convert – is that so much has happened, so much is involved, and there is so much to digest. We can now turn at once to this book in which Professor Ekins gives us the information that we need and does so in a clear, detailed, and authoritative way. It is the handbook that even the most informed will find invaluable but which, at the same time, is wholly accessible to the widest of audiences. Nobody, whether they are already seriously involved or just want to understand the issues for the first time, should be without *Stopping*

Climate Change. I will certainly have it close at hand as the battle to save the planet continues with increasing urgency.

The Rt Hon. the Lord Deben
Chairman of the Climate Change Committee, 2012–2023

Preface

Human-induced climate change is now under way. It has its own dynamic and will have impacts for millennia.

This book seeks to present in a scientifically robust way all the key issues and arguments around how to stop human activities contributing to, and increasing the rate of, climate change. 'Stopping human activities contributing to and exacerbating climate change' is how the title of this book, and any references in the book to 'stopping climate change' should be interpreted. As this book will make clear, all humans can do now to ameliorate the situation is to stop contributing to climate change and increasing its rate of change. They can also reduce the elevated concentrations in the atmosphere of the greenhouse gases that are driving it, in the hope that human societies will be able to adapt to the climate instabilities that they have unleashed.

Each chapter draws on some of the key scientific literature that informs the arguments that it presents. This literature is vast, as shown in the thousands of pages and references in the reports of the Intergovernmental Panel on Climate Change (IPCC), which has the task of assessing it in detail. Therefore this book's selection from the literature is very limited.

The chosen references are often to papers that have themselves sought to synthesise a body of other papers, so that the chapters are to some extent syntheses of syntheses, from which I hope that a coherent and digestible narrative emerges of the technological and policy options available to address climate change, and the major advantages and disadvantages, potentialities and limitations, of each. The scientific references in the text and the lists of references at the end of each chapter are supplemented by footnotes which give links to websites outside the scientific literature that refer to current issues and developments, or provide sources for some of the numbers I have quoted. Many of these footnotes refer to fast changing issues. With a few exceptions, the data cut-off date for these references was the end of April 2023. This means that, for example, the book dos not take comprehensive account of the highly regrettable policy changes, which weakened UK climate policy, made by UK Prime Minister Rishi Sunak in September 2023.

On the principle that very often an image can tell the story of a thousand words, I have liberally used or recreated (with permission) the many excellent graphics that other scientists have produced to aid comprehension and appreciation of their arguments and results. I am very grateful and much beholden to them for their visualisations.

Each chapter starts with a Summary of its content, and, where I thought necessary, an Introduction to the topics covered in the chapter. It ends with a Conclusion that seeks to draw together the points that it has made and contains my interpretation of the lessons that can be drawn from them in respect of reducing the emissions that contribute to climate change. A final chapter draws together all the conclusions from the rest of the report in what I hope are comprehensible summaries of each key issue discussed in the book.

Acknowledgements

My first acknowledgements are to Ida Ahmad and Zaichun Zhang, who recreated many of the figures in this book. Their visualisation skills are infinitely greater than mine, and I could not have created this book without them. Secondly, I am most grateful to Keith Bell and Callum MacIver, at Strathclyde University, Grant Wilson, at the University of Birmingham, and Michael Grubb, Serguey Maximov Gajardo and Omran Al-Kuwari at UCL, for sharing data with me that allowed us to make figures to illustrate important points in Chapter 5.

Thirdly, I would like to thank my colleagues at the Bartlett School of Environment, Energy and Resources at UCL for the many conversations and exchanges on these issues. In particular, I would like to note that part of Chapter 3 on energy efficiency was published online in 2021 as an Explainer as part of UCL's Bartlett Faculty's 'Together for Climate Action' campaign. I acknowledge the contribution to the Explainer of my Bartlett colleagues Jing Meng, Scott Orr and Rokia Raslan. Colleagues at the UK Energy Research Centre (UKERC) from universities all over the country have been just as important in sharing their insights and teaching me about these issues over the past ten or more years. All these discussions, and their academic papers, have enabled me to crystallise in my mind the thoughts, ideas and arguments that I share in the book.

Fourthly I would like to acknowledge the publisher's expert editorial team — Annabelle Harris, Jyotsna Gurung and Gillian Bourn – who put the book on a fast track, so that it could appear before COP28.

Finally, there are the many publishers who have given me permission to use material from their publications. These include the world's two premier scientific organisations in the fields of climate change and energy: the Intergovernmental Panel on Climate Change (IPCC) and the International Energy Agency (IEA). Other organisations which have kindly given me permission to reproduce or recreate their work are the UK's Climate Change Committee, the World Bank, the Food and Agriculture Organization of the United Nations, the United Nations Environment Programme (UNEP), the Organisation for Economic Co-operation and Development (OECD), the World Energy Council, Finland's innovation agency, Sitra, the Royal Institute of British Architects (RIBA), Carbon Brief, Climate Action Tracker, International Synergies, and the Energy Research Partnership. I have borrowed illustrations or figures from many Open Access sources, including the World Health Organization, the International Renewable Energy Agency (IRENA), the UN Economic Commission for Europe, the Energy Transitions Commission, Imperial College London's Energy Futures Lab, LSE's Grantham Institute for Climate Research and Harvard University's Science in the News blog series. Commercial publishers which have allowed me to recreate or tabulate work that they have published are Springer Nature, Taylor & Francis, Wiley, and Elsevier. I am enormously grateful to all of them. Full citations of and references to the works concerned are of course given in the relevant places.

A note on terminology and units

Terminology

Climate change and energy are technical subjects, and it is not possible to discuss them coherently without the use of some specialist terminology, some of which is briefly explained here, so as not to impede the flow of the book. In addition, such discussion inevitably involves numbers, which are expressed in a wide range of units. The main units used in the book are also explained here.

The human contribution to climate change, which is now the dominant cause of the phenomenon, is due to the emission to the atmosphere of so-called greenhouse gases (GHGs), which impede the incoming solar radiation being re-radiated out to space. This warms the atmosphere ('global warming') and changes the climate.

The main GHG is carbon dioxide (CO_2), but there are five others that are included in countries' GHG inventories: methane (CH_4), nitrous oxide (N_2O), the so-called F-gases (hydrofluorocarbons and perfluorocarbons) and sulphur hexafluoride (SF_6). The quantities and uses of these are described in the book.

These gases contribute differently to climate change and stay in the atmosphere for different periods of time. Their effect on the climate therefore changes over time. To compare their different effects with that of CO_2, a global warming potential (GWP) is calculated for each GHG, for a particular time after its emission, normally 100 years. The GWP of CO_2, which is a relatively long-lived gas (i.e. most of it remains in the atmosphere for over 100 years) is taken as 1. Methane, in contrast, is a much more potent GHG when it is emitted, but is removed by chemical reaction from the atmosphere on average after 12 years. However, even after 100 years, methane has a GWP of 34 (i.e. contributes 34 times as much to global warming per tonne as CO_2). Chlorofluorocarbons (CFCs) and hydrochlorofluorocarbons (HCFCs) are also powerful GHGs, but they are controlled under the Montreal Protocol and are being phased out, and are not discussed in this book.

Methane is one of three main fossil fuels (the other two being oil and coal) and is often called natural gas. However, methane is also produced by the anaerobic decomposition of organic matter, when it is called biogas, and this is starting to be injected into the gas network. For this reason, the fossil fuel in this book is called fossil methane (rather than natural) gas.

There is much uncertainty as to what GHG emissions will be in the future. Scientists tend to explore such uncertainties through constructing scenarios. Five scenarios for possible emission levels have been quite influential in thinking about this. They are called the Representative Concentration Pathways (RCPs), and will be encountered at various points in the book. The numbers associated with the RCPs (e.g. RCP1.9, RCP2.6, RCP4.5, RCP6, RCP8.5) are the radiative forcing that can be expected from certain atmospheric concentrations of GHGs. All that needs to be remembered for this book is that low RCPs derive from low emissions, and high RCPs from high emissions. In the book it is seen how these RCPs relate to both emission levels and atmospheric temperature increases.

Many factors affect emission levels, the most obvious of which are the level of the human population, economic activity, fossil fuel use, and government energy and climate policies. To construct emission scenarios, assumptions have to be made about these variables. Five standard sets of assumptions have been created, called the Shared Socioeconomic Pathways (SSPs). These are described in more detail in Chapter 2.

The main source of CO_2 from human activities is from the use of fossil fuels as energy. In their extracted form, before processing, such fuels, along with nuclear fuel and bioenergy, are called *primary* energy. After conversion into electricity, or the fuels that are actually used (such as petrol or diesel from oil), this energy is called *final* energy. Because some energy is always lost in any energy conversion, final energy is always less than the primary energy from which it is made. A further distinction is made between final energy and *useful* energy, which is that portion of final energy that actually provides the desired service. For example, much of the final energy in petrol and diesel is lost as heat in an internal combustion engine – only the energy that actually propels the car is useful energy in this case.

Different forms of energy contain different amounts of carbon per unit of energy. This is called their carbon intensity. Fossil fuels all emit carbon dioxide when they are burned, but different amounts per unit of energy produced. Wind, solar and nuclear energy all emit zero carbon when they generate electricity, as does electricity when it is used. The carbon intensity of wind, solar and nuclear is therefore zero. However, it is only zero across the full life cycle of these technologies if zero-carbon energy was also used to manufacture the equipment (e.g. wind turbines, photovoltaic panels, nuclear power stations). This is often not the case but the life cycle emissions of these zero-carbon energy sources are still much lower than those from fossil fuels.

In order to stop the human contribution to climate change, the net emissions of GHGs to the atmosphere from human activities will need to fall to zero, meaning that any continuing emissions from human activity to the atmosphere will need to be offset by removals of emissions from the atmosphere. This is what 'net-zero emissions', or just '*net zero*' means. This is a term which will often be encountered in the book. However, this book calls for '*real zero*', for reasons which will become apparent. This means ceasing completely the use of fossil fuels (and therefore CO_2 emissions from such use), while reducing other GHG emissions (e.g. from agriculture and deforestation) sufficiently so that the land and forests can remove more CO_2e (see below for what CO_2e means) from the atmosphere than is emitted.

Because energy use and carbon emissions are closely associated with economic activities, it is often useful to describe how much energy is used, or carbon is emitted, per unit of economic output. These are called the energy intensity and carbon intensity of the economy respectively. A related subject is energy efficiency (the subject of Chapter 3), which describes how much useful service is delivered by a unit of energy. Greater energy efficiency will therefore in itself lead to lower energy intensity. A greater share of zero-carbon energy (e.g. from renewables) in energy use will of itself lead to lower carbon intensity.

People, businesses and countries emit GHGs directly when they use fossil fuels, but emissions also arise from making the products, which may have been made in another country and imported, that they consume. The total emissions arising from a person's, business's or country's consumption is called their 'carbon footprint'. For a country this amounts to the emissions from their territory, plus the emissions associated with the manufacture of their imports, minus the emissions associated with the manufacture of their exports.

Finally, on terminology, there are three sorts of responses humans can make to climate change. They can seek to *reduce* the emissions which contribute to it, which is called *carbon abatement*, or *decarbonisation*. Or they can try to *remove* emissions from the atmosphere, and so have less climate change in the future. Both emission reduction and emission removal are called *climate mitigation*. Or humans can adapt to climate change to the best of their ability (e.g. install air-conditioning, grow crops that do better in a warmer climate), to reduce its negative impacts. This is called *climate adaptation*.

Whatever the mitigation or adaptation, climate change is already causing damage to human health, human lives and livelihoods, and human economies and societies, and will do more in the future. There are increasing calls that those who have contributed most to climate change through their cumulative emissions should compensate those who have contributed least to it but are suffering from it. This is called the *loss and damage* issue.

Climate mitigation, adaptation and damage all involve *costs*. It is intuitively obvious that public policy should now seek to stimulate action such that the sum of mitigation, adaptation and damage costs be kept to a minimum. Because of the uncertainties in all these costs, the practical approach that has been adopted globally is to try to keep global warming to 1.5 to 2°C above pre-industrial levels. These are targets which will be frequently encountered in this book.

Many other terms will be encountered in the book that may not be familiar. They will be explained as the book unfolds.

Units

This book adopts metric units throughout. Length is measured in metres (m) and mass in grams (g).

Factors of 10 of such units are expressed as follows.

- 1 thousand = kilo (k) = 1,000 = 10^3;
- 1 million = mega (M) = 1,000,000 = 10^6;
- 1 billion = giga (G) = 1,000,000,000 = 10^9;
- 1 trillion = tera (T) = 10^{12};
- other terms that will be encountered: peta (P) = 10^{15}; exa (E) = 10^{18}.

Using mass as an example, as this will often be encountered in the book, this translates into:

- 1,000 g = 1 kg = 10^3 g;
- 1,000 kg = 1 million (M) g = 10^6 g = 1 tonne (t);
- 1,000 t = 1 kt = 10^9 g;
- 1 million tonnes = 1 Mt = 10^{12} g;
- 1 billion t = 1 Gt = 10^{15} g.

CO_2 emissions are often measured in Gt. Other GHGs can be converted to 'CO_2-equivalents' (CO_2e) using the GWP concept described above. So when CO_2 is mentioned, it refers to just CO_2 emissions. When CO_2e is mentioned, it includes one or more of the other five GHGs (with or without CO_2) described above.

Electric power is measured in watts. An electric kettle is around 3 kW. An electric bar fire might be 1 kW. Large wind turbines are 3 MW. Fossil fuel power stations are hundreds of MW. Nuclear power reactors are 1–2 GW.

The unit of energy derived from watts of power is the kWh, which is the energy generated by a 1 kW fire used for 1 hour. It is common in this book to see energy measured in GWh or TWh.

However, there are many other units of energy. The main other one used in this book is joules (J). 1 MJ = 0.28 kWh. Many of the graphs in this book show EJ (10^{18} J). 1 EJ = 278 GWh.

Another unit of energy applied just to fossil fuels is barrels of oil equivalent (boe), which, as the name suggests, converts an amount of energy to the equivalent number of barrels of oil. This is not often used in this book, but, for the record, 1 boe = 1700 kWh.

Finally, concentrations of CO_2 or CO_2e in the atmosphere are measured as parts per million by volume (ppmv), or commonly shortened to just parts per million (ppm).

Graphs

There are many graphs in this book. Often the variables they measure have considerable uncertainty around their values, so they are measured as a range of possible values. This may be indicated by a shaded area on the graph around the central line.

For bar charts the range may be shown, with a mark towards the middle of the range, showing the mean (average) or median (the value at the middle of the frequency distribution), a *box* showing the 75th and 25th percentiles (with 25% of the values both below and above the box) and lines called 'whiskers' which span the whole distribution of the range. The text around a graph or a table always explains its key messages. But often conveying all the information in words would be tedious.

Acronyms and abbreviations

AFOLU	agriculture, forestry and other land use
BECCS	bioenergy with carbon capture and storage
BEV	battery electric vehicle
BF	blast furnace
BIM	Building Information Management
BNZP	Balanced Net Zero Pathway (of the CCC)
BOF	basic oxygen furnace
CAP	Common Agricultural Policy (of the European Union)
CCC	Climate Change Committee (UK)
CCP	Core Carbon Principles
CCS	carbon capture and storage
CCUS	carbon capture, use and storage
CDR	carbon dioxide removal
CE	circular economy
CFC	chlorofluorocarbon
CfD	contract for difference
CH_4	methane
CHP	combined heat and power
CO_2	carbon dioxide
COP	Conference of the Parties (of the UNFCCC)
CSP	concentrated solar power
DAC	direct air capture
DACCS	direct air carbon capture and storage
DEC	Display Energy Certificate
dLUC	direct land use change
DRI	direct reduced iron
EAF	electric arc furnace
ECO	Energy Company Obligation
EEO	Energy Efficiency Obligation
EPC	Energy Performance Certificate
ECT	Energy Charter Treaty
ETS	emissions trading system
EU	European Union

EV	electric vehicle
FAO	Food and Agriculture Organization of the United Nations
FCEV	fuel cell electric vehicle
FIP	feed-in premium
FIT	feed-in tariff
GDP	gross domestic product
GFANZ	Global Financial Alliance for Net Zero
GHG	greenhouse gas
GMSL	global mean sea level
GWP	global warming potential
GWP	gross world product
H-DRI	hydrogen direct reduced iron
H_2	hydrogen
H2FC	hydrogen fuel cell
HCE	hydrogen combustion engine
HCFC	hydrochlorofluorocarbon
HFC	hydrofluorocarbon
HFC	hydrogen fuel cell
ICAO	International Civil Aviation Organization
ICE	internal combustion engine
ICT	information and communication technologies
IEA	International Energy Agency
ILO	International Labour Organization
iLUC	indirect land use change
IoT	Internet of Things
IPCC	Intergovernmental Panel on Climate Change
IRA	Inflation Reduction Act (in the US)
IRENA	International Renewable Energy Agency
LA	local authority (UK)
LCA	life cycle analysis
LCOE	levelised cost of energy/electricity
LDV	light-duty vehicles (mainly cars and vans)
LED	Low Energy Demand (scenario)
LED	light-emitting diode
LNG	liquid natural (fossil methane) gas
LTS	long-term strategy
LULUCF	land use, land use change and forestry
MRV	monitoring, reporting and verification
MENA	Middle East and North Africa
NASA	National Aeronautics and Space Administration
NDC	Nationally Determined Contribution
NET	Negative Emissions Technology
NFGS	Network for Greening the Financial System
N_2O	nitrous oxide
NZE	Net Zero Emissions scenario (of the IEA)

OECD	Organisation for Economic Co-operation and Development
PCA	personal carbon allowance
PCC	per capita convergence
PFC	perfluorocarbon
PV	photovoltaics
R&D	research and development
RCP	Representative Concentration Pathway
R,D&D	research, development and demonstration
RPS	Renewable Portfolio Standard
RURR	remaining ultimately recoverable resources
SAI	stratospheric aerosol intervention
SBTi	Science-Based Targets Initiative
SCC	social cost of carbon
SDG	Sustainable Development Goal
SDS	Sustainable Development Scenario (of the IEA)
SFDR	Sustainable Finance Disclosure Regulation (of the European Union)
SF_6	sulphur hexafluoride
SMR	steam methane reforming
SRM	solar radiation management
SSA	sub-Saharan Africa
SSP	Shared Socioeconomic Pathway
STEPS	Stated Policies Scenario (of the IEA)
TCFD	Task Force on Climate-related Financial Disclosures
TRL	Technology Readiness Level
UK	United Kingdom
UNEP	United Nations Environment Programme
UNFCCC	United Nations Framework Convention on Climate Change
USA (or US)	United States of America
WAIS	West Antarctic ice sheet
WAM	West African Monsoon
WMO	World Meteorological Organization
ZEV	zero emission vehicle

Introduction

Summary

This chapter introduces the topic of climate change and the approach taken in the book to analyse it. To stop climate change, emissions of greenhouse gases (GHGs) will have to be reduced effectively to zero, and large quantities of GHGs will have to be removed from the atmosphere. How this can be done is then explored in subsequent chapters, the main topics of each of which are briefly introduced here.

This book aims to give a comprehensive overview, in a single coherent narrative, of how to stop climate change – or, at least, how to stop humans contributing further to it and worsening the situation, because the climate change that is already happening as a result of past emissions from human activity will work out in unpredictable ways for millennia to come.

Climate change is perhaps the defining issue of the current time, in that it is the major symptom of a period in human history when human numbers and activities are on such a scale as to disrupt global natural systems, such as the climate.

While the climate on Earth has always changed, the human addition to natural changes is now well understood in scientific terms, and its effects are already beginning to be felt around the world. The problem is easy to state. The solutions are both complex and more difficult to implement. They largely form the subject of this book.

The problem is that humans are changing the climate because of their large-scale emissions into the atmosphere of greenhouse gases (GHGs), most importantly carbon dioxide (CO_2) from the use of fossil fuels, but also other GHGs from deforestation, the conversion of land to agriculture, and the ways in which food is grown to feed large, and still growing, human populations.

In order to stop climate change, it is no longer sufficient for humans to stop emitting GHGs. The extra GHGs that humans have already emitted into the atmosphere will cause climate change for millennia to come, with very

DOI: 10.4324/9781003438007-1

uncertain effects on human societies in different parts of the world. Now, for humans to stop climate change, they not only have to stop emitting GHGs, they have also to remove very large quantities of past GHG emissions from the atmosphere.

This is a truly Herculean task. Stopping the emission of GHGs requires the transformation of two of the most fundamental human systems: the energy system and the food system. In respect of the former, the transformation involves the complete substitution of fossil fuels by non-carbon emitting energy sources, or the capture of remaining emissions from fossil fuels and their secure geological storage. In respect of the latter, the human use of land will need to grow sufficient nutritious food to feed at least the 9 billion people expected to be alive by the middle of this century, while at the same time allowing the land, through forests, other vegetation and new agricultural practices, to resume its historical role of acting as a net carbon sink, i.e. removing carbon from the atmosphere and storing it in trees and soils.

The following chapters describe how these changes can be brought about through technology and policy that fundamentally change how much and what kinds of energy and food are produced, and the incentives, regulations and other policies in the economy that could bring these changes about.

When faced with challenges of such enormity, it is tempting to opt for simplistic solutions and prescriptions. Why, it might be asked, can humans not simply adapt to using far fewer fossil fuels for non-essential purposes, for example those leisure activities related to driving and flying, and concentrate single-mindedly on decarbonising essential production and consumption? Doubtless decarbonisation will require some of such lifestyle adaptation. But there seems to be zero appetite among humans, even with the growing recognition of the potentially catastrophic implications of climate change, to revert on a large scale to simpler lifestyles, with lower consumption of material goods, far less use of energy, and a diet predominantly consisting of seasonal, home-grown, plant-based foods.

The assumption underpinning the decarbonisation strategies explored in this book is therefore that humans will continue to aspire to consumer lifestyles currently characteristic of more affluent industrial countries, and that the compromises that they are prepared to make to stop climate change are strictly limited. If technology, policy and economic incentives cannot stop climate change on these terms, then the evidence to date suggests that it will not be stopped.

The book starts with a brief synopsis of the climate science (Chapter 1) that puts in stark relief the imperative of decarbonisation. The next chapter (Chapter 2) then works through both the global context of the UN Framework Convention on Climate Change, with the size, nature and global distribution of current emissions, and three representative emission pathways (two global, one at the UK level) which reach net-zero emissions by 2050. The next eight chapters explore all the means by which these pathways may be brought about. These include greater energy efficiency (Chapter 3), the technologies of energy supply (Chapters 4, 5 and 6), carbon removal and storage (Chapter 7), the different sectors of energy demand (Chapters 8 and 9), and the transformation of agriculture and reduction of waste (Chapter 10). Chapter 11 is an in-depth exploration of the economics of decarbonisation, and Chapter 12 looks in detail at the policies that have been introduced, or will need to be introduced, to bring decarbonisation about at the necessary scale. Chapter 13 concludes with a realistic assessment of the prospects of stopping climate change before it becomes self-reinforcing and, therefore, unstoppable (the reasons this might happen are explored in Chapter 1).

The rest of this Introduction outlines the content of each chapter in more detail, to allow its underlying narrative to emerge more clearly as a whole, so that the entire journey on which the book seeks to take the reader can be appreciated before diving into the book itself.

Chapter 1: Why real zero?

This chapter gives a brief overview of the natural science of climate change, and what it is doing to Earth systems – the atmosphere, oceans, land (including land covered by ice) and ecosystems. With facts and figures drawn largely from the World Meteorological Organization's State of the World Climate, 2021 and 2022, it shows how the global average temperature has risen since the start of the industrial revolution, and how this is raising the level of the sea; making the world's oceans warmer and more acidic; melting glaciers, ice sheets and permafrost; putting ecosystems under stress; and making extreme weather events – including storms, floods, droughts and wildfires – more extreme and more frequent. The major impact of these events on human societies in the short term is likely to be in respect of disruptions to food supplies. This part of the chapter ends by outlining the dangers of climate change triggering 'tipping points', passing threshold conditions that cause various aspects of the Earth's natural systems – ocean currents, weather patterns, forests, permafrost – to change in fundamental ways, some of which will serve to intensify the global warming that caused the changes in the first place. Comparing the risks from climate change with those from aircraft accidents, it is shown that, even at current levels of global warming, humanity is taking risks with its future many times greater than it tolerates in other aspects of human life.

The second part of the chapter digs deeper into why climate change is happening, going back far before the industrial revolution to show how the atmospheric concentrations of heat-trapping GHGs, driven higher by human emissions of these GHGs, are at a level not seen for at least 3 million years, when sea levels were many metres higher than they are now. The GHG emissions mainly arise from burning fossil fuels, cutting down trees and growing food in ways that release GHGs from fertilisers, soils and livestock.

The chapter concludes with two blunt imperatives if human exacerbation of climate change is to be stopped. Humans must stop burning fossil fuels and transform the dominant methods of food production to reduce emissions to near-zero; and they must find ways to remove very large quantities of CO_2 from the atmosphere. The rest of the book describes how this can be done.

Chapter 2: The global context and pathways to net zero

Human societies recognised the need to reduce global GHG emissions in 1992, signing and later ratifying the UN Framework Convention on Climate Change (UNFCCC). This chapter briefly plots the evolution of negotiations at the annual Conference of the Parties (COP) meetings held under the auspices of the UNFCCC, through the Kyoto Protocol and Paris Agreement to the most recent COP26 and COP27. It shows how the Nationally Determined Contributions (NDCs) to global emission reductions, agreed under the terms of the Paris Agreement, and further commitments made at COP26, could stay within the 1.5–2°C temperature window of the Paris Agreement, if all countries' commitments and pledges were honoured. For this to happen, most countries of the world will need to reach 'net-zero' emissions by mid-century.

The chapter then zooms in on the emissions of different countries and notes the different ways in which they can be calculated: total annual emissions (as reported to the UNFCCC), emissions per person, cumulative emissions since the start of the industrial era, and emissions from consumption (including those from manufacturing imports), rather than those emitted on a country's territory (as reported to the UNFCCC). The different circumstances, emissions and national realities of the big emitters – the US, China, India, deforestation countries (e.g. Brazil and Indonesia), the fossil fuel exporters (e.g. Russia, Saudi Arabia and the Gulf States) – are explored through the different ways their emissions can be calculated. This leads to a brief consideration of the core issues of climate justice – historical responsibility for climate change, loss and damage arising from it, and climate finance – which are taken up in more detail in Chapter 12.

Limiting global warming to 1.5–2°C requires that only a certain amount of CO_2 and other GHGs can be released. These amounts are called 'carbon budgets' for the different temperature targets.

Reducing GHG emissions to net or 'real' zero by mid-century is a task of considerable complexity, as energy use and food production, the major causes of those emissions, are fundamental to the economy and human societies more broadly. There have been many exercises using complex climate-energy-economy models that show how this reduction in emissions, generically called 'decarbonisation scenarios', or 'decarbonisation pathways', might unfold. Four such modelling exercises are explored here. The first scenarios to be examined are those brought together by the Intergovernmental Panel on Climate Change (IPCC) in its 2018 report on how the 1.5°C temperature target of the Paris Agreement could be met, and how these scenarios differed from those that just met a 2°C target. The decarbonisation pathways of three further scenarios are explored: the Net Zero Emissions (NZE) global scenario of the International Energy Agency (IEA); a global scenario produced for the Finnish innovation agency, Sitra; and a scenario for the UK published by its Climate Change Committee (CCC). The global scenarios show broadly the decarbonisation pathways of different economic sectors, and which technologies are key for emission reduction, and they highlight the importance for most scenarios of those technologies that remove CO_2 from the atmosphere. The UK scenario allows the actual measures that bring about the emission reduction to be presented in more detail. Decarbonisation in other countries will necessarily have many similar measures, but tuned to their different contexts.

Chapter 3: Energy efficiency

Energy efficiency is sometimes described as the 'first fuel', because increasing energy efficiency means that the same energy service (kilometres driven, level of lighting, clothes washed, hours of entertainment watched) can be delivered with less energy use. This chapter looks in detail at the energy savings and CO_2 emission reductions that have been achieved through energy efficiency policies, and the further savings and reductions that could be achieved by such measures in the future. Largely drawing on UK and wider European experience, the chapter sets out the barriers to increasing energy efficiency, and how they can be overcome, in respect of industry, buildings and transport (vehicles).

Chapters 4, 5, 6: The energy supply technologies

These chapters explore all the various energy supply options, starting with fossil fuels, that will need to be phased out for climate stability, and then discuss the current status of and prospects for the energy sources that will need to replace them.

Chapter 4: Fossil fuels

Fossil fuels still supply 80% of the world's primary energy demand. For the Paris temperature targets to be met, a large proportion of fossil fuel reserves will need to remain unburned, and their use will need to fall precipitately, as shown by all net-zero scenarios. However, countries' and companies' plans to produce fossil fuels are wildly inconsistent with these targets. The great majority of emissions from the fossil fuel industry comes from the use (i.e. combustion) of their products, though the production of fossil fuels of some countries and companies is far more carbon-intensive than that of others. Many major fossil fuel producing countries are very dependent on fossil fuel revenues for both their exports and taxes, and the National Oil Companies they own show little desire to curb fossil fuel production. The major publicly owned oil and gas companies are more subject to pressure from shareholders and the public, and this has led some of them to claim that in some sense they are part of the 'clean-energy transition'. However, their businesses, and their investments, remain overwhelmingly focused on fossil fuels, and scientific papers have shown that their public utterances about their activities are at wide variance from the overall profile of their activities. Public pressure against them, through divestment campaigns, shareholder action and litigation, is growing. Although there are government-led organisations seeking to phase fossil fuels down or out, so far none of these have proved decisive in reducing fossil fuel supply.

Chapter 5: Electricity

All scenarios of deep decarbonisation see an important role for electricity provided by low- or zero-carbon sources (nuclear or renewables). Most such scenarios see a very great expansion of renewables (mainly solar and wind) and, to a lesser extent, nuclear. These sources of electricity are much less flexible than the fossil fuel generation that they are replacing. Renewables are dependent on time and place – they are not always available and different places have different amounts of sunlight and wind. This has important implications for many aspects of electricity provision, which are explored in this chapter. Energy storage becomes critical for those times when renewables are not available and for balancing the electricity system when there are fluctuations in supply or demand. Interconnection between the grids of different countries can help smooth out the different availabilities of renewables in different countries, and ambitious schemes have been planned to bring solar energy from North Africa to Europe at scale. Electricity networks may need extending to bring distant renewables to centres of demand, or to facilitate the local generation and consumption of renewable electricity. And electricity markets will need adjusting to ensure that consumers are able to benefit from cheap renewables, while still paying enough for their power to enable renewable generators to be profitable. The chapter discusses all these issues in some depth.

Chapter 6: Bioenergy and hydrogen

Bioenergy comprises biomass (trees, crops, crop residues, animal manure) that is used to produce power, heat or vehicle fuels ('biofuel'). The main constraint on the quantity of bioenergy that is available is land, given the need for land also to produce food for a growing population and to leave space for forests and wildlife. The chapter also touches on the various controversies around the production of biomass, such as the sustainability or otherwise of the production and use of biomass power generation or biofuels.

Zero-carbon energy scenarios also usually see a role for hydrogen, the production and potential uses of which are also discussed in this chapter. Elemental hydrogen for energy purposes has to be produced from hydrogen compounds, which is an energy-intensive process, and then more energy is needed to transport it to where it is needed. The most common processes for hydrogen production are steam reformation of methane gas ('grey' hydrogen, when the resulting CO_2 has to be captured and stored for the hydrogen to be 'low-carbon' or 'blue' hydrogen), or electrolysis of water with renewable electricity, to produce zero-carbon 'green' hydrogen. Hydrogen has a wide range of uses – for heat, power and transport – for which it can be burned, used in fuel cells for power generation, converted into synthetic fuels ('synfuels' or 'e-fuels') or used in the direct reduction of iron for steel-making. Hydrogen has been characterised over the last 25 years by cycles of enthusiasm and relative stagnation. However, the development of hydrogen-using products has advanced markedly in recent years, and there is now considerable investment going into the manufacture of electrolysers to produce 'green' hydrogen, so it may be that after a long gestation period, this energy source will become more widely used over the next 20 years.

Chapter 7: Carbon capture, use, storage and removal, geoengineering

Most emission scenarios that meet the Paris temperature targets make sometimes very extensive use of technologies that involve carbon capture and storage (CCS), some of which also use carbon (CCUS), or which may also, or separately, involve removing carbon dioxide (carbon dioxide removal, or CDR) from the atmosphere. In addition, there is growing interest, given already apparent effects of climate change, in techniques that involve 'geoengineering' the climate through direct, large-scale interventions in natural processes in the atmosphere, oceans or on land. This chapter discusses the technologies of CCS, CCUS, CDR, and geoengineering, their current deployment and possible effectiveness, and the use of some of them by companies and, potentially, governments, to 'offset' their continuing emissions. There is also some discussion, in respect of geoengineering, of how the use of these technologies should be governed and what their potential unintended or negative consequences might be.

Chapter 8: Digitalisation, the circular economy and critical minerals

New digital technologies already play an important role in the supply of electricity through, for example, optimisation of energy efficiency in generation, transmission and distribution, and smart grid sensing. Such uses will doubtless be extended. But the big new opportunities for digitalisation in energy relate to the use of electricity (the demand side) – not only by enabling more flexibility in the way electricity is used (demand response), but also through its application to electricity use in industry, transport and buildings, in everything from the increasing sophistication of smart meters to autonomous vehicles to the Internet of Things. This chapter discusses the many opportunities in this area and how the barriers to their realisation may be addressed.

The extraction of materials is responsible for 50% of global GHGs, and both the tonnage extracted and the associated emissions are projected on current trends to increase dramatically by 2060. The use of materials in industrial society has been characterised as following the model of 'take-make-dispose'. Materials are extracted from Earth, products are made, and they often last a relatively short time, before they are thrown away and end up on landfill sites or being burned. Large quantities of food are also wasted. Food waste, landfill sites and incineration all cause GHG emissions. An alternative model of material use is called 'the circular economy', where products are made to last longer and to be repaired, and at their end of life the materials they contain are recycled. Modelling shows that a combination of strong climate policy and policies to use resources more efficiently and move towards a circular economy can reduce both material extraction and GHG emissions, the former significantly below the projected 2060 levels, and the latter to levels consistent with a 2°C global warming temperature increase.

Certain materials, called 'critical minerals' have an especially important role in the transition to zero-carbon energy, also called the 'clean-energy transition'. This is because the technologies required for decarbonisation (e.g. batteries for electric vehicles, wind turbines, photovoltaic panels, fuel cells) require a wide range of metals and minerals, the use of which to date has been relatively small, and the production of which will need to be scaled up dramatically if these technologies are to replace fossil fuels over the next three decades. This chapter explores which materials are required for the clean-energy transition, how much of them might be needed, where they might come from, what the problems might be in supplying them in the required quantity, and how these problems might be overcome. The chapter concludes that it is likely that the mining industry will need to obtain a 'Sustainable Development Licence to Operate' before countries with these mineral resources will permit their extraction at the scale envisaged in net-zero scenarios.

Chapter 9: Buildings, transport, industry and business

Buildings, transport, industry and business are responsible for most global emissions. They will need to be transformed in order to shift from fossil fuels to clean energy. This chapter describes this transformation for each of these sectors.

Buildings globally account for around 40% of GHG emissions, the majority from their operation (e.g. their heating and cooling). Reaching real zero in many countries will require a transformation in both the energy efficiency of the building stock, and the systems through which they are heated and cooled. This chapter explores the extent to which greater energy efficiency could reduce the energy demand in buildings, and the technologies through which zero-carbon heating and cooling may be delivered. Because of the diversity of the building stock, detailed analysis of this issue is only possible at a country level, which in this chapter is undertaken for the UK. Issues explored include technologies, finance, energy advice and labelling, and policy.

The great majority of transport emissions come from the use of oil products, and the majority of those emissions come from road transport. Because there is no technology to capture tailpipe CO_2 emissions, the only way of eliminating them is to substitute the oil products with biofuels (which absorb CO_2 from the atmosphere before releasing it when they are burned) or synfuels, batteries or hydrogen, as an input to fuel cells or as a fuel to be burned. The chapter looks in detail at the implications of these alternatives in the different modes of transport – road passengers and freight, rail, aviation and shipping. Supplying adequate quantities of these alternative fuels would be made easier by reducing car ownership and use in urban areas. This could be achieved by encouraging a shift for short journeys in urban areas towards walking and cycling, the latter of which may need more infrastructure (e.g. more cycle lanes) for increased cycling, and by providing better public transport in compact, relatively densely populated cities. Car-sharing also has a role to play. Examples are given such as Freiburg in Germany, and various active travel initiatives in the UK.

With industry the second-largest direct emitter after the energy sector, its decarbonisation is clearly critical to real zero. The IEA shows how emissions in this sector can be reduced by 95% by 2050, with major roles being played by material efficiency, hydrogen, electrification and CCUS. After looking at the decarbonisation of industry in general, the chapter focuses in on the three most energy-intensive sectors – iron and steel, chemicals and cement – to look in more detail at how their emissions can be drastically reduced, including through the reduction of fluorinated gases (F-gases). While these industry sectors are the big emitters, the chapter also sketches out the important role in emission reduction for businesses more widely, both in their own operations and through their supply chains.

Chapter 10: Agriculture and waste

The world's current food system is a major contributor to climate change. Its food supply is also under significant threat from it. This chapter examines how food can both reduce its contribution to climate change, and become more resilient to the climate change that is in prospect. The main emission reduction strategies in agriculture centre on reducing emissions from livestock, by reducing meat consumption to healthy levels where it is currently excessive, and by producing their feed without associated deforestation. The main way to reduce the threat of climate change to agriculture and food security is to reduce the extent of climate change by reducing the emissions that contribute to it, and by reducing food waste.

Chapter 11: Economics of mitigation

Earlier chapters have shown that carbon emission reduction can be brought about by a wide range of technologies. This chapter explores how the costs of these technologies have changed and may change in the future, and what the effects of their large-scale introduction will be on economic growth.

Innovation is the process that creates new technologies and that leads to their cost reduction so that they can achieve large-scale deployment. This section explores the nature of this process throughout the journey from the laboratory to the mainstream economy, and presents the extraordinary cost reductions in low-carbon technologies which have already been brought about.

Reaching zero emissions by mid-century will require huge levels of investment in low-carbon technologies. This section shows that such investment has increased greatly in recent years, but is still well short of where it needs to be. The issue is not shortage of capital *per se*, but a shortage of projects with the right risk-return ratio for asset owners to invest in. There are now many initiatives in the financial sector to try to direct more capital into zero-carbon projects and technologies.

Investment is required not just in technologies, but also in the skills that are needed to manufacture and operate them effectively. The evidence suggests that the transition to clean energy will lead to a net increase in jobs, but there are challenges to ensure that those who lose their jobs in declining sectors are retrained to be able to take jobs in the low-carbon economy.

Economic growth is driven by investment in physical and human capital, and innovation and technical progress. Decarbonisation will require both investment and innovation, and it is therefore to be expected that it will lead to economic growth. This is borne out by the evidence which is explored in this chapter. It is surprising that a current of academic thought has arisen that posits that decarbonisation to zero emissions will require a reduction in economic output, in rich countries at least. This chapter argues that not only is this not necessary, but also that the suggestion itself is potentially very damaging politically in a world in which increased material living standards are still a major aspiration for most people.

There are a number of important 'co-benefits' from reducing GHG emissions. The largest of these is the health benefit from the improvement in outdoor air quality that would come from reducing fossil fuel use, and in indoor air quality from moving towards cooking and heating with cleaner fuels. Active travel in cities also has health benefits, as does reducing excessive meat and dairy consumption. Removing carbon dioxide from the atmosphere through afforestation or better land management can also deliver important ecosystem services (e.g. cleaner air and water) alongside carbon sequestration. This section explores the size and extent of these benefits.

Chapter 12: Climate justice, development, policy and delivery

This chapter is about the climate policies that will need to be adopted if the world is to have any hope of meeting the Paris temperature targets. To be acceptable globally, these policies will not only need to stimulate the adoption and deployment of clean-energy technologies. They will also need to be perceived as fair.

The UNFCCC is committed to a decarbonisation process that demands most from those who are best able to give, and seeks to protect the most vulnerable from climate impacts. This derives from and reflects perceptions that decarbonisation and the transition from fossil fuels should be just. However, there is a wide divergence of interpretation about both the principles of climate justice and the 'just transition', and how they should be implemented in practice. This chapter explores some of the key issues through an exploration of three relevant topics: the distribution between countries of the remaining carbon budget; a just distribution of the costs and benefits of the transition to clean energy; and which countries should produce the very limited quantity of fossil fuels which is consistent with achieving the Paris temperature targets.

For technology deployment, many of the relevant policies have already been introduced in earlier chapters. This chapter organises the policies according to various categories of policy instruments and other dimensions of policy: legislation and litigation; economic instruments; regulation; voluntary agreements; information; innovation instruments; circular economy policies; policy evaluation; policy mixes; policy evaluation; and sub-national (city-level, local government and community) policy. There is also a section on human behaviour change, including personal carbon calculators and personal carbon trading, and another on the institutions for policy delivery. The chapter ends with the key policy timeline identified by the IEA for decarbonisation to net-zero emissions globally, by 2050.

Chapter 13: Conclusion

This chapter draws together the threads of the book. It reiterates why the aim must be real (and not net) zero by 2050. It stresses the existence of the technologies and finance to achieve this, and the broadly positive economic implications of the clean-energy transition. It explains why, notwithstanding, politicians find it so difficult to introduce the necessary policies to bring this transition about. Overcoming these barriers will be the essential condition for humans to flourish in the decades and centuries ahead.

1   Why real zero?

Summary

The Earth's atmosphere is warming as a result of human emissions of greenhouse gases (GHGs), principally carbon dioxide (CO_2) from burning fossil fuels and deforestation, with methane (CH_4) from agriculture and leakages from the extraction and transport of fossil methane (sometimes called natural) gas also being important. These emissions have caused the atmospheric concentrations of GHGs, and global average atmospheric temperatures to increase beyond all human experience.

This global warming – currently 1.1°C above pre-industrial temperatures – is having major impacts on the weather and on all Earth's systems, on land and in the oceans. Extreme events are becoming more extreme and are projected to become more frequent. They have already caused loss of life and livelihoods and their future economic impacts could be catastrophic.

There are risks that global warming could cause Earth systems to cross several 'tipping points' that reinforce and cause climate change to accelerate out of human influence.

The human exacerbation of climate change can only be prevented by stopping the emission of GHGs from the human activities noted above, and removing CO_2 from the atmosphere. Subsequent chapters show the extent to which this must happen, and how it can be achieved.

Climate change and global warming

The climate is changing. The great majority of this change is being caused by human activities, past and present, as a result of emissions of carbon dioxide (CO_2), and some other so-called greenhouse gases (GHGs), into the atmosphere. These emissions are building up in the atmosphere and trapping the sun's heat in it, causing it to warm. So much we know from the publications of the Intergovernmental Panel on Climate Change (IPCC), the scientific

DOI: 10.4324/9781003438007-2

body set up by the United Nations Environment Programme (UNEP) and the World Meteorological Organization (WMO) in 1988, and which now has a membership of 195 governments. The IPCC "provides regular assessments of the scientific basis of climate change, its impacts and future risks, and options for adaptation and mitigation" with the objective to "provide governments at all levels with scientific information that they can use to develop climate policies").[1] So far according to the IPCC the average level of warming is 0.8°C to 1.3°C, with a best estimate of 1.07°C (IPCC, 2021a, A.1.3). The WMO estimate for 2021 is slightly higher: "The global mean temperature in 2021 was 1.11 ± 0.13°C above the 1850–1900 average" (World Meteorological Organization, 2022, p. 6). However, these numbers are already out of date. In 2023, Forster et al. (2023, Abstract) reported: "human-induced warming reached 1.14 [0.9 to 1.4]°C averaged over the 2013–2022 decade and 1.26 [1.0 to 1.6]°C in 2022 … Over the 2013–2022 period, human-induced warming has been increasing at an unprecedented rate of over 0.2°C per decade."

The WMO has confirmed that the last eight years (2015–2022) have been the warmest on record.[2] In its 2023 *Global Annual to Decadal Climate Update* the WMO estimate that the "chance of global near-surface temperature exceeding 1.5°C above preindustrial levels for at least one year between 2023 and 2027 is more likely than not (66%)", while "The chance of at least one year between 2023 and 2027 exceeding the warmest year on record, 2016, is very likely (98%)." Also at 98% is the probability of the five-year mean for 2023–2027 exceeding the last five years (2018–2022) (WMO, 2023, p.2).

The temperature increase is clearly shown in Figure 1.1: roughly 1.1°C between 1850 and 2019. The increase in temperature varies in different places, with "the highest increase in the temperature of the hottest days, at about 1.5 to 2 times the rate of global warming" occurring in mid-latitude (sub-tropical and temperate) regions' hottest days, and the Arctic projected "to experience the highest increase in the temperature of the coldest days, at about three times the rate of global warming" (IPCC, 2021a, B.2.3). The warming is already having dramatic effects on the climate, which will increase the hotter the atmosphere gets. The WMO meticulously documents these effects in its annual State of Climate reports. The data that follows is taken from the 2021 or 2022 reports unless stated to the contrary.

Long-term effects from global warming[3]

Some of the effects from global warming and the climate change it brings about are relatively slow but over time will have major effects on Earth's natural systems and the human societies they support.

---

1 www.ipcc.ch/about/. Accessed March 27, 2023.
2 https://public.wmo.int/en/media/press-release/past-eight-years-confirmed-be-eight-warmest-rec ord. Accessed April 22, 2023.
3 This and the next section contains information taken from (World Meteorological Organization, 2022), unless stated otherwise.

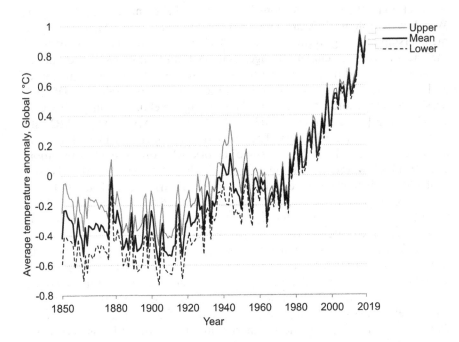

Figure 1.1 Global average temperature increase: 1850–2019.

Source: Our World in Data.[a]

[a] https://ourworldindata.org/co2-and-other-greenhouse-gas-emissions. Accessed April 22, 2023.

Ocean

The ocean takes up most of the extra heat in the Earth system caused by global warming, and this warms it, causing it to expand so that sea levels rise. Other contributors to sea level rise are the melting of land ice (see below), and land/ocean exchanges of water.

The global mean sea level rise between 1993 and 2002 was 2.1 mm/year, but between 2013 and 2021 this had risen to 4.5 mm/year, more than a doubling in the rate of increase, mainly due to an increased rate of melting of the ice sheets (World Meteorological Organization, 2022, Figure 5, p. 9).

The ocean absorbs around 23% of the annual emissions to the atmosphere of CO_2 from human activities. This causes a gradual acidification of the ocean (World Meteorological Organization, 2022, Figure 9, p. 12), which is now more acid (has a lower pH measurement) than it has been for at least 26,000 years. With rising acidity the ocean's ability to absorb CO_2 from the atmosphere declines. The WMO (World Meteorological Organization, 2022, p. 12) writes: "Ocean acidification threatens organisms and ecosystem services, and hence food security, tourism and coastal protection", but the effects vary greatly in different parts of the world. Coral reefs are especially vulnerable. Their area is projected to decline by 70–90% at 1.5°C of average global warming, and by over 99% at 2°C.

Cryosphere

The cryosphere comprises the frozen parts of Earth: sea ice, glaciers, ice sheets, snow, and permafrost. With global warming all these are experiencing profound change.

Sea ice is frozen water that floats on the sea (e.g. icebergs). When it melts it does not increase sea levels but is disruptive to the Earth system and organisms in other ways. The extent of sea ice in the Arctic has decreased dramatically, by between 1 and 2 million km² since 1980. The change in the Antarctic is much less pronounced (World Meteorological Organization, 2022, Figure 10, p. 12).

Glaciers are snow on land that has been compacted to form ice. Over 2000–2019 and excluding the Greenland and Antarctic ice sheets, global glaciers and ice caps lost an average of 267 Gt per year, but over 2015–2019 the average mass loss had increased to 298 Gt per year.

Ice sheets are glaciers that cover more than 50,000 km², and currently exist on Greenland and the Antarctic. They are also losing ice: on Greenland at an average of 326 Gt per year over 1986–2020, on the Antarctic much less (as with the loss of sea ice), with melting concentrated in low-lying and coastal areas. However, even this amounted to a total loss of around 1000 Gt since 2002.

Permafrost is defined as "ground that remains at or below 0°C for at least two consecutive years" (World Meteorological Organization, 2022, p. 19) and is beneath about one eighth of Earth's exposed land areas. Its thawing can lead to landscape instability, with areas of land sliding into rivers or the sea, the emission of greenhouse gases from previously frozen organic material (see the discussion of tipping points below), and, in Siberia, damage to whole cities as the earth under them thaws.[4,5]

Ecosystems

The world's ecosystems – including terrestrial, freshwater, coastal and marine ecosystems – are degrading at an unprecedented rate irrespective of climate change, but climate change is adding to the pressures on them, disrupting the cycles of temperature-sensitive plants, shifting the timing of marine and

---

4 The Guardian, October 14, 2016: www.theguardian.com/cities/2016/oct/14/thawing-permafr ost-destroying-arctic-cities-norilsk-russia. Accessed April 22, 2023.
5 www.reuters.com/world/europe/russias-remote-permafrost-thaws-threatening-homes-infrast ructure-2021-10-18/. Accessed April 22, 2023.

freshwater fish spawning events and animal migrations worldwide, and exacerbating other threats to biodiversity. The WMO reports:

> The number of species projected to go extinct increases dramatically as global temperatures rise – and is 30% higher at 2°C warming than at 1.5°C warming. … Species that are less able to relocate are projected to experience high rates of mortality and loss. … Rising temperatures heighten the risk of irreversible loss of marine and coastal ecosystems, including seagrass meadows and kelp forest. Coral reefs are … projected to lose between 70 and 90% of their former coverage area at 1.5°C of warming and over 99% at 2°C. Between 20 and 90% of current coastal wetlands are at risk of being lost by the end of this century, depending on how fast sea levels rise.
>
> (World Meteorological Organization, 2022, p. 39)

Increased concentrations of CO_2 in the atmosphere not only bring about global warming. They also enhance photosynthesis in plants (known as the 'CO_2 fertilisation effect'), and increase their water efficiency.[6] The effect of this on crops is discussed in Chapter 10 on agriculture.

Extreme events

The figures above show trends that are happening relatively slowly related to human timescales, though extremely fast compared to most changes in the Earth system. However, each year these trends are punctuated by more extreme events.

Coumou & Rahmstorf (2012) have showed the extent to which extreme weather events increased over the 20th century. The ratio of observed to expected extreme events increased by more than a factor of three over the 20th century (Coumou & Rahmstorf, 2012, Figure 2, p. 495).

Some of the most unsettling pictures on our television screens in 2021 and 2022 were of wildfires and floods. Major wild fires occurred in California and British Columbia, where the town of Lytton was effectively destroyed. In the Mediterranean Region there were major fires in Algeria, in which 40 people lost their lives, and in Turkey, Greece, France, Italy, North Macedonia, Lebanon, Israel, Libya, Tunisia and Morocco. Siberia experienced summer wildfires for the third year in succession. There were fires in the Amazon (though fewer than in 2019 and 2020), and extensive fires in other parts of Brazil, e.g. the Pantanal. Australian wildfires in 2020–2021 were thankfully well below the appalling levels of 2019–2020.

Unsurprisingly, many of these fires occurred simultaneously with or as a result of extreme heat waves. Many places recorded record temperatures: British

---

6 www.carbonbrief.org/guest-post-understanding-co2-fertilisation-and-climate-change/. Accessed April 22, 2023.

Columbia (49.6°C), California (54.4°C), Oregon (46.7°C), Sicily (48.8°C, the highest ever anywhere in Europe), Tunisia (50.3°C), Spain (47.4°C), Turkey (49.1°C), Georgia (40.6°C), and, further north, Northern Ireland (31.3°C).

Older people and very young children are the people most vulnerable to extreme heat. The Lancet Countdown on Global Health and Climate Change found that "children younger than 1 year were affected by 626 million more person-days of heatwave exposure and adults older than 65 years were affected by 3.1 billion more person-days of heatwave exposure in 2020 than in the 1986–2005 average" (Romanello et al., 2021, p. 1625). Extreme heat also makes it dangerous to work outside. In 2020, 295 billion hours of potential work were lost due to extreme heat exposure, with over half of all losses occurring in the agricultural sector in countries with low to medium human development, with obvious implications for food security (Romanello et al., 2021, p. 1626).

Despite average global warming, many places experienced abnormally cold conditions in 2021. A lengthy winter cold spell in Texas and Oklahoma (−19°C in Dallas, −25.6°C in Oklahoma City) disrupted power supplies for 10 million people, killed 172 people and caused economic losses of over US$20 billion, making it the most expensive winter storm ever in the US. Locations in many other countries had record or near record cold spells, including the Russian Federation, China, Japan, Spain, UK, France, Switzerland and Slovenia. Spring frosts caused economic damage to crops in France in excess of US$4.6 billion.

Warmer air is able to hold more moisture than colder air: for each 1°C increase, the air can hold 7% more moisture (Coumou & Rahmstorf, 2012, p. 491); that's about an extra 900 billion tonnes of water.[7] It is therefore not surprising that a recent paper in the journal *Nature* opened with the sentence: "Damaging floods are increasing in severity, duration and frequency, owing to changes in climate, land use, infrastructure and population demographics" (Tellman et al., 2021, p. 80). However, as with everything to do with climate and weather, the situation is not straightforward, with the IPCC writing: "A warmer climate will intensify very wet and very dry weather and climate events and seasons, with implications for flooding or drought" (IPCC, 2021a, B.3.2). So with climate change we can expect both more flooding *and* more droughts.

2021 certainly saw a lot of flooding; 302 people died, and losses of US$17.7 billion were reported, when in Henan over 200 mm of rain fell in one hour (a Chinese national record), 382 mm in 6 hours, and 720 mm over the full three-day event. In Western Europe exceptional floods killed 179 people in Germany and 36 in Belgium, with economic losses in Germany exceeding US$20 billion. Significant floods also occurred in France, the Netherlands, Luxembourg, the United Kingdom and Switzerland. Flash floods in 2021 killed 61 people in

---

7 USGS: "One estimate of the volume of water in the atmosphere at any one time is about 12,900 cubic kilometers (km³)" www.usgs.gov/special-topics/water-science-school/science/atmosphere-and-water-cycle. Accessed April 22, 2023. 1 cubic kilometre weighs 1 billion tonnes.

Afghanistan, and 77 people on Turkey's Black Sea coast. Record floods in New South Wales in Australia caused economic losses of US$2.1 billion. In South America, the Rio Negro at Manaus in Brazil reached its highest level on record, and there was also flooding in Guyana, Colombia and Venezuela. In South Asia there were 529 deaths from flooding in India, 198 in Pakistan (as of 30 September) and further deaths in Bangladesh and Nepal. In Africa flooding was reported in Niger, Sudan and South Sudan and Mali.

Flooding is also associated with hurricanes and cyclones, when it is accompanied by high winds. Hurricane Ida in the US had sustained 1-minute winds of 240 km/hour, with major wind damage and storm surge flooding, which caused damage estimated at US$63.8 billion. Hurricane Grace caused widespread flooding and storm damage in Mexico, killing 14 people, having earlier caused flooding in Haiti, the Dominican Republic, Jamaica, and Trinidad and Tobago. Cyclone Seroja in the southern hemisphere killed 272 people in Indonesia and Timor-Leste. In Africa, Cyclone Eloise caused damage and casualties in Mozambique, South Africa, Zimbabwe, and Madagascar. In Asia Cyclone Tauktae killed 144 people in India and 4 in Pakistan, while later in the year Cyclone Shaheen caused 39 deaths across India, Pakistan, Oman and Iran, mostly from flooding.

Drought was also a widespread problem in 2021. In South America, central and southern Brazil, Paraguay, Uruguay and northern Argentina all experienced well below average rainfall, leading, in Brazil, to crop losses, low river flows, reduced hydroelectric production and disrupted river transport. The River Paraguay fell to a record low and Chile, which has experienced drought for most of the last decade, had another dry year. In western North America widespread drought in 2020 spread and intensified in 2021. The 20 months from January 2020 to August 2021 were the driest on record for the south-western United States – the level of Lake Mead on the Colorado River fell in July to 47 m below full supply level, the lowest level on record since the reservoir was fully commissioned. Extreme to exceptional drought also extended eastwards on both sides of the United States-Canada border. Canada produced less wheat, canola, barley, soybeans and oats in 2021 compared with 2020, largely because of the drought conditions in Western Canada. Several major field crops grown mainly in Western Canada experienced their largest year-on-year yield decrease on record, falling to levels not seen in more than a decade.[8] In Asia, well below average rainfall occurred during the 2020–2021 cool season in most of Iran, Afghanistan, Pakistan, south-east Turkey, and Turkmenistan. Overall, the WMO writes: "Climate change may exacerbate water stress, especially in areas of decreased precipitation and where groundwater is already being depleted, affecting agricultural production, arable land, and the more than 2 billion people who are already experiencing water stress" (World Meteorological Organization, 2022, p. 38).

---

8 Statistics Canada, The Daily, December 3, 2021, www150.statcan.gc.ca/n1/daily-quotidien/211 203/dq211203b-eng.htm. Accessed April 22, 2023.

Sub-Saharan Africa (SSA) faces special challenges from droughts and floods (World Bank, 2021).[9] This is partly because of its relatively low incomes (SSA has 15% of the global population, and produces just 2% of world gross domestic product or GDP). It is also because, although SSA contributes about 7% to global GHG emissions, mainly from agriculture and land use change, the impacts of climate change in the region seem particularly intense. The World Bank reports:

> Africa's annual temperature has increased at an average rate of 0.13°C per decade since 1910; however, it has more than doubled to 0.30°C since 1981. … Rising temperature and changes in precipitation across many countries in Sub-Saharan Africa are leading to increasing frequency and intensity of extreme weather events – heatwaves, droughts, floods, and storms, among others. … Relative to 1970–79, the frequency of droughts in Sub-Saharan Africa nearly tripled by 2010–19, while it more than quadrupled for storms and increased more than tenfold in the case of floods.

Figure 1.2 shows the climate hazards that affect most people in SSA: drought in the north, east and south-west, floods in the west, centre and south-east, and storms also in the south-east. The loss of life from these events can be substantial: over 20,000 from the 2010 drought in Somalia, over 2,000 from Cyclone Idai in Mozambique. Madagascar, designated as in the storm area in Figure 1.2, is also experiencing a two-year, severe, localised drought in the southern part of the island, which resulted in the World Food Programme classifying over 1 million people as in urgent need of food assistance in August 2021 (World Meteorological Organization, 2022, p. 29).

Of course, extreme weather *per se* is not new. But even the limited climate change currently being experienced is intensifying it. Moreover, the IPCC projects that

> At 1.5°C global warming, heavy precipitation and associated flooding are projected to intensify and be more frequent in most regions in Africa and Asia, North America and Europe. Also, more frequent and/or severe agricultural and ecological droughts are projected in a few regions in all inhabited continents except Asia compared to 1850–1900.
>
> (IPCC, 2021a, C.2.2)

Impacts of extreme weather

The extreme events described in the previous section pose the most intense and immediate risks for agriculture and food security. According to the UN Food

---

9 Information about SSA is taken from World Bank 2021 'Africa's Pulse', October, https://documents1.worldbank.org/curated/en/786721633587328307/pdf/Africa-s-Pulse-No-24-October-2021-An-Analysis-of-Issues-Shaping-Africa-s-Economic-Future.pdf. Accessed March 27, 2023.

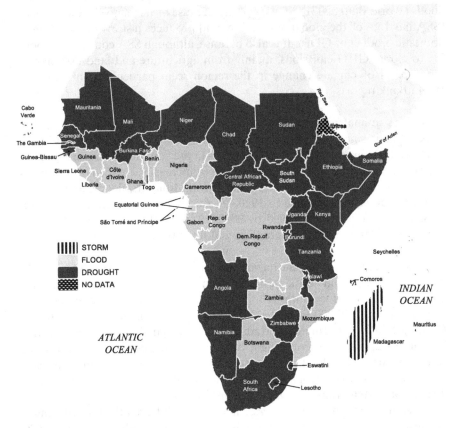

Figure 1.2 Weather events affecting sub-Saharan Africa.

Source: World Bank, 2021, Map 2.1, p. 58.

and Agriculture Organization (FAO), agriculture "underpins the livelihoods of over 2.5 billion people worldwide and up to 60 percent of those in LDCs" (Least Developed Countries) (Food and Agriculture Organization, 2021, p. 15). In the decade 2008–2018 the total disaster-related loss of crop and livestock production was US$108.5 billion in poorer countries (those classed as Least Developed Countries [LDCs], and Lower Middle Income Countries [LMICs]), and US$280 billion globally. While a small-sounding 4% of potential crop and livestock production, the FAO writes:

> This is a significant amount, capable of causing perceivable production disruptions with severe impacts on international markets and global food supply. Furthermore, disasters often occur within a limited geographical

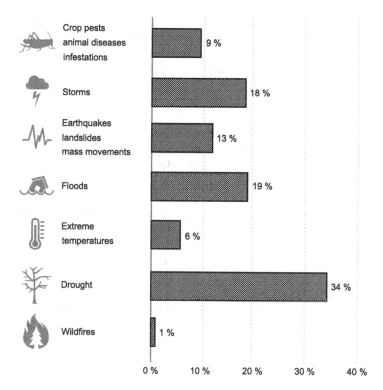

Figure 1.3 Total crop and livestock production loss per disaster type, LDCs and LMICs, 2008–2018.

Source: Food and Agriculture Organization, 2021, Figure 5, p. 33.

area, where they may cause the complete destruction of local production or infrastructure. … such impacts may fundamentally disrupt local livelihoods and food security in affected areas.

(Food and Agriculture Organization, 2021, p. 28)

Figure 1.3 shows that most of the food losses in the ten years to 2018 were from drought, floods and storms, three of the types of events that are projected to intensify with further climate change. One of the headline messages of the FAO report cited earlier is "The growing frequency and intensity of disasters, along with the systemic nature of risk, are jeopardizing our entire food system" (Food and Agriculture Organization, 2021, key message, final page).

Extreme weather in 2021 also led to the displacement of millions of people, with China, Vietnam, Indonesia and the Philippines recording the highest numbers. These millions join the at least 7 million people [who] were living in internal displacement following disasters related to natural hazard events in previous years. The highest numbers of these people were in Afghanistan, India and Pakistan, followed by Ethiopia, Sudan, Bangladesh, Niger and Yemen.

(World Meteorological Organization, 2022, p. 36)

Tipping points

One of the characteristics of some natural systems, such as ocean currents, forests, or established weather patterns is that they can remain seemingly stable and seem able to withstand changes for a considerable period of time, and then quite quickly 'flip' into a different state once a certain threshold of change has been passed. The name that has been given to such a threshold is 'tipping point'. This is a phenomenon that has been explored intensively in the scientific literature in recent years in respect of climate change.

Perhaps the first mention of tipping points in this context was by the climate scientist James Hansen in a 2005 lecture,[10] when he said:

I present multiple lines of evidence indicating that Earth's climate is nearing, but has not passed, a tipping point, beyond which it will be impossible to avoid climate change with far ranging undesirable consequences. ... we are on the precipice of climate system tipping points beyond which there is no redemption.

(Hansen, 2005, Abstract)

In 2005 the CO_2 atmospheric concentration was around 375 parts per million (ppm), since when it has risen by 40 ppm to 415 ppm, with increases in other GHGs as well. If humanity was on a tipping point precipice then, it is much closer to it now, if it has not already gone over the edge.

Subsequent work on climate tipping points defined the phenomenon as "a critical threshold at which a tiny perturbation can qualitatively alter the state or development of a system" (Lenton et al., 2008, p. 1786), and identified a number of elements in the Earth system (which were called 'tipping elements') that could lead to tipping points.

However, there are still many and extensive uncertainties around the concept, which include: the threshold beyond which the tipping points will be triggered; the impacts that will arise from doing so; the timescale over which they will occur; whether the change will be abrupt, or more a gradual adaptation to a warmer Earth; whether the tipping point will occur for the 'element' itself, or for the ecosystems that depend on it. And then, of course, because climatic 'elements' tend to be inter-connected, again sometimes in very complex

10 www.columbia.edu/~jeh1/2005/Keeling_20051206.pdf. Accessed March 27, 2023.

ways, there are further uncertainties as to how the evolution of these elements will affect each other. Notwithstanding these uncertainties, what scientists seem to agree on is that once triggered, there is unlikely to be a way back to the previous state in which these elements existed, over human timescales at least. In Hansen's language, there is no 'redemption'.

Table 1.1 and the brief subsequent explanations of tipping elements and the possible consequences if tipping points exist and are triggered are gross simplifications of scientific realities and uncertainties, but hopefully they provide a useful first overview of the much more complex explanations from which they are derived (Lenton et al., 2008; Carbon Brief, 2020 IPCC, 2021b, Box TS.9, p. 106).

Arctic sea ice: the loss of Arctic sea ice was described earlier in the chapter. The sea-ice extent seems to be subject to a gradual (on human timescales) decline, with year-to-year variations, rather than an abrupt change. But this is not true for the animals, ecosystems and human cultures that depend on the sea ice, which could more or less disappear in a period of very low or fragmented sea-ice extent. Less sea ice exposes more dark ocean, which absorbs more heat, so contributing further to the melting process.

Greenland ice sheet: The IPCC AR6 Technical Summary estimated that the Greenland ice sheet had lost around 5000 Gt of ice over 1992–2020, contributing 13.5 mm to global mean sea level rise (IPCC, 2021b, p. 77). It is still melting and its future rate of melting is highly dependent on future warming. Complete melting would lead to a rise in mean global sea levels of perhaps 7 metres,[11] but with warming of 3°C or less, this would take thousands of years.

West Antarctic ice sheet (WAIS): one of three regions making up Antarctica. The WAIS rate of ice loss tripled from 53 billion tonnes a year during 1992–1997 to 159 billion tonnes a year in 2012–2017. It is vulnerable to ice loss, and may already have passed a tipping point. Full melting of WAIS would raise sea levels by 3 metres or more.

Atlantic Thermohaline Circulation (also called Atlantic Meridional Overturning Circulation): an ocean current (that includes the Gulf Stream) bringing warm, salty water from the South Atlantic to the Arctic, freshwater melting from which reduces its propensity to sink to greater depths and travel back to the South Atlantic. Its shutdown would bring widespread changes in climate to Europe and elsewhere. Its weakening has already been recorded.

Indian monsoon: The monsoon rains – 70–90% of India's rainfall – are crucial for India's farm sector, which makes up about a sixth of India's economy and employs about half of the country's 1.3 billion population, so any shift towards drought or increased rainfall could have huge impacts. Weather processes are very complex, and made more so by the widespread air pollution.

---

11 This and subsequent numbers and quotes in the descriptions of tipping points come from Carbon Brief (2020).

Table 1.1 Potential tipping points

Tipping element	Current observed effect	Main reason(s) for concern	Trigger temperature/ timescale for major change	Effect on climate
Arctic summer sea ice	Decreasing in area	Loss of animals, ecosystems, human cultures	Already occurring/ steady loss of sea ice rather than tipping point	Exposed oceans absorb more heat, contributing to further warming
Greenland ice sheet (GIS)	Decreasing in volume	Sea level rise	2–3°C above pre-industrial/ thousands of years	Less reflective surfaces absorb more heat, contributing to further warming
West Antarctic ice sheet (WAIS)	Decreasing in volume	Sea level rise	1.5–2°C above present, 2–3°C above pre-industrial*/ after 2100	
Atlantic thermohaline circulation (THC)	Weakening	Substantial climate change, especially in Europe	3–4°C above pre-industrial/ after 2100	Cooling of western Europe and eastern US
Indian summer monsoon (ISM)	Uncertain but likely increased rainfall with significant warming	Significant change could lead to more floods or drought	3°C likely to lead to greater intensity of rainfall/ not much change before 2100	None
Amazon rainforest	Decreasing in area	Reduced rainfall, biodiversity loss, ecosystem and climate change, switch from sink to source of CO_2	3°C or more above pre-industrial, or 20–40% deforestation/ 15–20 years at current rates of deforestation	Extra CO_2 emissions, and reduced CO_2 absorption, cause of further warming
Boreal forest	Decreasing	Forest loss, ecosystem change, greater thawing of permafrost	3–4°C above pre-industrial/ decades	Extra GHG emissions cause of further warming

Table 1.1 (Continued)

Tipping element	Current observed effect	Main reason(s) for concern	Trigger temperature/ timescale for major change	Effect on climate
Permafrost	Thawing	CO_2, CH_4 emissions, infrastructure damage	Further warming would accelerate thawing/ Already occurring	Extra GHG emissions cause of further warming
Sahara/Sahel and West African monsoon (WAM)	Very uncertain, could make area drier or wetter	Increased droughts or floods (positive effect: perhaps northward shift of vegetation)	3°C, with negative heat impacts on agriculture in the region/ No significant change before 2100	None
Coral dieback	Coral bleaching and death	Loss of biodiversity, damage to ecosystems and economies	1.5–2°C warming could see loss of 90% of corals/ Loss already occurring	None

* These two temperatures are broadly equivalent because the poles warm faster than the average.
Source: Information derived from: Carbon Brief (2020); Lenton et al., 2008; IPCC, 2021b, Box TS.9, p. 106.

The outcome of further warming is very uncertain, but beyond 3°C seems likely to lead to more intense rainfall.

Amazon rainforest: rainforests are supported by high rainfall which the trees then cycle back into the atmosphere ('evapotranspiration'). At a certain level of deforestation this process could weaken such that the forest would die back and become savannah. Deforestation has currently reduced the Amazon rainforest by 17% of its original area. Some estimates suggest that tipping points in some parts of the Amazon have already been triggered.

Boreal forests: These lie in a cold climate, are often underlain by permafrost, and occur mainly in Russia and Canada, just south of the Arctic Circle. They comprise about 30% of the world's forests and store about one third of terrestrial carbon. They are susceptible to ecosystem shift (northwards) and change (e.g. from fires) as a result of global warming. Under some circumstances (e.g. frequent, severe fires) these changes could be quite quick (i.e. over decades). Overall the changes would add to global warming through forest loss and hastening thawing of permafrost.

Permafrost: Permafrost is ground that contains ice or frozen organic material that has remained at or below 0°C for at least two years. It covers around a quarter of the non-glaciated land in the northern hemisphere, in Siberia, Alaska, northern Canada and Tibet and, in the southern hemisphere, in Patagonia, Antarctica and New Zealand. The frozen ground holds a vast amount of carbon, from dead plants and animals over many thousands of years. As the climate warms, there is an increasing risk that permafrost will thaw, releasing CO_2 and methane. Thus, large-scale thawing of permafrost has the potential to cause further climate warming. There is evidence that thawing permafrost is already releasing a net 300–600 million tonnes of carbon into the atmosphere a year. Thawing permafrost destabilises infrastructure, and sometimes whole towns, that have been built on the permafrost in the expectation that it will remain solid. High-warming IPCC scenarios result in upwards of tens of billions of tonnes of CO_2 from thawing permafrost.

West African Monsoon (WAM): WAM brings rainfall to West Africa and the Sahel, a band of semi-arid grassland between the Sahara desert to the north and tropical rainforests to the south. In the 1960s and 1980s it was associated with a decline in rainfall and extended drought that killed many people. However, the effects of future warming on the WAM are still very uncertain.

Coral dieback: ocean warming can cause coral 'bleaching' (whitening of corals which can lead to their death), and corals are also damaged or killed by ocean acidification and various human activities. Mass bleaching events have become five times as common over the past 40 years. "Coral cover across the Caribbean declined by 80% from 1977 to 2001 and may completely disappear by 2035." Warming of 1.5–2°C could lead to the degradation or destruction of 90% of corals. Coral reefs currently support the livelihoods of over 500 million, mainly poor, people. "The widespread loss of coral reefs would be devastating for ecosystems, economies and people."

Risks from climate change

In discussing the potential impacts of climate change, the concept of risk has been frequently deployed. Climate change, and the impacts it brings in its train, are a substantial *risk* for human societies. Unfortunately it is well recognised that humans are not good at assessing and comparing the risks of different situations. This can be illustrated through the following example.

Table 1.2 shows the probabilities of a more than 6°C rise in average global temperatures at different atmospheric GHG concentrations, given the uncertainties in the relevant relationships. A rise of this level would certainly make large parts of the Earth uninhabitable and spell the end of human civilisation. Humans would revert to a global situation in which there were far fewer of them, and those still alive would be engaged in a desperate struggle to survive.

Table 1.2 shows that, at 700 ppm, which is a highly possible future CO_2 concentration on current emission pathways, as will be seen below, the median average temperature increase would be 3.4°C, but there is an 11% chance that it would reach a civilisation-ending 6°C.

Table 1.2 Probabilities of exceeding an average global temperature increase of 6°C at different atmospheric concentrations of GHGs

Probabilities of exceeding an average global temperature of 6°C at different atmospheric concentrations of GHGs

CO_2 concn. (ppm)	400	450	500	550	600	650	**700**	750	800
Median temp. increase (°C)	1.3	1.8	2.2	2.5	2.7	3.2	**3.4**	3.7	3.9
Chance of > 6°C (%)	0.04	0.3	1.2	3	5	8	**11**	14	17

Source: Wagner & Weitzman, 2016, Table 3.1, p. 54, cited in Ekins & Zenghelis, 2021, Table 1, p. 952.

To put this in perspective, it may be noted that in 2018 there were around 38 million aircraft flights per year,[12] with one fatal accident every 3 million flights,[13] a probability of 0.000033%. A 0.3% probability – the risk of civilisation-ending climate change at *current* GHG concentrations of 450 ppm (see Table 1.2) – of a fatal accident would mean over 300 fatal accidents *each day*. How many people would fly given that kind of accident rate reported daily on the news? Yet that is the risk human societies are currently taking in respect of catastrophic climate change.

Greenhouse gas emissions

There are six major GHGs which derive from human activities and which have contributed to human-induced global warming: carbon dioxide (CO_2), methane (CH_4), nitrous oxide (N_2O), the so-called F-gases (hydrofluorocarbons [HFCs] and perfluorocarbons [PFCs]) and sulphur hexafluoride (SF_6). As noted in the Preface, it is possible to express the 'global warming potential' (GWP) of each of these GHGs in the same unit, which is 'CO_2-equivalents', or CO_2e, which means the amount of CO_2 that would cause the same amount of global warming over 100 years as 1 tonne of the GHG. The specification of the timeframe is necessary because the different GHGs stay in the atmosphere for different amounts of time. For example, methane is a much more potent greenhouse gas than CO_2, but it only stays in the atmosphere for around 12 years, whereas CO_2 can stay there for centuries. Over 20 years methane has a GWP of 84–86, whereas over 100 years its GWP is 28–34 (IPCC, 2013, Ch.8, pp. 714, 731). Reducing methane emissions therefore has a relatively large short-term effect compared to reducing CO_2. N_2O has a 100-year GWP of 265, and stays

---

12 ICAO (International Civil Aviation Organization) (2018) Annual Report 2018: the World of Air Transport in 2018, ICAO, www.icao.int/annual-report-2018/Pages/the-world-of-air-transport-in-2018.aspx. Accessed April 22, 2023.

13 https://edition.cnn.com/2019/01/02/health/plane-crash-deaths-intl/index.html. Accessed April 22, 2023.

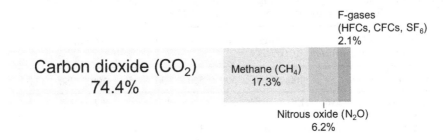

Figure 1.4 Relative emissions of the main GHGs in CO_2e in 2016.

Source: Our World in Data, https://ourworldindata.org/greenhouse-gas-emissions. Accessed May 5, 2023.

in the atmosphere for 121 years.[14] F-gases have 100-year GWPs that differ between the gases, but are very high and often in the thousands, e.g. over 6,500 for PFC14, and 23,600 for sulphur hexafluoride.

Figure 1.4 shows the relative emissions of the different GHGs in 2016, in terms of their GWP. CO_2 comprised nearly three quarters of CO_2-equivalent emissions that year, which is broadly still the situation.

GHG emissions have increased fairly steadily since 1990, which is the reference year for the UN Framework Convention on Climate Change (UNFCCC), as shown in Figure 1.5. From 1990 to 2000 the rate of emissions increase was 0.7% per year, increasing in the following decade to 2.1% per year, before falling back in 2010–2019 to 1.3% per year. Total annual emissions reached 59 $GtCO_2e$ in 2019. To keep global warming to 1.5°C, they will need to fall by 40–50% over 2020–2030, an average annual emissions reduction of 1.2 Gt per year. In 2022, they were still rising, though the rate of CO_2 increase had slowed to 0.9%.[15]

The great majority of the emissions are CO_2 from the burning of fossil fuels (coal, oil and fossil methane gas), and industrial processes (mainly cement manufacture). Methane is the second biggest source, from agriculture (rice cultivation and livestock, especially cattle, goats and sheep), leakage from oil and gas extraction ('fugitive emissions'), coal mining, and waste (decomposition of organic matter in landfills). Then comes emissions from land use change, mainly deforestation, but also from other land use changes (e.g. the draining of peat). The main source of N_2O is agriculture (from fertiliser use). F-gases are used in industry. These gases are now briefly discussed in turn.

14 https://ourworldindata.org/greenhouse-gas-emissions. Accessed April 22, 2023.
15 www.iea.org/reports/co2-emissions-in-2022. Accessed April 22, 2023.

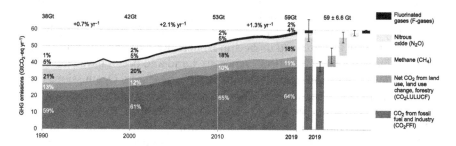

Figure 1.5 Global anthropogenic GHG emissions, 1990–2019.

Source: IPCC, 2022b, Figure SPM1, p.SPM-6.

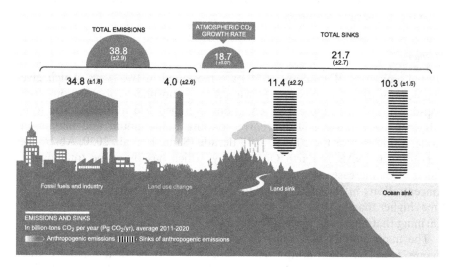

Figure 1.6 Representation of the sources and sinks of human emissions of CO_2 (global annual average for the decade 2011–2020, $GtCO_2$/year).

Note: The budget imbalance is the difference between the estimated emissions and sinks.

Source: Global Carbon Project, www.globalcarbonatlas.org/en/content/global-carbon-budget. Accessed April 22, 2023.

Carbon dioxide

The cycling of carbon dioxide through land, vegetation and oceans is one of the fundamental characteristics of the Earth system. Figure 1.6 gives a diagrammatic representation of the human-induced flows, the quantity taken up by the land and oceans and, therefore, the amount that remains in the atmosphere (18.7 Gt CO_2 annually, on average, over 2011–2020).

Over 2011–2020 humans emitted an average of 35 GtCO$_2$ from burning fossil fuels and 4 GtCO$_2$ from land use change per year. Of that, 11 GtCO$_2$ was taken up by the biosphere (soils and vegetation), 10 GtCO$_2$ was taken up by the ocean, and 19 GtCO$_2$ remained in the atmosphere. Remarkably, even as human emissions have increased, the IPCC reports: "Land and ocean have taken up a near-constant proportion (globally about 56% per year) of CO$_2$ emissions from human activities over the past six decades, with regional differences" (IPCC, 2021a, A.1.1). However, the IPCC projects that, as shown in Figure 1.7, the higher cumulative emissions rise over this century, the lower will be the proportion of those emissions taken up by the land and ocean, so that a greater proportion is left in the atmosphere.

Figure 1.8 shows the relentless rise in global CO$_2$ emissions from fossil fuels (the main source of CO$_2$, as seen in Figure 1.5) since 1960, punctuated only by major economic disruptions (oil crises, end of the Soviet Union, financial crisis, Covid). And of course they have been rising well before 1960, in fact since the industrial revolution.

Methane

Global emissions of methane (CH$_4$) are measured in two ways, which give significantly different answers. Measured by atmospheric observations (the 'top-down' view in Figure 1.9), emissions averaged 576 million metric tons (Mt) per year (range 550–594 corresponding to minimum and maximum model estimates) for the 2008 to 2017 decade (Saunois et al., 2020, Abstract). According to NASA's Earth Observatory, which has the same total figure, this is a 9% increase compared to the previous decade, which took methane concentrations in the atmosphere to more than 1,875 parts per billion, 2.5 times higher than in 1850, with methane responsible for a quarter of the global warming that has taken place since then.[16]

The main sources and sinks of methane over 2008–2017 are shown in Figure 1.9. The 'bottom-up' view (estimates from particular natural and human processes) in Figure 1.9 shows total emissions to be about 30% higher (737 Mt, range 594–880 Mt), with the majority of the difference coming from the estimated 'Other natural emissions'. About 60% of methane emissions in the top-down view are the result of direct human activities. The emissions have caused the amount of methane in the atmosphere to grow by an average of 18.2 Mt per year.

Other GHGs

Nitrous oxide: Human-induced emissions of N$_2$O increased by 30% over 1980–2020 (Tian et al., 2020, p. 248).

---

16 https://earthobservatory.nasa.gov/images/146978/methane-emissions-continue-to-rise. Accessed April 22, 2023.

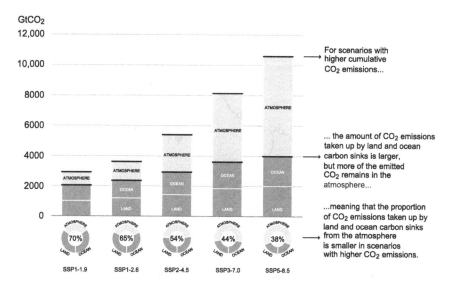

Figure 1.7 Cumulative anthropogenic CO_2 emissions taken up by land and ocean sinks by 2100 under five illustrative scenarios.

Note: the SSP (Shared Socioeconomic Pathways) scenarios (SSP1–SSP5) have different assumptions about the state of the world and climate change policies that lead to very different levels of emissions. They are explained in more detail in the next chapters. The numbers after the SSP1 etc. are the Representative Concentration Pathways (RCPs) (1.9 being the lowest emissions, 8.5 being the highest emissions, as explained in the Preface).

Source: IPCC, 2021a, Figure SPM.7, p. 20.

F-gases: F-gas emissions mainly arise from hydrofluorocarbons (HFCs) being used in refrigeration. They have increased strongly this century, from around 750 MtCO2e in 2005 to 4,000 MtCO2e in 2050, and are projected to increase much further, mainly from refrigeration, unless controlled.[17]

Figure 1.10 shows the projected effects of these GHG emissions on average global temperatures, with the horizontal axis showing the ranges of projected temperature increase above the pre-industrial level, and the vertical axis giving three different timescales (Near [2000s], Mid [2050s], and Long [2090s]) for different Shared Socioeconomic Pathways (SSPs) (socio-economic conditions, as discussed in more detail in Chapter 2) and emission levels (low [RCP1.9] to high [RCP8.5] – the RCPs [Representative Concentration Pathways] are explained in the Preface). The dots show the best estimates in each case, and the bars the 90% confidence intervals. Thus, by the end of the century, the average global temperature increase under SSP1–1.9 (lowest emissions) stays

17 Umweltbundesamt, 2011, Figure 2, www.umweltbundesamt.de/sites/default/files/medien/publ ikation/short/k3866.pdf. Accessed April 22, 2023.

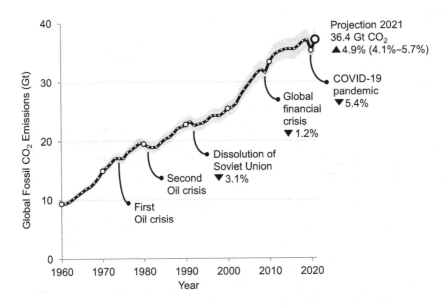

Figure 1.8 Global CO$_2$ emissions from fossil fuel use, 1960–2020.

Source: Global Carbon Project, https://robbieandrew.github.io/GCB2021/PNG/s18_2021_FossilFuel_and_Cement_emissions_1959.png. Accessed April 22, 2023.

below 1.5°C, while with the best estimate for SSP5–8.5 (highest emissions) it is close to 4.5°C.

The emissions increase the concentrations of GHGs in the atmosphere, and it is this that is driving the temperature increase shown in Figure 1.1.

Greenhouse gas concentrations

CO$_2$ concentrations in the atmosphere have changed dramatically over the very long term. 400 million years ago atmospheric CO$_2$ concentrations were around 2,000 ppm. The concentration went down and up, but overall declined through various natural processes until around 800,000 years ago. Under RCP8.5 they could reach 2,000 ppm again round about the year 2200 (Foster et al., 2017, Figure 4, p. 5).

Over the last two millennia, up until around 1750, atmospheric concentrations of CO$_2$ were relatively stable at 270–285 ppm. With the industrial revolution they rose precipitately to their current level of over 400 ppm, as shown in Figure 1.11. It can be seen that the current rate of increase of atmospheric CO$_2$ concentrations far exceeds anything that can be seen in the past 800,000 years. Changes that previously unfolded over thousands or millions of years in the historic record are now being driven in decades by human emissions.

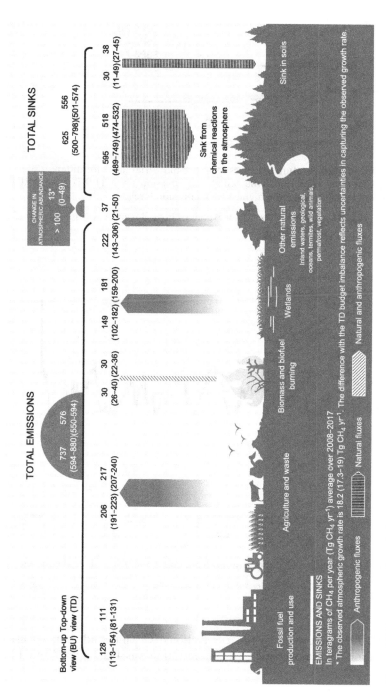

Figure 1.9 Representation of the sources and sinks of human emissions of methane (CH₄) (global annual average for the decade 2008–2017, Tg CH₄/year) (Note: 1 Tg = 1 million tonnes).

Note: The budget imbalance is the difference between the estimated emissions and sinks.

Source: Saunois et al., 2020, Figure 6, p. 1594, data from Global Carbon Project.

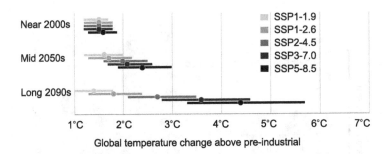

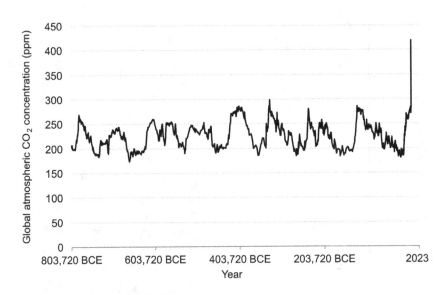

Figure 1.10 Global temperature increases from pre-industrial times for different socio-economic conditions (the SSPs) and emission levels (1.9 is lowest RCP emission level, 8.5 the highest RCP emission level).

Source: IPCC, 2022a, Chapter 16, Cross Sector Working Group Box ECONOMIC, Figure 1 (f), p. 2497.

Figure 1.11 Atmospheric CO_2 concentration from about 800,000 BCE to the present.

Source: Our World in Data, https://ourworldindata.org/atmospheric-concentrations. Accessed April 22, 2023.

In April 2023 according to the US National Oceanic and Atmospheric Administration the atmospheric CO_2 concentration was 422.9 ppm,[18] and it is

18 https://gml.noaa.gov/ccgg/trends/weekly.html. Accessed April 22, 2023.

currently rising by around 2 ppm every year. This is considered to be its highest atmospheric concentration in the past 3 million years.[19] 2022 was the first year in which the average global atmospheric CO_2 concentration was 50% above the 1750–1800 average of 278 ppm.[20]

Atmospheric concentrations of methane have risen from 719 parts per billion (ppb) in 1750 to 1,858 ppb in 2018. Atmospheric concentrations of N_2O have risen from 270 ppb in 1750 to 329 ppb in 2016.[21] Adding the global warming potential of all GHGs to that of CO_2 gives a total atmospheric concentration of GHGs in 2019 of 460 ppm CO_2e[22] (410 ppm of CO_2 alone), with undoubtedly further increases since then.

Economic damages from climate change

Economists have struggled with estimating the economic damages from climate change. The key metrics in which such damages are usually expressed are as a percentage of GDP in a certain year, or as the damage caused at a certain date by the emission of a single tonne of CO_2. This latter metric is called the 'social cost of carbon' (SCC).

In 2007 the Stern Review on the economics of climate change set out its conclusions thus:

> Climate change will affect the basic elements of life for people around the world – access to water, food production, health and the environment. Hundreds of millions of people could suffer hunger, water shortages and coastal flooding as the world warms.
>
> Using the results from formal economic models, the Review estimates that if we don't act, the overall costs and risks of climate change will be equivalent to losing at least 5% of global GDP each year, now and forever. If a wider range of risks and impacts is taken into account, the estimates could rise to 20% of GDP or more.
>
> In contrast, the costs of action – reducing greenhouse gas emission to avoid the worst impacts of climate change – can be limited to around 1% of global GDP each year.
>
> (Stern, 2007, p. xv)

On this analysis the benefit-cost calculus, giving a minimum ratio of 5:1, seemed clear. However, this estimate was well out of line with many other model-based projections from economists, dating back to 1994. A 2009 review

19 Our World in Data, https://ourworldindata.org/atmospheric-concentrations
20 www.carbonbrief.org/met-office-atmospheric-co2-now-hitting-50-higher-than-pre-industrial-levels. Accessed April 22, 2023.
21 https://ourworldindata.org/atmospheric-concentrations. Accessed April 22, 2023.
22 www.eea.europa.eu/ims/atmospheric-greenhouse-gas-concentrations. Accessed April 22, 2023.

paper of the extant estimates of climate impacts found that the damages of doubling the GHG concentration in the atmosphere from the pre-industrial level, taking the concentration to 560 ppm and resulting in global warming of around 3°C are "relatively small – a few percentage points of GDP" (Tol, 2009, p. 33), and that at relatively low levels of warming (e.g. 1°C) the damages are actually negative, i.e. that level of climate change actually delivers benefits, because of the CO_2 fertilisation effect. However, the paper also found that uncertainty around these estimates is high, and that high-income countries have caused most of the warming, but low-income countries will suffer most from it. The SCC estimates consistent with these impacts on GDP are similarly low – half are below US$30/tC (about US$8/tCO$_2$), but again with much uncertainty – 5% of the estimates are above US$360/tC (about US$98/tCO$_2$) (Tol, 2009, Table 2, p.31).

Tol's conclusion in 2009 from the evidence of economic modelling at that date was: "The quantity and intensity of the research effort on the economic effects of climate change seems incommensurate with the perceived size of the climate problem, the expected costs of the solution, and the size of the existing research gaps."

A review of the estimates five years later identified the continuing uncertainties in estimates of climate damages as:

> how the climate responds to carbon dioxide concentrations; positive and negative feedback loops in the climate system; emissions growth rates for various socio-economic scenarios; the completeness and accuracy of damage functions (especially with regard to catastrophic harms, migration and conflict, weather variability and feedbacks on economic growth); the ability of future generations to adapt to climate change; and the economic 'discount rate' used to translate future costs to current dollars.
>
> (Revesz et al., 2014, p. 174)

This latter factor, the discount rate, is particularly important in these calculations, because most climate damages will occur well into the future and a high discount rate reduces the cost estimate of such damages when expressed in present value terms. The Stern Review was widely criticised by some economists for using what they thought to be an inappropriately low discount rate.

However, in contrast to Tol, who did not believe that projected climate damages justified the research effort at that time, Revesz et al. (2014)'s conclusion from the literature is "The leading economic models all point in the same direction: that climate change causes substantial economic harm, justifying immediate action to reduce emissions" (Revesz et al., 2014, p. 174). They also conclude that even these estimates probably underestimate the economic damages from climate change, because the models "omit some major risks associated with climate change, such as social unrest and disruptions to economic growth", may overestimate the resilience of societies and economies to changes in the climate, and "omit damages to labour productivity, to productivity growth, and to the value of the capital stock, including buildings and infrastructure" which reduce annual economic growth rates (Revesz et al., 2014, p. 174).

Three years later the analysis of climate damages by Howard & Sterner (2017, p. 222) found "non-catastrophic damages are likely between 7 and 8% of GDP for a 3 °C increase (from the pre-industrial period)", a significant uptick on the 'few percentage points' of Tol in 2009.

In 2015 Burke et al. (2015) pioneered a new methodology to estimate climate damage costs, by seeking to determine the effect of past temperature changes on economic growth. They found that unmitigated global warming, the RCP8.5 (i.e. broadly the continuation of current levels of emissions) would reduce global income (GDP) by 23% by 2100, very similar to the estimate of the Stern Review 18 years before.

A later paper (Burke et al., 2018) estimated that, depending on different values for the various uncertainties, RCP8.5 would lead to global warming of between 3°C and 5°C by 2100 and reductions in global GDP of 30–35% (Burke et al., 2018, Figure 4, p. 552). They also suggest that moving from RCP8.5 to RCP2.6 would keep the average global temperature increase by 2100 to below 3°C, and there is some chance that average global temperatures would increase little beyond today's level (around 1.1°C relative to a world without warming). However, even at this level of warming, on the RCP2.6 trajectory, damages from global warming are close to 10% of global GDP by 2100, and a 1.5–2°C level of warming increases that to around 15% of global GDP. A loss of global GDP of 30–35% and its associated damages would be experienced as nothing less than catastrophic.

Just as more recent estimates of losses in global GDP from climate change have tended to increase over the last 20 years, so too the same trend is shown with SCC. The median estimate in Rennert et al. (2022) is US$185, with the big changes from earlier estimates due to increased agriculture and mortality costs, as well as a lower discount rate. The press release that launched this study stated:

The **social cost of carbon** is a critical metric that measures the economic damages, in dollars, that result from the emission of one additional ton of carbon dioxide into the atmosphere. A high social cost of carbon can motivate more stringent climate policies, as it increases the estimated benefits of reducing greenhouse gases.

And "**The study**, published today in the journal *Nature*, finds that each additional ton of carbon dioxide emitted into the atmosphere costs society **$185 per ton** – 3.6 times the current US federal estimate of $51 per ton"[23] (all

---

23 www.rff.org/news/press-releases/social-cost-of-carbon-more-than-triple-the-current-federal-estimate-new-study-finds/. Accessed March 28, 2023.

emphases in the original). However, far from charging polluters more than three times the official rate of the US Government for their carbon pollution, governments round the world are still *subsidising* fossil fuel use, i.e. paying polluters to pollute, as will be seen in Chapter 11.

In its most recent assessment of estimates of the GDP cost of climate damage, the IPCC concluded: "The wide range of estimates, and the lack of comparability between methodologies, does not allow for identification of a robust range of estimates with confidence (*high confidence*)" (IPCC, 2022a, Chapter 16, Cross Sector Working Group Box ECONOMIC, p. 2498).

Given such confusion over the economics of climate damages, with some estimates suggesting that they are "relatively small", and others that they are potentially catastrophic, and although I am an economist myself, I consider that policymakers are best guided by the weight of evidence from climate science, as expressed through the IPCC Working Group I reports, that the loss of climate stability is far too great a risk for humans to take, and that, even at this late stage, we must do what is necessary to rein it back by reducing both GHG emissions and the concentration of GHGs in the atmosphere.

What is required for climate stability?

The maximum global warming temperature targets that have been adopted by the global community are 'well below 2°C and as close as possible to 1.5°C'. The rationale for and process by which these targets were adopted are discussed in the next chapter. This chapter concludes by considering the atmospheric concentration of CO_2 and other GHGs at which these targets might be achieved.

There is no straightforward answer to this question, as with so many in climate science. The answer depends on the timescale that is being considered, the sensitivity of the climate response to the higher atmospheric concentrations, about which there is still considerable uncertainty, and the responses of the land and ocean CO_2 sinks to the higher concentrations, among other factors.

The European Environment Agency's analysis of this topic[24] has concluded:

> According to the IPCC's most conservative peak and 2100 concentration levels – those corresponding to a 67% chance of staying below target values – global greenhouse gas concentrations must not exceed 465 (range 445–485) ppm CO_2e and should return to 411 (390–430) ppm by 2100 to limit the increase to 1.5°C; for the 2°C limit, the corresponding values are 505 (470–540) ppm and 480 (460–500) ppm CO_2e, respectively. Given these numbers, at the present decadal growth rate of about 5 ppm per year, the peak concentration for limiting the increase to 1.5°C was exceeded around 2020. In the case of the 2°C limit, the peak concentration will be reached around 2028. Taking into account uncertainty ranges, peak concentrations

24 www.eea.europa.eu/ims/atmospheric-greenhouse-gas-concentrations. Accessed April 22, 2023.

will be reached within 0–5 years (for +1.5°C) or from 2–15 years (for +2°C) (compared to 2019).

This is broadly consistent with the conclusion in the IPCC's 1.5°C report: "In 2100, the median CO_2 concentration in 1.5°C-consistent pathways is below 2016 levels, whereas it remains higher by about 5–10% compared to 2016 in the 2°C-consistent pathways" (IPCC, 2018, Ch.2, p. 101).

In other words, current atmospheric CO_2e concentrations (460 ppm) are *already* above the median estimate (411 ppm) of those which would give the world a 67% chance of keeping the 2100 average global warming to 1.5°C. Considering CO_2 alone, the atmospheric CO_2 concentration consistent with this maximum temperature rise of 1.5°C is about 50 ppm lower, i.e. about 361 ppm. This is remarkably consistent with the statement in an article by the climate scientist James Hansen and colleagues in 2008 (Hansen et al., 2008, Abstract), which stated:

If humanity wishes to preserve a planet similar to that on which civilization developed and to which life on Earth is adapted, paleoclimate evidence and ongoing climate change suggest that CO_2 will need to be reduced from its current 385 ppm to at most 350 ppm, but likely less than that.

Since then, as has been seen, atmospheric CO_2 concentrations have risen another 30 ppm, and are still going up by 2 ppm per year.

The conclusion of Foster & Rohling (2013), in terms of projected sea level rise, is even more radical. They estimate that at 400–450 ppm of CO_2 (i.e. the range of current levels) the long-term sea level rise is likely to be more than 9 metres above its present level. They write: "[O]ur results imply that to avoid significantly elevated sea level in the long term, atmospheric CO_2 should be reduced to levels similar to those of preindustrial times." These levels were, as has been seen, around 280 ppm CO_2.

The long-term implications for sea level rise of not reducing CO_2 concentrations to this kind of level are simply mind boggling, as Dumitru et al. (2019, p. 233) write in their *Nature* article:

Here we show that during the mid-Piacenzian Warm Period [around 3 million years ago], which was on average two to three degrees Celsius warmer than the pre-industrial period [more or less the temperature range we can expect with current policies – PE], the GMSL [Global Mean Sea Level] was about 16.2 metres higher than today owing to global ice-volume changes, and around 17.4 metres when thermal expansion of the oceans is included. ... The mid-Piacenzian Warm Period has been used as an analogue for future anthropogenic warming since atmospheric CO_2 conditions were comparable to present-day values (~400 ppm) and estimated global mean temperatures were elevated by 2–3°C relative to the pre-industrial period.

Such rises in sea level would completely redraw the land maps of Earth, and would drown most of the great cities of human civilisation. And yet humanity continues to pump around 50 Gt CO_2e annually into the atmosphere.

Conclusion

These numbers lead to two main conclusions, if the consequences of the previous paragraph, and all the other worst impacts of climate change described in this chapter, are to be avoided:

1 Humanity must stop emitting GHGs into the atmosphere. This means a shutdown of the fossil fuel industry as quickly as possible.
2 Humanity must put in place more or less immediately a massive programme of removing GHGs from the atmosphere.

Both these issues are addressed in detail in later chapters of this book, the main message underlying which is: 'Every tonne of GHG counts'. Every tonne contributes to climate change and its impacts. Every tonne emitted now will have to be removed later if its impacts are to be avoided. Do we really want to lay this burden and responsibility on our children and grand-children simply because we found the task of kicking our addiction to fossil fuels too difficult?

References

Burke, M., Davis, W. M., & Diffenbaugh, N. S. (2018). Large potential reduction in economic damages under UN mitigation targets. *Nature, 557*(7706), 549–553. https://doi.org/10.1038/s41586-018-0071-9

Burke, M., Hsiang, S. M., & Miguel, E. (2015). Global non-linear effect of temperature on economic production. *Nature, 527*(7577), 235–239. https://doi.org/10.1038/nature15725

Carbon Brief. (2020). Nine tipping points that could be triggered by climate change. www.carbonbrief.org/explainer-nine-tipping-points-that-could-be-triggered-by-climate-change. Accessed March 27, 2023.

Coumou, D., & Rahmstorf, S. (2012). A decade of weather extremes. *Nature Climate Change, 2*(7), 491–496. https://doi.org/10.1038/nclimate1452

Dumitru, O. A., Austermann, J., Polyak, V. J., Fornós, J. J., Asmerom, Y., Ginés, J., Ginés, A., & Onac, B. P. (2019). Constraints on global mean sea level during Pliocene warmth. *Nature, 574*(7777), 233–236. https://doi.org/10.1038/s41586-019-1543-2

Ekins, P., & Zenghelis, D. (2021). The costs and benefits of environmental sustainability. *Sustainability Science, 16*(3), 949–965. https://doi.org/10.1007/s11625-021-00910-5

Food and Agriculture Organization. (2021). *The impact of disasters and crises on agriculture and food security: 2021*. Food and Agriculture Organization, Rome.

Forster, P. M., Smith, C. J., Walsh, T., Lamb, W. F., Lamboll, R., Hauser, M., Ribes, A., Rosen, D., Gillett, N., Palmer, M. D., Rogelj, J., von Schuckmann, K., Seneviratne, S. I., Trewin, B., Zhang, X., Allen, M., Andrew, R., Birt, A., Borger, A., Boyer, T., Broersma, J. A., Cheng, L., Dentener, F., Friedlingstein, P., Gutiérrez, J. M., Gütschow, J., Hall, B., Ishii, M., Jenkins, S., Lan, X., Lee, J.-Y., Morice, C., Kadow,

C., Kennedy, J., Killick, R., Minx, J. C., Naik, V., Peters, G. P., Pirani, A., Pongratz, J., Schleussner, C.-F., Szopa, S., Thorne, P., Rohde, R., Rojas Corradi, M., Schumacher, D., Vose, R., Zickfeld, K., Masson-Delmotte, V., & Zhai, P. (2023) Indicators of Global Climate Change 2022: annual update of large-scale indicators of the state of the climate system and human influence. *Earth System Science Data*, 15, 2295–2327. https://doi.org/10.5194/essd-15-2295-2023

Foster, G. L., & Rohling, E. J. (2013). Relationship between sea level and climate forcing by CO_2 on geological timescales. *Proceedings of the National Academy of Sciences*, *110*(4), 1209–1214. https://doi.org/10.1073/pnas.1216073110

Foster, G. L., Royer, D. L., & Lunt, D. J. (2017). Future climate forcing potentially without precedent in the last 420 million years. *Nature Communications*, *8*(1), 14845. https://doi.org/10.1038/ncomms14845

Hansen, J. (2005). Is There Still Time to Avoid 'Dangerous Anthropogenic Interference' with Global Climate? www.columbia.edu/~jeh1/2005/Keeling_20051206.pdf. Accessed April 23, 2023.

Hansen, J., Sato, M., Pushker, K., Beerling, D., Berner, R., Masson-Delmotte, V., Pagani, M., Raymo, M., Royer, D., & Zachos, J. (2008). Target Atmospheric CO_2: Where Should Humanity Aim? *The Open Atmospheric Science Journal*, 2, 217–231. https://openatmosphericsciencejournal.com/contents/volumes/V2/TOASCJ-2-217/TOASCJ-2-217.pdf

Howard, P. H., & Sterner, T. (2017). Few and not so far between: A meta-analysis of climate damage estimates. *Environmental and Resource Economics*, *68*(1), 197–225. https://doi.org/10.1007/s10640-017-0166-z

IPCC. (2013). *Climate Change 2013: The Physical Science Basis. Contribution of Working Group I to the Fifth Assessment Report of the Intergovernmental Panel on Climate Change* [Stocker, T. F., D. Qin, G.-K. Plattner, M. Tignor, S. K. Allen, J. Boschung, A. Nauels, Y. Xia, V. Bex and P. M. Midgley (eds.)]. Cambridge University Press, Cambridge, United Kingdom and New York, NY, USA.

IPCC. (2018). *Mitigation Pathways Compatible with 1.5°C in the Context of Sustainable Development*. [J. Rogelj, D. Shindell, K. Jiang, S. Fifita, P. Forster, V. Ginzburg, C. Handa, H. Kheshgi, S. Kobayashi, E. Kriegler, L. Mundaca, R. Séférian, and M. V. Vilariño] Chapter 2. In: Global Warming of 1.5°C. An IPCC Special Report on the impacts of global warming of 1.5°C above pre-industrial levels and related global greenhouse gas emission pathways, in the context of strengthening the global response to the threat of climate change, sustainable development, and efforts to eradicate poverty [Masson-Delmotte, V., P. Zhai, H.-O. Pörtner, D. Roberts, J. Skea, P. R. Shukla, A. Pirani, W. Moufouma-Okia, C. Péan, R. Pidcock, S. Connors, J. B. R. Matthews, Y. Chen, X. Zhou, M. I. Gomis, E. Lonnoy, T. Maycock, M. Tignor, and T. Waterfield (eds.)]. Cambridge University Press, Cambridge, UK and New York, NY, USA, pp. 93–174. https://doi.org/10.1017/9781009157940.004

IPCC. (2021a). *Summary for Policymakers*. In: *Climate Change 2021: The Physical Science Basis. Contribution of Working Group I to the Sixth Assessment Report of the Intergovernmental Panel on Climate Change* [Masson-Delmotte, V., P. Zhai, A. Pirani, S. L. Connors, C. Péan, S. Berger, N. Caud, Y. Chen, L. Goldfarb, M. I. Gomis, M. Huang, K. Leitzell, E. Lonnoy, J. B. R. Matthews, T. K. Maycock, T. Waterfield, O. Yelekçi, R. Yu, and B. Zhou (eds.)]. Cambridge University Press, Cambridge, United Kingdom and New York, NY, USA, pp. 3–32, doi:10.1017/9781009157896.001.

IPCC. (2021b). *Technical Summary*. [Arias, P. A., N. Bellouin, E. Coppola, R. G. Jones, G. Krinner, J. Marotzke, V. Naik, M. D. Palmer, G.-K. Plattner, J. Rogelj, M. Rojas,

J. Sillmann, T. Storelvmo, P. W. Thorne, B. Trewin, K . Achuta Rao, B. Adhikary, R. P. Allan, K. Armour, … K. Zickfeld]. In *Climate Change 2021: The Physical Science Basis. Contribution of Working Group I to the Sixth Assessment Report of the Intergovernmental Panel on Climate Change* [Masson-Delmotte, V., P. Zhai, A. Pirani, S. L. Connors, C. Péan, S. Berger, N. Caud, Y. Chen, L. Goldfarb, M. I. Gomis, M. Huang, K. Leitzell, E. Lonnoy, J. B. R. Matthews, T. K. Maycock, T. Waterfield, O. Yelekçi, R. Yu, and B. Zhou (eds.)]. Cambridge University Press, Cambridge, United Kingdom and New York, NY, USA, pp. 33–144. doi: 10.1017/9781009157896.002.

IPCC. (2022a). *Key Risks Across Sectors and Regions.* [O'Neill, B., M. van Aalst, Z. Zaiton Ibrahim, L. Berrang Ford, S. Bhadwal, H. Buhaug, D. Diaz, K. Frieler, M. Garschagen, A. Magnan, G. Midgley, A. Mirzabaev, A. Thomas, and R. Warren. In: *Climate Change 2022: Impacts, Adaptation, and Vulnerability. Contribution of Working Group II to the Sixth Assessment Report of the Intergovernmental Panel on Climate Change* [H.-O. Pörtner, D. C. Roberts, M. Tignor, E. S. Poloczanska, K. Mintenbeck, A. Alegría, M. Craig, S. Langsdorf, S. Löschke, V. Möller, A. Okem, B. Rama (eds.)]. Cambridge University Press, Cambridge, UK and New York, NY, USA, pp. 2411–2538, doi:10.1017/9781009325844.025.

IPCC. (2022b). *Summary for Policymakers.* In: *Climate Change 2022: Mitigation of Climate Change. Contribution of Working Group III to the Sixth Assessment Report of the Intergovernmental Panel on Climate Change* [P. R. Shukla, J. Skea, R. Slade, A. Al Khourdajie, R. van Diemen, D. McCollum, M. Pathak, S. Some, P. Vyas, R. Fradera, M. Belkacemi, A. Hasija, G. Lisboa, S. Luz, J. Malley (eds.)]. Cambridge University Press, Cambridge, UK and New York, NY, USA. doi: 10.1017/9781009157926.001.

Lenton, T. M., Held, H., Kriegler, E., Hall, J. W., Lucht, W., Rahmstorf, S., & Schellnhuber, H. J. (2008). Tipping elements in the Earth's climate system. *Proceedings of the National Academy of Sciences*, *105*(6), 1786–1793. https://doi.org/10.1073/pnas.0705414105

Rennert, K., Errickson, F., Prest, B. C., Rennels, L., Newell, R. G., Pizer, W., Kingdon, C., Wingenroth, J., Cooke, R., Parthum, B., Smith, D., Cromar, K., Diaz, D., Moore, F. C., Müller, U. K., Plevin, R. J., Raftery, A. E., Ševčíková, H., Sheets, H., … Anthoff, D. (2022). Comprehensive evidence implies a higher social cost of CO_2. *Nature*, *610*(7933), 687–692. https://doi.org/10.1038/s41586-022-05224-9

Revesz, R. L., Howard, P. H., Arrow, K., Goulder, L. H., Kopp, R. E., Livermore, M. A., Oppenheimer, M., & Sterner, T. (2014). Global warming: Improve economic models of climate change. *Nature*, *508*(7495), 173–175. https://doi.org/10.1038/508173a

Romanello, M., McGushin, A., Di Napoli, C., Drummond, P., Hughes, N., Jamart, L., Kennard, H., Lampard, P., Solano Rodriguez, B., Arnell, N., Ayeb-Karlsson, S., Belesova, K., Cai, W., Campbell-Lendrum, D., Capstick, S., Chambers, J., Chu, L., Ciampi, L., Dalin, C., … Hamilton, I. (2021). The 2021 report of the Lancet Countdown on health and climate change: code red for a healthy future. *The Lancet*, *398*(10311), 1619–1662. https://doi.org/10.1016/S0140-6736(21)01787-6

Saunois, M., Stavert, A. R., Poulter, B., Bousquet, P., Canadell, J. G., Jackson, R. B., Raymond, P. A., Dlugokencky, E. J., Houweling, S., Patra, P. K., Ciais, P., Arora, V. K., Bastviken, D., Bergamaschi, P., Blake, D. R., Brailsford, G., Bruhwiler, L., Carlson, K. M., Carrol, M., … Zhuang, Q. (2020). The Global Methane Budget 2000–2017. *Earth System Science Data*, *12*(3), 1561–1623. https://doi.org/10.5194/essd-12-1561-2020

Stern, N. (2007). *The Economics of Climate Change.* Cambridge University Press, Cambridge/New York. https://doi.org/10.1017/CBO9780511817434

Tellman, B., Sullivan, J. A., Kuhn, C., Kettner, A. J., Doyle, C. S., Brakenridge, G. R., Erickson, T. A., & Slayback, D. A. (2021). Satellite imaging reveals increased proportion of population exposed to floods. *Nature, 596*(7870), 80–86. https://doi.org/10.1038/s41586-021-03695-w

Tian, H., Xu, R., Canadell, J. G., Thompson, R. L., Winiwarter, W., Suntharalingam, P., Davidson, E. A., Ciais, P., Jackson, R. B., Janssens-Maenhout, G., Prather, M. J., Regnier, P., Pan, N., Pan, S., Peters, G. P., Shi, H., Tubiello, F. N., Zaehle, S., Zhou, F., … Yao, Y. (2020). A comprehensive quantification of global nitrous oxide sources and sinks. *Nature, 586*(7828), 248–256. https://doi.org/10.1038/s41586-020-2780-0

Tol, R. (2009). The economic effects of climate change. *Journal of Economic Perspectives, 23*(2), 29–51.

Wagner, G., & Weitzman, M. L. (2016). *Climate Shock: The Economic Consequences of a Hotter Planet.* Princeton University Press, Princeton, NJ. https://doi.org/10.2307/j.ctv7h0rzq

World Bank. (2021). *Africa's Pulse: An Analysis of Issues Shaping Africa's Economic Future.* World Bank, Washington, DC.

World Meteorological Organization. (2022). *The State of the Global Climate 2021.* WMO, Geneva.

World Meteorological Organization. (2023). *Global Annual to Decadal Climate Impact Update.* WMO, Geneva. https://library.wmo.int/doc_num.php?explnum_id=11629

2 The global context and pathways to net zero

Summary

In order to address the challenge of climate change, countries agreed the United Nations Framework Convention on Climate Change (UNFCCC) in 1992. So far, despite annual Conferences of the Parties (COPs) they have failed to stop global greenhouse gas (GHG) emissions increasing. The commitments made by countries under the UNFCCC are nowhere near enough to keep global warming well below the temperature target of 2°C, let alone reach the more ambitious target limit of 1.5°C, agreed at the Paris COP meeting in 2015. Meeting this target requires global 'net-zero' emissions – whereby remaining GHG emissions are offset by atmospheric removals of them – by 2050.

Different countries have very different levels of emissions, which can be counted in many different ways: total, per person, cumulative since 1750, territorial or those from consumption. This makes assignment of 'responsibility' for climate change and what to do about it – expressed in the UNFCCC's phrase 'common but differentiated responsibility' – both complex and disputed. Further issues of climate justice arise from the fact that the poorest countries that have done least to cause climate change through their emissions seem likely to suffer most from it.

The total amount of GHGs that can now be emitted while keeping within the Paris temperature targets – the so-called 'carbon budgets' – is now very small, and at current emission levels will be exhausted in less than 10 years for the 1.5°C target. While global emissions have not yet peaked, some countries have now experienced reductions in emission even as their economies have continued to grow. However, the rate of emissions reductions even in these countries will need to be increased if the Paris temperature targets are to be met.

Numerous scenarios of emission reduction have been produced in order to show what technologies of energy efficiency, energy supply, energy demand and agriculture will be needed, and the extent to which

DOI: 10.4324/9781003438007-3

they will need to be implemented, to reach 'net-zero' emissions by 2050. Three of these scenarios – two global and one for the UK – are described in some detail, before considering an overview of the kinds of policies that will need to be implemented in the UK context for the UK scenario to be realised.

Introduction

Chapter 1 explored what is causing climate change and what its impacts are likely to be. Its conclusion was that climate change can only be effectively addressed by reducing the emissions of GHGs, especially CO_2 from burning fossil fuels, to zero, and by bringing about the massive removal of past emissions of CO_2 from the atmosphere. The chapter set out why humanity should exert every best effort to stop climate change, and identified the level of atmospheric concentration of CO_2 and other GHGs that would be required for this, which is well below today's level. It showed that continuing GHG emissions are currently increasing these concentrations, when they actually need to be reduced.

This chapter focuses principally on the emissions reduction part of the challenge. It looks first at the global institutional arrangements that have been put in place to reduce emissions, as part of the global effort that is being made, if not to stop global warming, at least to keep it within a 1.5–2°C temperature bound. It also breaks these emissions down to country level and shows which countries are contributing most to climate change, on both a current total and per capita basis, and which have contributed most on the basis of their cumulative past emissions. These are the countries which will need to do most if climate change is to be effectively addressed. This chapter also shows how much more carbon can be emitted if the global temperature targets for the end of the century are to be met, and the extent to which the commitments made by different countries so far fall short of these so-called 'carbon budgets'.

The chapter then goes on to describe various possible pathways, or scenarios, through which GHG emissions can be drastically reduced. This will require the implementation of a wide range of technologies of energy efficiency, energy supply and energy demand, which are the subjects of Chapters 3, 4, 5, 6, 8 and 9. They also require technologies that will capture and store CO_2 before it is emitted into the atmosphere, and measures to remove from the atmosphere GHGs that have already been emitted into it. These are the subject of Chapter 7. The 'greenhouse gas removal' measures will be particularly important to reduce the atmospheric concentrations of GHGs to the levels identified in Chapter 1 as required to *stop* further global warming from human activities. This chapter also contains a preliminary consideration of the governance, policies and changes in the behaviours of people and organisations that will be necessary actually to achieve the pathways that are set out in the first part. This subject is explored in much more detail in Chapter 12.

The UN Framework Convention on Climate Change

Climate change first became an issue of global concern at the Earth Summit – the UN Conference on Environment and Development – in Rio de Janeiro in 1992, two years after the publication of the First Assessment Report of the Intergovernmental Panel on Climate Change (IPCC). This conference agreed the UN Framework Convention on Climate Change (UNFCCC), the website of which states: "The ultimate objective of the Convention is to stabilize greenhouse gas concentrations at a level that would prevent dangerous anthropogenic (human induced) interference with the climate system."[1]

Given the evidence of the impacts of climate change which was briefly surveyed in the previous chapter, some of which we can see for ourselves on our television screens almost daily, there is little doubt that current atmospheric GHG concentrations are already 'dangerous'. To fulfil the objective of the Convention, these concentrations need to be brought down to at most 350 ppm of CO_2 alone[2] (as seen in Chapter 1), or around 400 ppm of all GHGs. A level of 350 ppm of CO_2 alone is still around 80 ppm more than before the industrial revolution, but hopefully 350 ppm would maintain the climatic conditions in which humans and other life forms have prospered for the past 10,000 years or more, assuming that the 'tipping points' discussed in Chapter 1 have not already been passed.

The UNFCCC was ratified by enough nations to come into operation in 1994 and the first annual Conference of the Parties (COP1) was held in 1995 in Berlin, setting in train the negotiations that in 1997 were to lead to the Kyoto Protocol at COP3.

The Kyoto Protocol split the world into two parts with a 'common but differentiated responsibility' to seek to reduce emissions: Annex 1 (broadly industrialised countries), which had supposedly binding emission reduction targets for the period 2008–2013, and the rest of the world, which had no such targets. In the event the US never ratified the Kyoto Protocol; Canada did, but then withdrew when it was clear that it would not meet its target. At COP15 in 2009 in Copenhagen, attempts were made to extend binding targets beyond 2013, but these failed, leading in due course to the Paris Agreement at COP21 in 2015. The Paris Agreement abandoned both binding targets and the Annex 1/Non-Annex 1 distinction, in favour of all countries making voluntary commitments to reduce their emissions. It was understood that countries' commitments would differ "in the light of different national circumstances".[3]

---

1 https://unfccc.int/files/essential_background/background_publications_htmlpdf/application/pdf/conveng.pdf. Accessed April 23, 2023.
2 www.eea.europa.eu/highlights/climate-change-targets-350-ppm-and-the-eu-2-degree-target. Accessed April 23, 2023.
3 https://unfccc.int/process-and-meetings/the-paris-agreement/the-paris-agreement/key-aspects-of-the-paris-agreement. Accessed April 23, 2023.

These commitments are called Nationally Determined Contributions (NDCs), and the Paris Agreement envisaged that countries would initially submit their NDCs for emission reductions for 2030, and update them every five years, and that in aggregate these NDCs would keep the global average temperature increase from climate change "well below 2 degrees Celsius above pre-industrial levels and (to) pursue efforts to limit the temperature increase even further to 1.5 degrees Celsius".[4] Many countries, but not all, did indeed update their NDCs, as expected, prior to the COP26 meeting in Glasgow in November 2021. However, India and China did not submit new NDC commitments for 2030, and some countries (most notably Australia, Brazil, Indonesia and Mexico) submitted NDCs that were *less* ambitious than those they had submitted earlier. The most recent COP (COP27 in Egypt in November 2022) was a further disappointment in this respect as very few countries submitted new updates of their NDCs, as they had promised to do at COP26.

It was understood at the Paris COP15 that in order to stabilise global warming at any specific level, net emissions of GHGs to the atmosphere (i.e. emissions to the atmosphere minus those removed from it by the land or oceans, see Figure 1.5) would need to fall to zero, and that to keep within the Paris temperature target of 1.5°C, this global 'net zero' would need to be achieved by 2050. The UK in 2019 was the first major economy to set a 'net zero by 2050' target, and many other countries have now followed suit, though China's net zero date is 2060, and that of India 2070.

The emissions gap

The United Nations Environment Programme (UNEP) conducts an analysis each year to calculate the 'emissions gap' between what countries have promised through their NDCs (some of which are conditional on the actions or provision of finance by other countries), and what would need to be achieved in order for the temperature targets in the Paris Agreement (2°C or 1.5°C) to be met. The UNEP Emissions Gap Report for 2021 (United Nations Environment Programme, 2021) found that on the basis of the NDCs submitted before COP26, the global average temperature would rise by 2.7°C over pre-industrial levels, but that this could be reduced to 2.2°C if all those countries that had submitted 'net-zero' commitments actually achieved them within the time stated. However, in order to be on track to reach the targets agreed in the Paris Agreement, 2030 emissions would need to be reduced beyond the NDC commitments and pledges made before COP26 by a further 11 $GtCO_2e$ (for the 2°C target) or 25 $GtCO_2e$ (for the 1.5°C target) (United Nations Environment Programme, 2021, Figure ES6, p. XXV).

---

4 https://unfccc.int/process-and-meetings/the-paris-agreement/the-paris-agreement/key-aspects-of-the-paris-agreement. Accessed April 23, 2023.

The global average temperature increase would be reduced further if all the various extra pledges made at COP26 were also achieved.[5] These included the 'phase down' of the use of coal in power generation; an effort to reduce methane (CH_4) emissions by 30% from 2020 levels by 2030; a commitment to stop deforestation by 2030; and a phasing out of 'inefficient fossil fuel subsidies'. Climate Action Tracker estimated that, if all countries' pledges, NDCs and 'net-zero' commitments, and long-term strategies (LTS) were met, the global average temperature increase from pre-industrial times might be kept to $1.8°C$[6] (see Figure 2.1). Given the lack of further emission reduction commitments at COP27, this is broadly still the global picture, with the UNEP 2022 Emissions Gap Report estimating the $2°C$ and $1.5°C$ emission gaps as 12 $GtCO_2e$ and 20 $GtCO_2$ respectively (United Nations Environment Programme, 2022, p. 33).

Country differences in emissions

One of the features of global climate negotiations that makes agreement between countries very difficult is that different countries have very different profiles in respect of their CO_2 and wider GHG emissions, depending on their level and date of industrialisation. These emissions can be measured in three very different ways, according to: total current emissions of CO_2/GHGs, emissions of CO_2/GHGs per head of population, and cumulative emissions of CO_2/GHGs since the start of the industrial era. This last is relevant because CO_2, in particular, can stay in the atmosphere for hundreds of years, so that those countries that industrialised early, and predominantly are now relatively rich, have contributed considerably more to climate change than those which industrialised later.

Figure 2.2 shows that the ranking of seven major emitters comes out rather differently depending on which of the three means of measurement are used, and which GHGs they include. China is easily the top emitter in total terms, with emissions of nearly 15 $GtCO_2e$, followed by the USA at slightly less than 6 $GtCO_2e$. In per capita terms, China drops below the USA and Russian Federation, the former of which has emissions close to 15 tCO_2e per person. However, China now has significantly higher emissions per person than the European Union. The deforestation emissions of Brazil and Indonesia significantly increase their totals, while land use, land use change and forestry (LULUCF) actually absorb CO_2 emissions from the atmosphere in China, USA, India, EU and Russian Federation. Emissions from international transport are around 1 $GtCO_2e$. India is the only country shown with per capita emissions below the world average.

---

5 See www.ucl.ac.uk/bartlett/news/2021/nov/cop26-what-happened-and-where-next for a slightly longer description of what happened at COP26 (accessed April 23, 2023).

6 https://climateactiontracker.org/documents/997/CAT_2021-11-09_Briefing_Global-Update_Glasgow2030CredibilityGap.pdf. Accessed April 23, 2023.

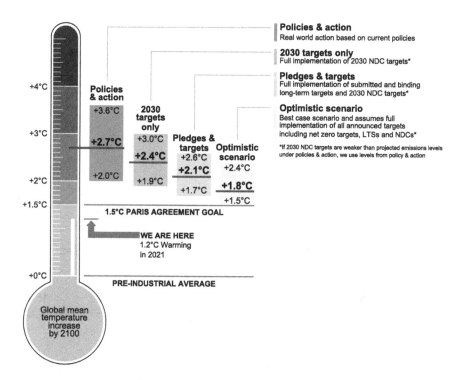

Figure 2.1 Climate Action Tracker's projection of global warming on the basis of different policies and commitments.

Source: Climate Action Tracker 2021, https://climateactiontracker.org/documents/997/CAT_2 021-11-09_Briefing_Global-Update_Glasgow2030CredibilityGap.pdf. Accessed April 23, 2023.

Figure 2.3 shows the per capita territorial emissions from fossil fuel use and cement manufacture of the top 20 country emitters. In per capita terms Saudi Arabia is the top emitter, China has dropped from top place in terms of total emissions (Figure 2.2) to number 11, and India has dropped from 3rd place to 20th.

The order again looks different when cumulative emissions since the industrial revolution are plotted, as shown in Figure 2.4. The USA is now in top place by a large margin, followed by China, again very well ahead of Russia, Germany and UK in third, fourth and fifth places respectively. Figure 2.5 shows some country shares of cumulative emissions since 1750.

There is one other difference in the way emissions can be calculated. *Territorial or production emissions* are those from production or consumption emitted in the country in question, and is the way emissions are calculated under the UNFCCC. *Consumption emissions* are those deriving from consumption in the country in question, including those associated

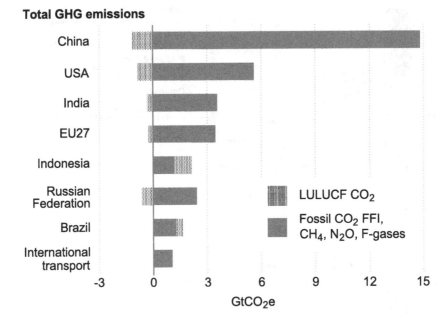

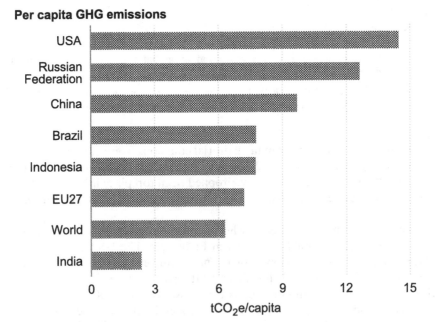

Figure 2.2 GHG emissions of seven major emitters in 2020, in total (top panel) and per capita (bottom panel), distinguishing (in top panel) between fossil fuel and industry (FFI) emissions, and those from land use, land use change and forestry (LULUCF).

Source: United Nations Environment Programme, 2022, Figure ES1, p. XVII.

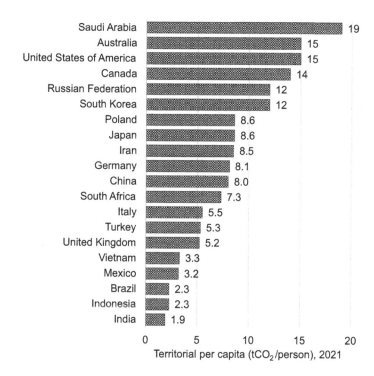

Figure 2.3 The top 20 CO_2-emitting countries in 2021 in terms of emissions per capita (person) from fossil fuel use and cement manufacture.

Source: Data from Global Carbon Project, Global Carbon Atlas, www.globalcarbonatlas.org/en/ CO_2-emissions. Accessed April 23, 2023.

with the manufacture of imports, and excluding those associated with the manufacture of exports. They are different to the extent that the country in question imports more or less energy-intensive goods and services than it exports.

The difference between territorial and consumption emissions for China, the USA, EU27 and India is shown in Figure 2.6. For the USA and EU27 their consumption emissions are above their territorial emissions, indicating that they import more energy-intensive goods and services than they export. The reverse is true for China and India, whose territorial emissions are above their consumption emissions. Both the consumption and territorial emissions of the USA and EU27 have been falling since about 2007, but both sets of emissions from India and China are still growing strongly.

These different ways of calculating emissions, and the different orders in which they rank countries, highlight the fundamentally different circumstances and interests of different countries in terms of who should reduce emissions and by how much. There are the old industrial countries, including North America, Western Europe, Australia and Japan, with relatively high

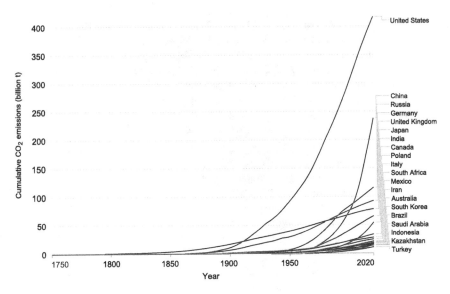

Figure 2.4 Cumulative emissions since 1750, for the 20 countries with the highest emissions in 2020.

Source: Our World in Data, https://ourworldindata.org/co2-and-other-greenhouse-gas-emissions#co2-and-greenhouse-gas-emissions-country-profiles. Accessed April 23, 2023.

cumulative emissions and high emissions per person. There are the so-called 'emerging economies', of which China is easily the most important, which is now the world's largest emitter and has per person emissions higher than some countries in Western Europe. There are the so-called 'petrostates', epitomised by oil-producing Middle East countries, which tend to have high subsidies for fossil fuel use, and high per person emissions, and the economies of which are almost totally dependent on fossil fuels continuing to be the world's major energy source. There are the countries in which most deforestation is occurring, including Brazil, Indonesia, Malaysia and some countries in Africa, from which CO_2 emissions per person from energy use are relatively low, but which have high emissions from tree loss and land use change. And then there are low-income countries, with low total and per person emissions, the main priority of which is industrialisation and development, even if this means a significant increase in their GHG emissions. The annual COP meetings have to find a way to reconcile the interests of these very different countries, in such a way that all countries feel not only that their economic and other interests are being adequately taken into account (COP decisions have to be unanimous), but also that basic norms of justice are being upheld.

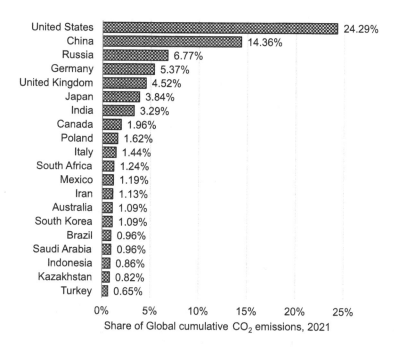

Figure 2.5 Share of cumulative emissions since 1750, for the 20 countries with the highest emissions in 2021.

Source: Data from Our World in Data, https://ourworldindata.org/co2-and-other-greenhouse-gas-emissions#co2-and-greenhouse-gas-emissions-country-profiles. Accessed April 23, 2023.

Climate justice

The differing contributions of different countries to climate change is a major source of contention in the COP negotiations. Countries with high current, per capita and cumulative emissions, which are most responsible for climate change, and which also tend to be richer, are expected to contribute most to help poorer countries, which tend to have much lower emissions in all categories, both to adapt to the climate change that is occurring and to develop their economies in a low-carbon way. And they are also being asked to make further contributions for the 'loss and damage' that climate change is causing, and will cause in the future, to poor countries. A mechanism to address 'loss and damage' was put in place at COP19 in Warsaw, and reaffirmed at COP21 in Paris, but there was little progress in the years following. Only further 'dialogue' on the subject, rather than a dedicated fund, was promised at COP26 in 2021. However, at COP27 something of a breakthrough was achieved, with countries finally agreeing to set up a specific 'loss and damage' fund, with the details to be agreed at a subsequent COP.

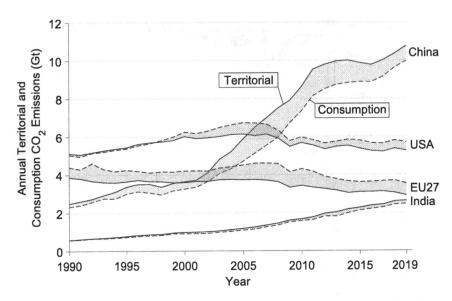

Figure 2.6 Territorial and consumption emissions for China, USA, EU27 and India

Source: Data from Global Carbon Project, www.globalcarbonproject.org/carbonbudget/archive/2021/GCP_CarbonBudget_2021.pdf. Accessed April 23, 2023.

In respect of the costs of adaptation and low-carbon development, countries agreed at COP15 in 2009 that by 2020 'climate finance' of US$100 billion per year would be made available to poorer countries. One of the causes of anger in poor countries at COP26 in 2021 was that this climate finance is not yet in place, and is now only envisaged to be available by 2023.

Climate finance and provision for loss and damage are recognised as the principal means of addressing issues of 'climate justice'. That this is now perceived as a justice issue derives from the fact that, as noted above, some countries have developed their economies through the extensive use of fossil fuels over a long period, have therefore contributed most to climate change and have become relatively rich. In contrast, relatively poor countries, with low per capita and cumulative emissions, and which therefore have contributed least to climate change, are in many cases currently suffering most from its impacts, and may do so more intensely in the future (especially some small island states which seem likely to disappear under rising sea levels). Considerations of climate justice suggest that the relatively rich countries, with high current per capita and cumulative emissions, should give substantial support to poorer countries both to adapt to the climate change that occurs and develop their economies in a low-carbon way, i.e. without the use of fossil fuels. Such support is explicitly envisaged in the UNFCCC and its agreements, but so far the relatively rich countries have come nowhere

Table 2.1 Remaining carbon budgets from 2020 for different temperature limits and likelihoods of keeping within them

Temperature limit (°C)	Estimated remaining carbon budgets from the beginning of 2020 ($GtCO_2$) to give stated likelihood of keeping temperature within limit				
	17%	33%	50%	67%	83%
1.5	900	650	500	400	300
1.7	1450	1050	850	700	550
2.0	2300	1700	1350	1150	900
From beginning of 2023					
1.5	500	300	250	150	100
1.7	1100	800	600	500	350
2.0	2000	1450	1150	950	800

Note: budgets refer to global CO_2 emissions while accounting also for non-CO_2 emissions.

Source: 2020 figures from IPCC, 2021, Table SPM2, p. 29; 2023 figures from Forster et al., 2023, Table 7.

near meeting the demands of poorer countries in this respect. This issue is discussed in more detail in Chapter 12.

Carbon budgets

As noted above, countries have said that they wish to limit global warming to below 2°C above pre-industrial levels, and preferably to 1.5°C, by 2100. These temperature targets imply a strict limit to the quantity of net emissions over the rest of this century. This limit is called the 'global carbon budget' for the rest of the century for whichever temperature limit is under consideration.

The IPCC estimates that cumulative CO_2 emissions of 2,390 $GtCO_2$ from 1850 to 2019 caused average global warming of 1.07°C. Table 2.1 shows the remaining carbon budgets that give a certain probability of keeping average global warming within the 1.5–2°C range in the Paris Agreement. The top right number means that in order to have an 83% chance of keeping global warming within the 1.5°C limit, CO_2 emissions from 2020 should be limited to 300 Gt or less. If emissions from 2020 total 650 Gt, then the chance of staying within this temperature limit falls to 33%. With global CO_2 emissions in 2020 around 36 Gt (Figure 1.8), and still rising, the 83% 1.5°C target is set to be used up by 2030 unless countries engage in unprecedented emissions reduction.

Table 2.1 also shows that every year, with every tonne of CO_2 emitted, the budget for remaining within any given temperature is reduced. The 2023 estimates from Forster et al. (2023) show that, between the 2020 estimate from the IPCC's AR6 report and the Forster et al. estimate, using the same methods as the IPCC, the remaining carbon budget for a 67% chance of remaining within global warming of less than 1.5°C shrank from 450 to 150 $GtCO_2$. At the 2022 CO_2 emission rate of 36.8 GtCO2 from just energy use and industrial processes (International Energy Agency, 2023), humans will have burned through this budget in a little over 4 years.

Decarbonisation scenarios

As already noted, Figure 2.6 shows that, unlike China and India, USA and EU27 are now on a declining emissions path, with regard to both territorial and consumption emissions. Lamb et al. (2022) find that in 24 countries (22 of them European, plus USA and Jamaica), CO_2 and GHG emissions seem to have peaked and are now declining. However, the total reductions in these countries since their peak years, which may have been some time ago, represent only approximately 9% of 2018 emissions, and have been far outweighed by the growth of emissions in other countries.

This section explores in more detail the kinds of emissions reductions that will be required by 2030 and thereafter in order for the temperature limits agreed at the Paris COP21, and shown in Table 2.1, not to become completely out of reach.

Global greenhouse gas emissions by sector in 2016 are shown in Figure 2.7. Energy use – essentially burning fossil fuels, plus leakage from energy production (mainly methane gas) – accounted in 2016 for nearly three quarters of this. Fossil fuels, and low- or zero-carbon energy sources, are the subject of Chapters 4, 5 and 6. The big energy demand sectors, that currently emit most CO_2, are industry (24.2% CO_2, plus a further 8.4% CO_2e from cement and waste), transport (16.2% CO_2) and buildings (17.5% CO_2), and each of these will be examined in Chapter 9, while agriculture (18.4% CO_2e), the greenhouse gas emissions from which are mainly non-CO_2, is the subject of Chapter 10. As noted above, Chapter 7 looks at ways in which carbon emissions can be captured and stored before they get into the atmosphere, or removed from it.

This section is concerned with overall pathways, or scenarios, of decarbonisation, and how, in general terms, these can be brought about.

Projecting how energy systems might develop in the future with drastically reduced GHG emissions requires the use of complex energy system models. It also requires many assumptions to be made about the global context of emissions reduction (e.g. levels of economic and population growth), what technologies will be available and how quickly they can be deployed. For people to understand what level of investment will be required to make these technologies available, and what the broader economic impact of introducing them will be, assumptions also have to be made about how much these investments will cost in the future, over the period that is being projected.

Given the differences in these assumptions, and the technologies that can reduce emissions, the scientific community has engaged intensively in producing scenarios with much reduced GHG emissions, and a number of these will be explored in this chapter, in order to see which of the emission-reducing assumptions and technologies turn out to be particularly

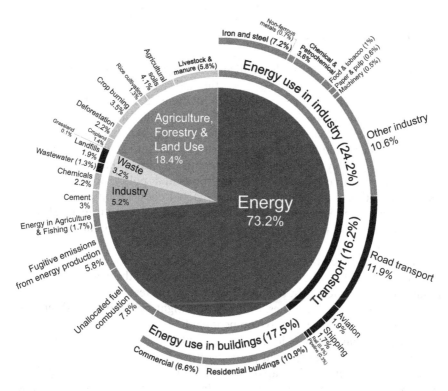

Figure 2.7 Global greenhouse gas emissions by sector, 2016.

Source: Data from Our World in Data, https://ourworldindata.org/emissions-by-sector. Accessed April 23, 2023.

important for emission reduction, though the technologies themselves will be described in much more detail in later chapters, as noted above. The first scenarios to be explored are those reviewed by the IPCC in the report that it produced (IPCC, 2018) to compare the differences between the 1.5°C and 2°C targets, once COP21 in Paris had identified these as the temperature targets which it did not want global warming to exceed. First, however, it is necessary briefly to describe the 'Shared Socioeconomic Pathways' that have been developed to describe the non-energy ways in which the world might develop.

The Shared Socioeconomic Pathways (SSP) scenarios

In order to characterise different states of the world, which might make emissions reduction easier or more difficult, researchers developed the so-called 'Shared Socioeconomic Pathways' (SSPs) (O'Neill et al., 2014; O'Neill et al., 2017; Riahi et al., 2017). Figure 2.8 provides a diagrammatic representation of these different ways in which the world might develop, and some of their different characteristics.

The five SSPs in Figure 2.8 are arranged across two axes. The horizontal axis illustrates the extent of the challenge of *adapting* to any climate change that occurs, while the vertical axis shows the challenge of emissions reduction, or *mitigating* climate change. On the right of Figure 2.8 are the two elements of any scenario: the storyline that describes and justifies in words the various choices that have been made between the different characteristics; and the elements that are quantified when the scenario is put into a numerical model.

It may be readily seen that SSP1 presents the least challenge for both mitigation and adaptation. Strong environmental policy, low population growth, rapid technological development, global cooperation and the other characteristics mentioned for SSP1 in Figure 2.8 mean that it is relatively easy to reduce emissions in this scenario, so less climate change occurs that needs to be adapted to.

At the other extreme, in the top right-hand corner, the opposite characteristics deliver the worst of both worlds: in SSP3 the world is unable to reduce emissions effectively, but also doesn't have the resources to adapt to the climate change that then ensues. In the bottom right, SSP4, the world remains very divided, with economic development eluding poor countries. This means that emissions stay relatively low, and rich countries can reduce their emissions, so climate change is limited, but poor countries find it hard to adapt to even that. In the top left (SSP5) rapid global economic development using fossil fuels means that emissions rise fast, and climate change with it, but the world has the resources to spend on adaptation. SSP2, the 'Middle of the Road' scenario, has characteristics between those of the other SSPs.

People will have their own views on which scenario is more likely or more desirable. Adjudicating that is not the point of a scenario. What is important is that all the scenarios are plausible and that the differences between them illuminate the implications and consequences of the situation that they are investigating, in this case the extent of emissions reductions.

The IPCC 1.5°C scenarios

Figure 2.9 shows some scenarios in the IPCC database that meet the temperature limit of 1.5°C in 2100.

The five solid, dotted and dashed lines show scenarios from different models, based on four different SSPs, plus one extra Low Energy Demand (LED) scenario from Grubler et al. (2018) that is discussed further below. It can be seen that all the scenarios cross the line of zero CO_2 emissions by or fairly soon after 2050, and then for the rest of the century have *negative emissions* (in other words, they suck more emissions out of the atmosphere than are emitted into it). In the SSP5 scenario, the most challenging for mitigation, as seen above, these net negative emissions total around 20 $GtCO_2$ per year by the end of the

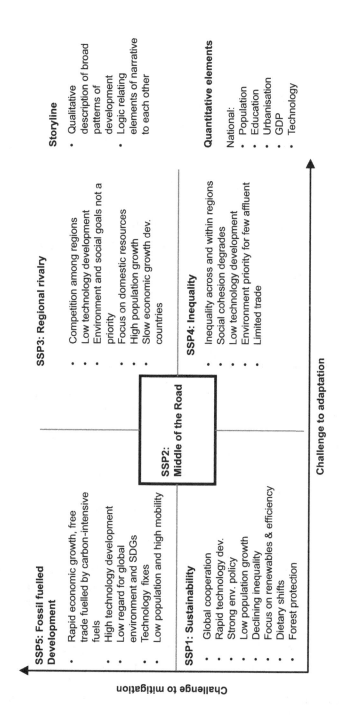

Figure 2.8 Outline of the Shared Socioeconomic Pathways (SSPs).

Source: UNECE, https://unece.org/fileadmin/DAM/energy/se/pdfs/CSE/PATHWAYS/2019/ws_Consult_14_15.May.2019/supp_doc/SSP2_Overview.pdf. Accessed April 23, 2023.

Note: 'SDGs' stand for the UN Sustainable Development Goals.

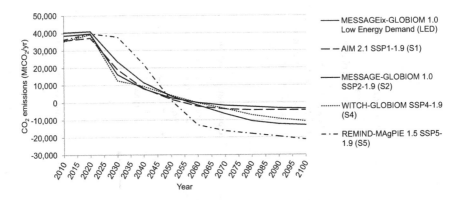

Figure 2.9 1.5°C-compliant scenarios in the IPCC database.

Note: The 1.9 following the SSP numbers refers to RCP1.9, i.e. the lowest of the RCP emission scenarios.

Source: Data from Rogelj et al., 2018, and Huppmann et al., 2019, adapted from Ekins et al., 2022, Figure 6, p. 6.

century. The feasibility or otherwise of this is discussed in Chapter 7. The LED scenario has the least requirement for negative emissions in order to meet the temperature constraint, but both this and the SSP1 scenario achieve this by very great emissions reduction in the 2020s, such that by 2030 emissions are around 50% of what they were in 2020. Such a rate of emissions reduction is absolutely unprecedented and, at the time of writing (2023), shows no sign of being achieved.

It will be noted from Figure 2.9 that there is no SSP3 scenario that meets the 1.5°C temperature limit. Given the assumptions in the SSP3 scenario, and the other assumptions in the models, none of them was able to reduce emissions enough, or generate enough negative emissions, for this temperature limit to be met. The relevant model assumptions in this and the other scenarios are discussed further below.

Figure 2.10 shows that one of the major differences between the scenarios shown is in their final energy demand – the amount of electricity, oil, gas and coal that is used by 'final consumers', namely households, government and businesses (this does not include the energy that goes actually to make the electricity, or to refine crude oil or other energy products). In scenarios S2 and S5 final energy demand rises quite strongly from the level in 2020, because of the assumptions in SSP2 and SSP5. In S1 and LED, in contrast, final energy demand stays flat, or falls, from 2020 levels. This makes it much easier to reduce carbon emissions in the near term, so that S1 and LED require much less use of negative emissions later in the century, as seen in Figure 2.9. The measures through which final energy demand might be reduced in this way are discussed later in this chapter and in Chapter 3.

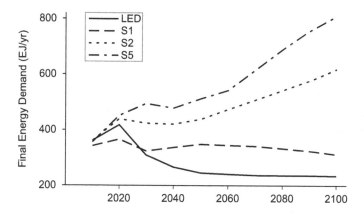

Figure 2.10 Final energy demand in different 1.5°C scenarios.
Source: IPCC, 2018, p. 111.

Whatever the final energy demand in the scenario, reaching 'net-zero' emissions around the middle of this century requires the complete transformation of the energy system, with the implementation of quite different energy technologies to those dominant today. To show the range and scale of the technologies required, three more scenarios (two global and one for the UK) will be explored in a little more detail: the Net Zero Emissions (NZE) scenario published by the International Energy Agency (IEA) in 2021, a scenario devised for Sitra, the Finnish innovation agency, which also provides important results for Chapter 11 on the economics of this transformation, and the Balanced Pathway Scenario for the UK developed by the UK Climate Change Committee. These scenarios provide the context for the more detailed discussion of the technologies that follows in Chapters 3, 4, 5, 6, 7, 8, 9 and 10.

The IEA's Net Zero Emissions (NZE) scenario

The publication of the IEA's NZE scenario in 2021 (International Energy Agency, 2021) was a milestone for the global energy industry and policy community. It was the first time that the IEA, the world's pre-eminent energy institution, had explored the implications of the net-zero policy position in such detail. The focus of NZE is the energy sector, which is currently responsible for around 75% of global GHG emissions, as has been seen. The IEA considers "Alongside corresponding reductions in GHG emissions from outside the energy sector, this [the NZE scenario] is consistent with limiting the global temperature rise to 1.5 °C without a temperature overshoot (with a 50% probability)" (International Energy Agency, 2021, p. 47).

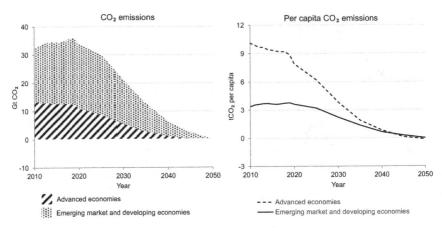

Figure 2.11 CO$_2$ emissions from the energy sector in the IEA's NZE.

Source: International Energy Agency, 2021, Figure 2.2, p. 53.

Figure 2.11 shows how emissions in the NZE, divided between the so-called 'Advanced economies' and 'Emerging market and developing economies' fall to net zero by 2050 (or 2045 in the case of the Advanced economies). This fall in emissions is despite the global economy in 2050 being around three times the size in money terms than it was in 2000.

Figure 2.12 shows that emissions fall rapidly in all sectors, but most rapidly in electricity. The need for negative emissions (brought about by BECCS and DACCS, which will be explained in Chapter 6) by 2050 is relatively small (less than 2 GtCO$_2$ per year).

Global energy consumption in 2050 in the NZE is at about the same level as it was in 2010, despite the population having grown by 3 billion people and the economy being over three times larger. The main reason for this is the strong uptake in energy efficiency measures across all sectors, meaning that the increase in emissions from increased economic activity is substantially outweighed by emission reduction from changes in the behaviours that use energy, efficiency increases in buildings, industry and transport, and radical substitution of fossil fuels by zero-carbon energy sources (as shown in International Energy Agency, 2021, Figure 2.4, p. 56, not reproduced here). The details of all these changes will be discussed further in later chapters.

The IEA estimates that the NZE scenario has cumulative CO$_2$ emissions between 2020 and 2050 of about 500 Gt CO$_2$, a carbon budget that, according to the IPCC, provides a 50% chance of limiting warming to 1.5°C. In addition to reducing CO$_2$ emissions, NZE reduces methane emissions from fossil fuel production and use from 115 million tonnes (Mt) methane in 2020 (3.5 Gt CO$_2$e) to 30 Mt in 2030 and 10 Mt in 2050 (International Energy Agency, 2021, p. 54).

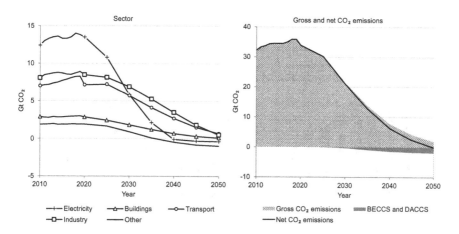

Figure 2.12 Sectoral net CO_2 emissions (left panel), and gross and net CO_2 emissions (right panel) in the IEA's NZE scenario.

Note: BECCS and DACCS stand for 'bioenergy with carbon capture and storage', and 'direct air carbon capture and storage' respectively. These technologies will be explained in Chapter 7.

Source: International Energy Agency, 2021, Figure 2.3, p. 55.

The Sitra net zero and 1.5°C scenario

Figure 2.13 shows the CO_2 and temperature pathways in the Sitra scenario. As with the NZE scenario, CO_2 emissions drop sharply from 2020. However, in common with many of the IPCC 1.5°C scenarios, the average global temperature (the solid line that rises and then falls) overshoots the 1.5°C target, reaching nearly 1.9°C shortly after 2060, before falling back to 1.5°C by 2100 because of the sizeable negative emissions (the bars below the 0-line of the horizontal axis), which start in 2030, and reach 10 GtCO2 per year by 2100. One of the reasons for more negative emissions in this scenario than in the NZE scenario is that CO_2 emissions fall less steeply, so that more CO_2 needs to be sucked out of the atmosphere later in the century. Again the various contributors to the emissions reduction and the negative emissions will be explained in more detail in later chapters.

The Balanced Net Zero Pathway of the UK Climate Change Committee

In 2008 the UK Government passed into law the Climate Change Act. Three of the provisions in the Act were the setting of an emissions reduction target for 2050, the establishment of the Climate Change Committee (CCC) to advise on how the target should be reached, and a duty on the CCC to recommend 5-yearly 'carbon budgets' to the UK Parliament in the period to 2050 that would enable the 2050 target to be reached in a cost-effective way.

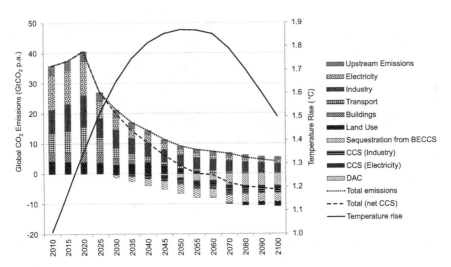

Figure 2.13 CO_2 emissions and temperature pathways to 2100 in the Sitra scenario.

Note: DAC (direct air capture), CCS (carbon capture and storage) and BECCS (bioenergy with CCS) will all be explained in Chapter 7.

Source: Drummond et al., 2021, Figure 15, p. 63, www.sitra.fi/en/publications/growth-positive-zero-emission-pathways-to-2050/

In 2019 the UK Government increased the ambition of the 2050 target to net zero, something which a number of other countries, and the European Union, have now also done.

The CCC is obliged to describe in detail the measures which it thinks could meet the carbon budgets and the 2050 target, and as part of this process it published its Sixth Carbon Budget report in 2020,[7] covering the period 2033–2037. This explored a number of scenarios for getting to net zero by 2050, but the CCC's recommended scenario was the one it called the Balanced Net Zero Pathway (BNZP).

Figure 2.14 shows the sources of UK GHG (i.e. not just CO_2) emissions in 2018. The largest source is transport, mainly from the use of petrol and diesel road vehicles, but there are significant emissions from all the main sources. This pattern of emission sources is quite typical of other industrialised countries, except that the UK's use of coal for electricity generation is relatively small, and therefore other countries might have a larger share of emissions from electricity (the third band up in Figure 2.14).

Figure 2.15 shows how the emissions are reduced to net zero from their 2018 level. The initial thick black line is historical emissions to 2020, and the

---

7 www.theccc.org.uk/publication/sixth-carbon-budget/. Accessed April 23, 2023.

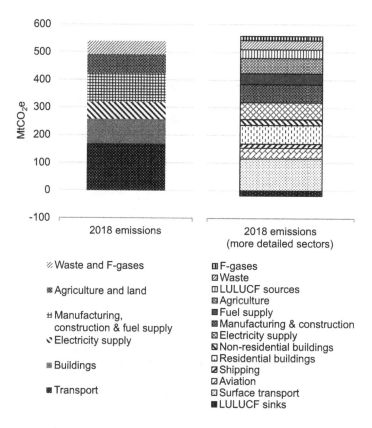

Figure 2.14 Sources of UK GHG emissions in 2018.

Note: LULUCF stands for land use, land use change and forestry, and is distinct from agriculture in the right-hand chart, but included with it in the left-hand chart. What is involved in LULUCF will be explained in Chapter 10.

Source: Climate Change Committee, 2020, Figure 2.1, p. 61.

line continues along the top of the graph to show what the projected emission pathway would be if no further emission reduction policies are implemented. It shows that, without further measures, UK GHG emissions would rise by about 20% by 2050 from their 2018 level.

Each of the differently hatched wedges of carbon emission reductions comes from a different technological option, that will be described in more detail in later chapters. The exception is 'reduced demand', which involves people changing their habits and behaviours towards low-carbon consumption options, which will be discussed later in this chapter. Notable in the figure is the very important role played by electricity in emissions reduction (the largest of the hatched areas), as more and more electricity is generated by renewable and other zero-carbon sources, and electricity substitutes for fossil fuels in much

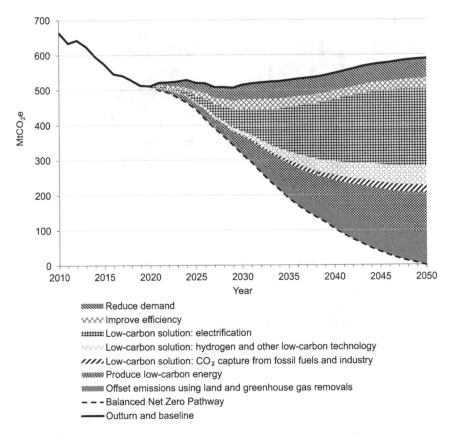

Figure 2.15 Sources of GHG emissions reduction in the Balanced Net Zero Pathway.

Source: Climate Change Committee, 2020, Figure 2.6, p. 69.

transport and home heating. This is a common feature of scenarios of deep decarbonisation.

Finally, it should be stressed that, as its name implies, the BNZP (and most other such scenarios) is a 'net zero' pathway. In 2050 there are still substantial GHG emissions which, to reach net zero, have to be 'offset' by sucking emissions out of the atmosphere into the land or using machines, as described in Chapter 7. If the whole world reached net zero at this time and in this way, then the atmospheric concentrations of GHGs would stabilise at their then level, as would temperatures, probably in 2100 at around 1.5°C above their pre-industrial level. But the effects of climate change, as described in Chapter 1, would continue, at a considerably greater intensity than currently (when the average global temperature increase is still only 1.1°C). To *stop* climate change requires '*real* zero' – a complete cessation of the use of fossil fuels and emissions of other GHGs, *plus* the use of land and other means of sucking CO_2 out of the atmosphere in order to reduce atmospheric CO_2 concentrations to the 350 ppm recommended by James Hansen back in 2005. Net zero is a necessary start, but real zero is what stopping climate change demands.[8]

Getting onto the net zero pathway

Climate change from human activities is the result of human aspirations for a better life. However, getting onto a net zero pathway in any country will require a great many people and organisations to do many things differently. At present politicians in relatively rich countries (e.g. North America, Europe, Japan, Australia and New Zealand) interpret human aspirations for a better life largely in terms of a desire for increased incomes, and more of the goods and services that they can buy. In 'less developed' countries, the aspiration is for 'development', of which an essential component is increased incomes for their populations.

However, there is increased awareness of the damage that current patterns of production and consumption are doing to the natural world, and hence to human societies and quality of life, including through climate change. There is a growing desire to reduce human impacts on the natural world, but little desire as yet to do this by doing less of the activities that cause these impacts. Rather the approach has been to use technology to try to 'decouple' the activities themselves from the environmental damage they cause – to keep doing the things people want to do, but enable them to do it with a small fraction of the environmental impact.

This approach has not been wholly unsuccessful. As was seen earlier in this chapter (Figure 2.6), a number of relatively rich countries have managed to reduce their CO_2 emissions while their economies have kept growing, i.e. as their money-based activities have increased. The trouble in the current context is that none of these countries have managed to reduce their emissions at anything like the rate required to reach net zero (let alone real zero) by 2050. And the emissions in pretty well all other countries, including those with very large populations, like China, India, Indonesia and Nigeria, are still increasing rapidly.

There are two possible approaches to the challenge of getting to the dramatic emissions reductions this decade that are required to hit the 1.5°C target in 2100. One approach is to double down on technological innovation. This approach is based on a presumption that neither relatively rich nor less rich countries will be prepared to forgo their aspirations for activities that require relatively high energy use, but that this energy use can be greatly reduced by determined implementation of energy efficiency, and the energy that is still

8 As explained in the Preface, in this book 'stopping climate change' means stopping further human contributions to climate change. The climate change already engendered by human emissions is likely to continue for millennia with unpredictable results.

required can be provided by sources that emit no carbon. Energy efficiency plays a prominent role in decarbonisation especially in the LED scenario, which has very low energy use through to 2050 (see Figure 2.10), but all the scenarios include it to some extent. Reducing the demand for energy in this way makes decarbonisation easier – energy that is not used emits no carbon, and avoiding energy use also means avoiding the cost of building the zero-carbon technologies that would otherwise be necessary to supply it. Chapter 3 discusses the whole issue of energy efficiency in detail.

The other approach is to say that 'the good life' in rich countries does not need these countries to get richer, and that in fact everyone in these countries could still live a good life if they were to get quite a bit poorer (O'Neill et al., 2018). In order to achieve this there would need to be redistribution from rich to poor, so that social inequalities became considerably less in these countries than they currently are. This is broadly the approach to decarbonisation advocated by the 'degrowth' literature, where degrowth means a reduction in economic activity, especially high-energy activities, in the rich world (Kallis et al., 2018).

It can be seen from the scenarios above that they do incorporate some assumed changes in behaviour away from greenhouse gas emitting activities. This is what causes the 'reduce demand' wedge of decarbonisation in Figure 2.15. The IEA has a similar cause of reduction through 'Behaviour and avoided demand' in the NZE scenario (see International Energy Agency, 2021, Figure 2.4, p. 56). But both these scenarios assume growth in economic activities overall, measured by gross domestic product (GDP) in the UK in the BNZP and globally in the NZE. And it is clear from these scenarios that the great majority of emissions reduction comes from the implementation of low-carbon technologies. These are not degrowth scenarios based on dramatic changes in behaviour, although the behaviour changes that they do comprise are discussed further below.

The issue of what constitutes 'a good life' is picked up again later in this book, and the relative economics of the technological and degrowth options is explored in Chapter 11. For the moment it need only be noted that no country in the world has a strong political movement in favour of degrowth, with the result that all governments are still overwhelmingly taking the technological approach to emission reduction, and it is mainly governments that need to introduce the required policies for decarbonisation. So if countries are to get anywhere near the 50% reduction in emissions by 2030 that 2050 net-zero targets require, it is on technologies that they will need to rely.

Scenario exercises such as those in the BNZP for the UK, in Figure 2.15, are the result of many detailed numerical assumptions about how energy supply and demand, and the economic activities and behaviours that drive them, will change through to 2050. These numerical assumptions are shown in Table 2.2.

Of course, different countries will reduce their emissions in different ways according to their contexts and cultures, and the numerical assumptions in Table 2.2 are wholly appropriate only to the context of the UK, for which they were generated, but all industrial countries will need to make broadly similar decisions across their economies.

Some of the assumptions stand out in terms of their ambitions. Net GHG emissions, of course, fall to zero by 2050. Demand for meat and dairy products falls significantly, by 34% and 20% respectively. The land freed up by this reduction in livestock allows significant growth in the land area used for woodland (13% to 18%), bioenergy (10,000 to 720,000 ha), peatland restoration (25% to 79%) and carbon sinks (18 to 39 $MtCO_2$), all over 2019–2050. This means that by 2050 UK land is removing large amounts of carbon from the atmosphere (39 $MtCO_2$), as well as growing food.

Car-km per driver fall by a modest 9%, but from 2035 100% of new sales are battery electric vehicles (BEVs), and the fuel for new lorries and trucks (HGVs) is also carbon-free by 2050. By this time, too, the carbon intensity of the UK power system has fallen almost to zero, driven by a huge growth in offshore wind (from 10 to 95 GWe), which supplies near-zero carbon electricity to the millions of BEVs on the roads and to the nation's heat pumps, which have largely replaced gas boilers and more than 1 million of which are installed each year.

Each year 225 TWh of low-carbon hydrogen are produced and mainly used to substitute for fossil fuels in industrial processes that cannot be electrified. The economy has become much less wasteful, producing nearly 40% less waste per person by 2050, with computers and other electronic goods lasting more than twice as long. Carbon capture and storage (CCS) of emissions from burning bioenergy or fossil methane gas, is stopping over 100 $MtCO_2$ per year from going into the atmosphere, and new technologies that suck CO_2 from the air are sequestering another 58 $MtCO_2$ per year. By 2050 most remaining fuel use of oil products, together with some synthetic fuels, is in aviation, which has grown 17% over 2019 levels in terms of plane-km flown – but even this represents a significant reduction from the 65% growth that takes place in the baseline, which is broadly a scenario based on assumptions of the continuation of current policies.

Later chapters in this book go into considerable detail about the challenges and opportunities entailed by this enormous expansion of low-carbon technologies, but without exception, for the changes in Table 2.2 to come about, governments at all levels will need to introduce stringent policies. Many of these policies are explored in the chapters on different economic sectors which follow. Chapter 12 brings these policies together with others to provide a comprehensive palette of measures of climate action which could stop global warming, and the climate change it causes, from getting worse.

Conclusion

The use of energy is fundamental to nearly all economic activities, whether of production or consumption. The great majority of this energy is still provided by the combustion of fossil fuels, with its associated emissions of CO_2. Modern

Table 2.2 Numerical assumptions for the CCC's Balanced Net Zero Pathway

		2019	*2025*	*2030*	*2035*	*2050*
UK greenhouse gas emissions	UK greenhouse gas emissions (MtCO$_2$e)	522	445	316	191	0
	UK greenhouse gas emissions per person (tCO$_2$e/capita)	7.8	6.5	4.5	2.7	0
Demand reduction	Weekly meat consumption (g) (includes fresh and processed meat)	960	880	770	730	630
	Weekly dairy consumption (g)	2,020	1,840	1,620	1,620	1,620
	Plane-km per person	11,700	11,000	11,000	11,400	13,700
	Car-km per driver	12,900	12,600	12,400	12,200	11,700
	Remaining waste per person, after prevention & recycling (kg)	490	400	310	280	300
Efficiency	Carbon intensity of a new HGV (gCO$_2$/km)	680	580	420	20	0
	Increase in longevity of electronics	0%	30%	80%	120%	120%
Electrification, hydrogen and carbon capture and storage	Carbon intensity of UK electricity (gCO$_2$e/kWh e)	220	125	45	10	2
	Offshore wind (GWe)	10	25	40	50	95
	Share of BEVs in new car sales	2%	48%	97%	100%	100%
	Heat pump installations (thousand per year)	26	415	1,070	1,430	1,480
	Manufacturing energy use from electricity or hydrogen	27%	27%	37%	52%	76%
	Low-carbon hydrogen (TWh)	<1	1	30	105	225
	CCS in manufacturing (MtCO$_2$)	0	0.2	2	5	8

Table 2.2 (Continued)

		2019	2025	2030	2035	2050
	CCS in rest of the economy ($MtCO_2$)	0	0.1	20	48	96
Land	UK woodland area	13%	14%	14%	15%	16%
	Energy crops (kha)	10	23	115	266	720
	Peat area restored	25%	36%	47%	58%	79%
	Land-based carbon sinks ($MtCO_2$)	18	18	20	23	39
Removals	Greenhouse gas removals ($MtCO_2$)	0	<1	5	23	58

Source: Climate Change Committee, 2020, Table 2.2, p. 68.

agriculture, too, has developed in a way that produces large quantities of GHG emissions. To meet the challenge of climate change, all modern energy-using systems – industry, transport and the built environment, as well as the energy system itself, along with agriculture – will have to be fundamentally transformed, by reducing GHG emissions from human activities to zero, and below, by mid-century. So far, this transformation has hardly begun, and even in those countries that have started to reduce their emissions, the emissions reduction is too slow for the Paris temperature targets to be met. The rest of this book spells out in some detail what needs to happen for real-zero GHG emissions to be met by mid-century, with atmospheric emissions removals thereafter; what the economic implications of this transformation are likely to be; and the policies that will be necessary to bring it about.

References

Climate Change Committee. (2020). *The Sixth Carbon Budget: The UK's path to net zero*. CCC, London.

Drummond, P., Scamman, D., Ekins, P., Paroussos, L., & Keppo, I. (2021). *Growth-positive zero-emission pathways to 2050*. Sitra, Helsinki. www.sitra.fi/en/publications/growth-positive-zero-emission-pathways-to-2050/

Ekins, P., Drummond, P., Scamman, D., Paroussos, L., & Keppo, I. (2022). The 1.5°C climate and energy scenarios: impacts on economic growth. *Oxford Open Energy, 1*. https://doi.org/10.1093/OOENERGY/OIAC005

Forster, P. M., Smith, C. J., Walsh, T., Lamb, W. F., Lamboll, R., Hauser, M., Ribes, A., Rosen, D., Gillett, N., Palmer, M. D., Rogelj, J., von Schuckmann, K., Seneviratne, S. I., Trewin, B., Zhang, X., Allen, M., Andrew, R., Birt, A., Borger, A., … & Zhai, P. (2023). Indicators of Global Climate Change 2022: annual update of large-scale indicators of the state of the climate system and human influence. *Earth System Science Data, 15*, 2295–2327. https://doi.org/10.5194/essd-15-2295-2023

Grubler, A., Wilson, C., Bento, N., Boza-Kiss, B., Krey, V., Mccollum, D. L., Rao, N. D., Riahi, K., Rogelj, J., De Stercke, S., Cullen, J., Frank, S., Fricko, O., Guo, F., Gidden, M., Havlík, P., Huppmann, D., Kiesewetter, G., Rafaj, P., … Valin, H. (2018). A low energy demand scenario for meeting the 1.5 °C target and sustainable development goals without negative emission technologies. *Nature Energy*. https://doi.org/10.1038/s41560-018-0172-6

Huppmann, D., Kriegler, E., Krey, V., Riahi, K., Rogelj, J., Calvin, K., Humpenoeder, F., Popp, A., Rose, S. K., Weyant, J., Bauer, N., Bertram, C., Bosetti, V., Doelman, J., Drouet, L., Emmerling, J., Frank, S., Fujimori, S., Gernaat, D. … Zhang, R. (2019). IAMC 1.5°C scenario explorer and data hosted by IIASA. https://doi.org/10.5281/ZENODO.3363345

International Energy Agency. (2021). *Net Zero by 2050: A Roadmap for the Global Energy Sector*. IEA, Paris.

International Energy Agency. (2023). CO_2 *emissions in 2022*. March, IEA, Paris. www.iea.org/reports/co2-emissions-in-2022

IPCC. (2018). *Mitigation Pathways Compatible with 1.5°C in the Context of Sustainable Development*. [Rogelj, J., D. Shindell, K. Jiang, S. Fifita, P. Forster, V. Ginzburg, C. Handa, H. Kheshgi, S. Kobayashi, E. Kriegler, L. Mundaca, R. Séférian, & M. V. Vilariño] Chapter 2. In: Global Warming of 1.5°C. An IPCC Special Report on the impacts of global warming of 1.5°C above pre-industrial levels and related global greenhouse gas emission pathways, in the context of strengthening the global response to the threat of climate change, sustainable development, and efforts to eradicate poverty [Masson-Delmotte, V., P. Zhai, H.-O. Pörtner, D. Roberts, J. Skea, P. R. Shukla, A. Pirani, W. Moufouma-Okia, C. Péan, R. Pidcock, S. Connors, J. B. R. Matthews, Y. Chen, X. Zhou, M. I. Gomis, E. Lonnoy, T. Maycock, M. Tignor, & T. Waterfield (eds.)]. Cambridge University Press, Cambridge, UK and New York, NY, USA, pp. 93–174. https://doi.org/10.1017/9781009157940.004

IPCC. (2021). *Summary for Policymakers*. In: *Climate Change 2021: The Physical Science Basis. Contribution of Working Group I to the Sixth Assessment Report of the Intergovernmental Panel on Climate Change* [Masson-Delmotte, V., P. Zhai, A. Pirani, S. L. Connors, C. Péan, S. Berger, N. Caud, Y. Chen, L. Goldfarb, M. I. Gomis, M. Huang, K. Leitzell, E. Lonnoy, J. B. R. Matthews, T. K. Maycock, T. Waterfield, O. Yelekçi, R. Yu, & B. Zhou (eds.)]. Cambridge University Press, Cambridge, United Kingdom and New York, NY, USA, pp. 3–32, doi:10.1017/9781009157896.001

Kallis, G., Kostakis, V., Lange, S., Muraca, B., Paulson, S., & Schmelzer, M. (2018). Research On Degrowth. *Annual Review of Environment and Resources, 43*(1), 291–316. https://doi.org/10.1146/annurev-environ-102017-025941

Lamb, W. F., Grubb, M., Diluiso, F., & Minx, J. C. (2022). Countries with sustained greenhouse gas emissions reductions: an analysis of trends and progress by sector. *Climate Policy, 22*(1), 1–17. https://doi.org/10.1080/14693062.2021.1990831

O'Neill, B. C., Kriegler, E., Riahi, K., Ebi, K. L., Hallegatte, S., Carter, T. R., Mathur, R., & van Vuuren, D. P. (2014). A new scenario framework for climate change research: the concept of shared socioeconomic pathways. *Climatic Change, 122*(3), 387–400. https://doi.org/10.1007/s10584-013-0905-2

O'Neill, B. C., Kriegler, E., Ebi, K. L., Kemp-Benedict, E., Riahi, K., Rothman, D. S., van Ruijven, B. J., van Vuuren, D. P., Birkmann, J., Kok, K., Levy, M., & Solecki, W. (2017). The roads ahead: Narratives for shared socioeconomic pathways describing world futures in the 21st century. *Global Environmental Change, 42*, 169–180. https://doi.org/10.1016/j.gloenvcha.2015.01.004

O'Neill, D. W., Fanning, A. L., Lamb, W. F., & Steinberger, J. K. (2018). A good life for all within planetary boundaries. *Nature Sustainability, 1*(2), 88–95. https://doi.org/10.1038/s41893-018-0021-4

Riahi, K., van Vuuren, D. P., Kriegler, E., Edmonds, J., O'Neill, B. C., Fujimori, S., Bauer, N., Calvin, K., Dellink, R., Fricko, O., Lutz, W., Popp, A., Cuaresma, J. C., KC, S., Leimbach, M., Jiang, L., Kram, T., Rao, S., Emmerling, J., … Tavoni, M. (2017). The Shared Socioeconomic Pathways and their energy, land use, and greenhouse gas emissions implications: An overview. *Global Environmental Change, 42*, 153–168. https://doi.org/10.1016/j.gloenvcha.2016.05.009

Rogelj, J., Popp, A., Kalvin, K., Luderer, G., Emmerling, J., Gernaat, D., Fujimori, S., Strefler, J., Hasegawa, T., Marangoni, G., Krey, V., Kriegler, E., Riahi, K., van Vuuren, D. P., Doelman, J., Drouet, L., Edmonds, J., Fricko, O., Harmsen, M. … Tavoni, M. (2018) Scenarios towards limiting global mean temperature increase below 1.5oC. *Nature Climate Change.* **8**: 325–332. https://doi.org/10.1038/s41558-018-0091-3

United Nations Environment Programme. (2021). *Emissions Gap Report 2021: The Heat is On – a world of climate promises not yet delivered.* UNEP, Nairobi.

United Nations Environment Programme. (2022). *Emissions Gap Report 2022: The Closing Window – Climate crisis call for transformation of societies.* UNEP, Nairobi.

3 Energy efficiency, the 'first fuel'

Summary

Energy efficiency is sometimes described as the 'first fuel', because increasing energy efficiency means that the same energy service (kilometres driven, level of lighting, clothes washed, hours of entertainment watched) can be delivered with less energy use. This chapter looks in detail at the energy savings and CO_2 emission reductions that have been achieved through energy efficiency policies, and the further savings and reductions that could be achieved by such measures in the future. Largely drawing on UK and wider European experience, the chapter sets out the barriers to increasing energy efficiency, and how they can be overcome, in respect of industry, buildings and transport (vehicles).

Introduction

The energy used in our homes, industry, vehicles and appliances has been through several stages of transformation. Natural resources (sunlight, wind, uranium, crude oil and other fossil fuels – so-called 'primary energy') are converted into useful forms: petrol, diesel, electricity, piped methane gas (so-called 'final energy'). This is then transported to where it is needed to deliver an 'energy service': journeys travelled, clothes washed, homes heated and entertainment watched, kettles boiled, streets and rooms lit.

At each stage energy is wasted, so that the useful energy that actually delivers the desired services is a fraction, often a small fraction, of the primary energy input. Money is wasted along with the energy. Low energy efficiency requires larger investments in energy supply to be made than are strictly necessary, making energy services more expensive. Where the primary energy is a fossil fuel, more CO_2 is emitted than the delivery of the energy services requires.

Increasing energy efficiency along the chain of energy production, transport and use reduces the fraction of wasted energy, the negative consequences that go along with it, and the amount of investment required in energy supply. It also saves money from the use of energy by consumers. With increased energy

DOI: 10.4324/9781003438007-4

efficiency, more 'energy services' (kilometres driven, level or hours of lighting experienced, weight of clothes washed, hours of entertainment watched) can be delivered with less energy use. For this reason, energy efficiency is sometimes called the 'first fuel'.

Across the economy energy efficiency is measured by the primary energy intensity of economic output (total primary energy/gross domestic product (GDP)). Over 2011–2016 energy efficiency increased by about 2% per year, which was only about half of the reduction required in the Net Zero Emissions scenario of the International Energy Agency (IEA) discussed in Chapter 2. In the years since 2016 the increase in energy efficiency has been lower still, falling below 2% each year (International Energy Agency, 2021, p. 8). Substantially greater investment in energy efficiency will be required in the 2020s if the Paris temperature targets discussed in Chapter 2 are to be met.

Barriers to increases in energy efficiency

Because energy is priced (has to be bought), it might be thought that people would save energy by increasing the energy efficiency with which they use it, up to the economically efficient level (i.e. where the cost of saving energy equals the cost of the energy saved). Of course, this happens to some extent and such financial incentives do result in energy efficiency increases. But there are also many reasons why such motivations alone do not lead to as much of an increase in energy efficiency as would be financially desirable from an individual's and organisation's point of view, or as socially desirable.

For example, in many sectors, energy use represents a relatively small proportion of costs, so there is little incentive to use it efficiently. Many buildings were constructed before the 1970s, when energy was relatively cheap and little thought was given to energy efficiency. Buildings and much machinery are long-lasting, and retrofitting them for greater energy efficiency can be expensive up front, even if it does prove cost-effective in the long term. People and businesses may not have the ready money to make the investments, or be able to borrow it easily. Energy prices are volatile, so that energy efficiency investments that make financial sense at times of high prices, such as in 2022, may not do so at lower prices, and people considering such investments may fear price falls in the future. All these factors mean that the economic incentives to increase the efficiency of energy use in these sectors are weak.

These weak incentives are accompanied by market failures that further inhibit energy efficiency measures. Fossil energy, in particular, is cheaper than it should be because its prices in the market all too often do not account for the pollution it causes. In fact, many countries actually subsidise fossil fuels, making them even cheaper than they would be in unregulated markets. In some markets energy efficiency investments are further discouraged because the investor finds it difficult to benefit from energy savings from greater energy efficiency (e.g. in buildings where the landlord has to make the investment, but the tenant pays the energy bills).

There can be further problems related to energy efficiency technologies. The measures required to increase energy efficiency are different across different sectors and regions, which prevents a widespread understanding of what needs to be done. Many people and businesses do not know what technologies are available to increase energy efficiency, and engineers and installers of energy technologies, especially in buildings, are often not trained in how to install them effectively. Perhaps because of low general awareness of energy efficiency, energy efficiency tends to have a low priority in people's purchasing decisions, whether of buildings, vehicles or appliances, and therefore energy efficiency labels, where they are available, have a limited influence.

As a result of these and other factors, the delivery of energy efficiency has not been able to reach its potential for reduced energy use and emissions for a range of reasons: older buildings, particularly traditional and heritage buildings, perhaps in conservation areas, may require more complex and expensive refurbishments to increase their energy efficiency than their contemporary counterparts; in vehicles substantial gains in the energy efficiency of engines has not been converted into increased miles or kilometres per gallon because of the increase in the size and weight of the vehicles purchased; in energy-intensive industry pressures from global competition have caused resistance to efforts to tax fossil energy for the pollution it causes; and in other industry energy efficiency remains a low priority for both analysis and investment.

Policy-driven increases in energy efficiency

For these reasons, many governments have introduced energy efficiency targets, standards and policies to increase energy efficiency, and have reaped rich rewards in terms of energy savings. The IEA has estimated the energy savings from the introduction of energy efficiency standards for appliances across nine jurisdictions, including the USA, the EU and China. In the USA and EU, with the longest-running appliance programmes, around 15% of total electricity generation has been saved. A similar 15% improvement across all the countries would have cut electricity consumption by 3,500 TWh – roughly half of China's current electricity consumption (International Energy Agency, 2021, p. 17).

Energy efficiency was one of the three pillars of the European Union (EU)'s 20/20/20 by 2020 legislation, requiring a 20% CO_2 reduction, 20% of final energy demand to come from renewables, and energy efficiency to be increased by 20% (energy demand to be reduced by 20% below its projected level in 2020), all by 2020.[1] The targets were increased for 2030: 32.5% increase in energy efficiency, 32% renewables in final energy demand[2] and a 55% CO_2 emissions

[1] https://ec.europa.eu/clima/eu-action/climate-strategies-targets/2020-climate-energy-package_en#ecl-inpage-904. Accessed April 23, 2023.
[2] For energy security reasons, following the Russian invasion of Ukraine, the European Commission proposed to increase this to 45%, https://energy.ec.europa.eu/topics/renewable-energy/renewable-energy-directive-targets-and-rules/renewable-energy-targets_en. Accessed April 23, 2023.

reduction from the 1990 level.[3] According to the IEA, the new energy efficiency targets amount to a 36% and 39% reduction in final energy and primary energy demand from the level in 2007 respectively, and new savings of 1.5% of final energy consumption a year between 2024 and 2030, twice the 2020 rate of 0.8% (International Energy Agency, 2021, p. 31).

Energy efficiency policy in the EU has been implemented through three directives: the Ecodesign Directive, which covers the energy efficiency of products; the Energy Performance of Buildings Directive, which covers aspects of energy use in buildings; and the over-arching Energy Efficiency Directive, which requires more efficient energy use from production to consumption.

As usual with EU directives, Member States of the EU implement them in different ways to suit their national circumstances, but, of course, countries learn from each other, and a successful policy in one country may be implemented in others. One particularly successful policy, pioneered in the UK, but which then spread to many other countries, is the Energy Efficiency Obligation (EEO). EEOs require energy suppliers or other energy companies to implement energy efficiency measures that will help their customers save energy or reduce carbon emissions, normally according to some target set by the government. Fifteen European countries had implemented EEOs by 2019, but in many different ways (Fawcett et al., 2019).

The UK EEO was introduced in 1994, and, after a slow start, it saved around 120 TWh in the 2008–2012 period, when the targets were at their highest, before being cut back by the UK Government (Fawcett et al., 2019, Figure 1, p. 61). The current UK EEO, called the Energy Company Obligation (ECO) is targeted exclusively on low-income, fuel-poor and vulnerable households. Its fourth iteration is set to run from 2022–2026.[4]

Many policies have been used to increase energy efficiency, often together in what is called a 'policy mix'. Apart from EEOs, the individual policies have included: appliance and equipment efficiency standards and building regulations; information and labelling about the energy use of buildings and appliances; economic instruments, including subsidies, loans, taxes, rebates, performance contracting, on-bill financing schemes, and tradable certificates; providing feedback such as through smart meters; and pledges, commitments and rewards (such as tax credits) (Saunders et al., 2021).

As with EEOs different countries have implemented their own mix of these policies, designed in their own ways, but it is clear that increased energy efficiency has played a major role in reducing carbon emissions. In the UK, Lees & Eyre (2021) calculated that between 1989 and 2019 increasing energy efficiency saved 224 million tonnes of CO_2, over three times as much as the switch

---

3 https://ec.europa.eu/clima/eu-action/climate-strategies-targets/2030-climate-energy-framework _en. Accessed April 23, 2023.

4 www.ofgem.gov.uk/environmental-and-social-schemes/energy-company-obligation-eco. Accessed April 23, 2023.

from coal to gas in power generation in the 1990s, or the more recent growth of power generation from renewables. They identify the key changes responsible for these savings as major increases in building energy efficiency (through the UK EEO), and higher efficiency boilers (condensing gas boilers became mandatory from 2005 when boilers were being replaced, leading to efficiency gains of 25%), vehicles and appliances. The UK Climate Change Committee has also identified product standards set at an EU level as an important driver of increased energy efficiency, and consequent emissions reductions in lighting, appliances and vehicles (Climate Change Committee, 2020, p. 41). Lees & Eyre (2021) also note that effective energy efficiency policies have generally focused on individual sectors and even technologies, and use a wide range of policies, including regulation, incentives and information.

Increased energy efficiency in households has so far largely been driven by government policy, though it will be interesting to see whether the large increase in gas and electricity prices because of Russia's invasion of Ukraine will lead to increased private uptake of energy efficiency measures. Another source of behaviour change in this respect may be 'peer effects'. These refer to changes "when the attitudes, values or behaviours of an individual are influenced by the behaviours of members within a peer group" (Wolske et al., 2020, p. 202). Clearly peer effects relating to energy behaviours are likely to be stronger in respect of more visible energy technologies, such as solar panels or electric vehicles. However, word of mouth communication has also been shown to influence energy behaviour, including with regard to the adoption of home energy efficiency measures.

Different factors affect the strength of peer effects in relation to energy behaviours. This will obviously depend to some extent on the perceived difficulty of the behaviour, with home retrofits being at the difficult end of that spectrum. Influence through communication may occur through passive or active channels, with the former involving observation of the behaviour of others and perceptions of existing or changing social norms. And the information itself may be absorbed largely intuitively on the basis of 'gut-level opinions' or require some processing and deliberation. Peer influence is likely to become greater the more people become aware of the threats of climate change and the need to respond to it, and the more the technologies to do so become familiar and experienced in their communities (Wolske et al., 2020, Figure 2, p. 209).

High-efficiency energy pathways

Chapter 2 briefly described the Low Energy Demand (LED) scenario, as one of the scenarios that achieved the 1.5°C temperature limit by 2050. As shown in Figure 2.10, this has considerably lower energy demand than the other 1.5°C-compliant scenarios discussed there, or indeed elsewhere in the literature. One of the ways it achieves this very low energy demand, despite very significant increases in end-user energy service provision both in developed and developing countries, is through maximal incorporation of energy efficiency measures and technologies.

The LED scenario considers how energy demand might develop in four important areas of energy consumption: heating and cooling buildings, consumer goods, mobility (i.e. travel and transport) and food. The scenario also considered the energy demand of the business activities that supply these consumer demands: public and commercial buildings, industry, freight transport, energy supply, agriculture and land use. The main reason for the very low energy demand in this scenario, despite a great increase in the demand for energy services, is an increase in the efficiency with which the end-use service is delivered. In this scenario, people have more thermal comfort in buildings, far more consumer goods, increased mobility and access, and sufficient (but not excess) food. In addition, developing countries catch up dramatically with developed countries in these areas of consumption. The scenario also distinguished between two kinds of processes, both of which lead to less energy use: increased material efficiency, whereby the same product is made with less material (e.g. through light-weighting) and therefore requires less energy to produce; and 'dematerialisation', whereby products are used more intensively (e.g. through being shared, such as in car clubs) and therefore fewer products are required to satisfy the same service demands. This chapter only considers the more dramatic improvements in end-user efficiency. The more detailed aspects of energy efficiency in different sectors are discussed in Chapters 8 and 9.

Table 3.1 shows the increase in end-user service demands in the LED scenario, and the dramatic reductions in energy use that accompany these because of the increased efficiency with which the energy to deliver them is used. It must be emphasised that each of these increased efficiencies is derived through detailed consideration of the technologies concerned and realistic, though ambitious, assumptions of how these technologies will be developed and delivered over the next three decades.

It can be seen that all the end-use activities show a robust increase over 2020–2050, but only the 175% increase in consumer goods in the Global South, broadly developing countries, results in an increase in energy demand (54%). All the rest show substantial reductions because of increased energy efficiency: a 122% increase in mobility in the South is accompanied by a decrease of 59% in energy demand; a 79% increase in consumer goods in the North is accompanied by a 25% decrease in energy demand. These dramatic reductions in energy use mean that much less zero-carbon energy supply needs to be built, and no carbon capture and storage or negative emission technologies are required, as will be seen in Chapters 5, 6 and 7.

The authors of the LED scenario are clear that policy "plays a critical role to drive and enable the change depicted by LED" (Grubler et al., 2018, p. 522). Strict and tightening standards and regulations for energy efficiency are needed both for retrofitting and for new buildings, in both developed and developing countries. The same applies for appliances and equipment globally. Innovation policies, and policies to increase market demand for energy

Table 3.1 Percentage changes in activity levels and energy demands in the LED scenario

		Region	% change in activity levels (2020–2050)	% change in energy demand (2020–2050)
End-use services	Thermal comfort	North	6	−74
		South	63	−79
	Consumer goods	North	79	−25
		South	175	54
	Mobility	North	29	−60
		South	122	−59
Upstream	Public and commercial buildings	North	49	−64
		South	77	−82
	Industry	North	−42	−57
		South	−12	−23
	Freight transport	North	109	−28
		South	75	−12
Total		North		−53
		South		−32

Source: Grubler et al., 2018, Table 2, p. 520.

Note: North and South refers broadly to developed and developing countries respectively.

efficiency technologies and their installation, are also required, to achieve the widespread cost reductions that will make these products affordable, and to enable them to be diffused at scale globally. The actual technologies and policies concerned, and the new business models that will be required for them to dominate the market, will differ from sector to sector. The rest of this chapter will concentrate on some of the main efficiency issues in buildings, transport and industry, which will be picked up again when these sectors are discussed in more detail in Chapter 9. As with the other chapters, the issues will be illustrated where necessary with examples from the UK, but many of these are relevant more widely.

Industry and business

Industry can contribute to increasing energy efficiency in two broad ways: through increasing the energy efficiency of its processes of production, and through increasing the energy efficiency of the products it creates when they are used.

With regard to industrial processes, these differ enormously by sector and their energy efficiency needs to be considered sector by sector and process by process. Chapter 9 will do this for some of the most energy-intensive sectors and processes. However, in general terms the means of improving the energy efficiency of industrial processes are well known, but all too often not adopted in businesses that could benefit from them. They include: support at Board level, especially from the finance director; appointing an internal energy management team to draw up a plan to monitor and reduce energy usage and implement energy-saving ideas, with incentives and bonuses based on performance; carrying out energy audits to identify the business's hotspots of energy use; controlling and managing the use of machinery and electric motors; and increasing the use of energy-efficient lighting.

Benchmarking energy efficiency in different sectors can show up the difference in business performance and highlight the potential for improvement, and governments can require companies over a certain size to report on their energy efficiency actions each year, and promote industrial energy efficiency through regional targets, financial support, and specific regulation and standards. The widespread use of benchmarking for different sectors, within and between countries, would provide a better understanding of energy performance and opportunities for improvement in industry and companies. This will require substantial improvements in the quantity and quality of data collected and reported, especially in developing countries. Government policy could then focus on the areas with most potential for energy efficiency increase.

In respect of their products, the major means of increasing the energy efficiency of electrical goods is through minimum performance standards and labelling programmes. Figure 3.1 shows the energy saving of the average appliance in use due to such programmes – it can be seen that a more than 50% improvement in appliance energy efficiency is commonplace through these programmes. It may also be noted from Figure 3.1 that more savings are delivered by longer-running programmes (some have run for more than 20 years) with stronger standards, as there is more time for inefficient appliances and equipment to be replaced.

However, despite this evidence that government policies in some countries are playing a role in accelerating the development and adoption of energy-efficient appliances and equipment, the policies on energy efficiency are not equally distributed across all industrial and business sectors. There is a lack of mandatory policies and energy efficiency obligations in industry, at least partly due to concerns in some sectors about the possible impacts on international competitiveness. It is especially important that emerging and developing economies, which are experiencing a rapid increase in their economic growth and energy consumption, and often have a relatively high energy intensity, update their energy efficiency policy frameworks according to best practice.

Buildings

Increasing energy efficiency in homes (the domestic sector) and commercial and public buildings (the non-domestic sector) provides many opportunities for energy saving and CO_2 emissions reductions. It is also associated with a wide range of other positive impacts or 'co-benefits', such as the improved

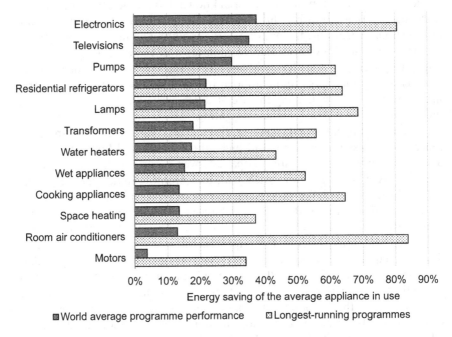

Figure 3.1 Energy savings from energy efficiency standards and labels over the life of programmes.

Note: Electronic devices include: external power supply units, monitors, DVD/VCR units and other personal electronics.

Source: International Energy Agency, 2021, p. 16 (based on reviews of over 400 published reports covering energy efficiency standard and label programmes in a wide range of countries).

health, productivity and social welfare of building occupants, in addition to increased employment opportunities in the energy efficiency industry. Energy efficiency measures for buildings include: cavity wall, solid wall and loft insulation; draught proofing and improved glazing; replacement of old gas boilers (with either more efficient condensing gas boilers or heat pumps); and more energy-efficient appliances, air conditioning systems, hot water tanks, heating systems and light bulbs, as discussed in the previous section; and, for commercial buildings, energy management systems.

Achieving increased energy efficiency in buildings involves the detailed consideration together of all the complex factors which influence energy consumption, including the building fabric, the systems and equipment/appliances installed in it and, most importantly, the people who occupy it. Effective energy efficiency policies have included minimum standards and labelling, and a wide range of regulations and voluntary certification schemes for both new and existing buildings. Retrofitting existing buildings is particularly important given that 75–85% of buildings that will exist in 2050 already exist today (Dowson

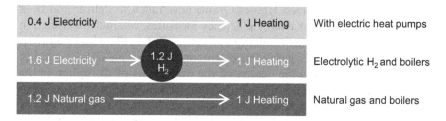

Figure 3.2 Energy efficiency of different means of heating buildings.
Source: Lowe & Oreszczyn, 2020.

et al., 2012). Heat pumps have huge potential to deliver more energy-efficient heating to buildings, as shown in Figure 3.2, where it can be seen that they can deliver two and a half times as much heat as their electricity input. However, as noted by Lowe & Oreszczyn (2020), and discussed further in Chapter 9, many obstacles to their installation at scale in the UK still need to be overcome.

Retrofitting traditional or heritage buildings, for which energy retrofit strategies for more modern constructions are not suitable and can be damaging, requires particular care, both to provide a healthy, cost- and resource-effective environment, and to take account of heritage values or significance.

While UK building energy efficiency has increased markedly in the last 30 years, as has been seen, often attempts to increase the energy efficiency of buildings have not been as effective as anticipated. Although policies are in place, enforcement is inconsistent, and policy messaging is often contradictory. A study by Zero Carbon Hub in 2010 showed that, of 16 new-build houses studied, not one achieved its performance standard for energy use predicted at the design stage, only five failed to meet the standard by less than 20%, and several exceeded predicted heat loss by more than 100%.[5] While the number of houses studied was small, and the study makes no claim to representativeness, such results are deeply worrying given the challenges of the net-zero target.

There are similar concerns about the UK Government's commitment to, and the quality of, energy efficiency retrofits to existing buildings. Fawcett et al. (2019, Figure 1, p. 61) showed that there has been a 10-year decline in

5 Zero Carbon Hub (2010) Carbon compliance for tomorrow's new homes: A review of the modelling tool and assumptions. Topic 4: Closing the gap between designed and built performance. August 2010. London, Zero Carbon Hub, p.12. Zero Carbon Hub was a body advising how the UK Government could meet its target of all new homes having to be 'zero-carbon' by 2016. Following the scrapping of the target in 2015, the Zero Carbon Hub was shut down in 2016. This publication was available on the internet in 2022, while this section of the book was being drafted, but has since been removed. It may still be available from NHBC House, Davy Avenue, Milton Keynes, MK5 8FP, UK, see www.thenbs.com/PublicationIndex/documents/details?Pub=ZCH&DocID=301152. Accessed April 23, 2023.

policy support for insulation installations, and several recent high-profile policies (e.g. Green Deal, Green Homes Grant) have failed to deliver and been withdrawn. The uptake of energy efficiency measures, whether of insulation of the building fabric or the installation of new equipment like heat pumps, is still hindered by high costs, lack of incentives and appropriate finance options, concerns about safety and the quality of the work, fears of disruption in the case of retrofit, institutional issues (e.g. different freeholder/leaseholder responsibilities, or landlord/tenant mismatches, where the landlord pays the costs of increased energy efficiency and the tenant gets the lower fuel bills), and the lack of independent advice on the complex process of upgrading buildings for higher energy efficiency. Even when energy efficiency measures are installed, a substantial difference between the expected and achieved energy savings is common – Dowson et al. (2012, p. 301) cited one study that reported that actual monitoring following the installing of cavity wall insulation and loft insulation showed that only 10–17% energy savings were achieved, when theoretical calculations suggested that the refurbishment would save 49% of fuel consumption. This 'performance gap' has been attributed to poor on-site installation quality and incorrect assumptions in the models and methods used to formulate energy efficiency strategies and estimate potential energy savings.

For the future, the energy efficiency potential of buildings can only be realised, in the UK and elsewhere, via long-term, large-scale and comprehensive measures that are supported by a standards framework which ensures that the right products are fitted to the right properties in the right way, and a monitoring and enforcement system which ensures that poor-quality work is dealt with effectively and safety is guaranteed. Financial incentives even for those with the means to pay will need to be provided to improve the energy efficiency of the building fabric, and to stimulate the installation of more efficient heating equipment such as heat pumps. Institutional innovation will be required to overcome freeholder/leaseholder bottlenecks or landlord/tenant mismatches.

Vehicles

Internal combustion engine vehicles (ICEs) are very inefficient and in the case of petrol cars only convert about 12–30% of the energy stored in the fuel to power at the wheels.[6] Figure 3.3 indicates the losses that result in only achieving the lower end of that range.

Improvements in engine technology and vehicle design, often driven by regulatory standards, have increased the fuel efficiency (km travelled/litre of fuel use) of internal combustion engine (ICE) cars in Europe by around 20%. This means that, despite increases in the number of cars (over 50% since 1990) and the total distance driven (over 30% increase since 1990, broadly in

---

6 www.fueleconomy.gov/feg/evtech.shtml. Accessed April 23, 2023.

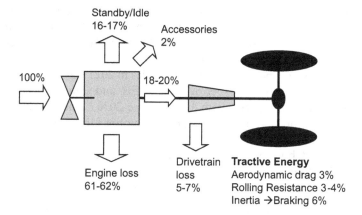

Figure 3.3 Energy losses in a petrol engine (light-duty, mid-size, urban driving mode). The 100% refers to the energy in the fuel input.

Note: Diesel cars are around 17% more fuel-efficient than petrol cars.[a]

[a] www.which.co.uk/reviews/new-and-used-cars/article/petrol-vs-diesel-cars-which-is-better-az4UV9R1twEE. Accessed April 23, 2023.

Source: Kobayashi et al., 2009, Figure 3, p. 128.

line with the growth of GDP), energy consumption and CO_2 emissions have increased by less than 10% over 1990–2015.[7]

Had cars not been getting bigger and more powerful over the last few decades, the increase in vehicle efficiency would have been considerably greater than the 20% figure cited above. The US Environmental Protection Agency has estimated that the fuel economy of the year 2004 fleet of cars and light trucks would have been 5.5 mpg (22%) higher had it remained at the same average weight and acceleration performance of the 1987 fleet (Kobayashi et al., 2009, p. 126). The trend continues: SUVs, which emit about a quarter more CO_2 than a medium-size car, accounted for 21.2% of the UK's new car market in 2018, up from 13.5% in 2015 (UKERC, 2019).

While the fuel efficiency of ICEs may be expected to improve further in the future, the really big efficiency gains will come from the shift to battery electric vehicles (BEVs), which typically convert 77–80% of the electrical energy from the grid to power at the wheels.[8]

Sales of BEVs are growing fast. In 2020 new BEVs accounted for just 2.7% of UK new car sales, but in 2022 this grew to 12.5%.[9] A continuing drag on that growth comes from a combination of their current cost, and 'range anxiety' due

---

7 www.eea.europa.eu/data-and-maps/figures/fuel-efficiency-and-fuel-consumption. Accessed April 23, 2023.
8 www.fueleconomy.gov/feg/evtech.shtml. Accessed August 8, 2023.
9 www.smmt.co.uk/vehicle-data/evs-and-afvs-registrations/. Accessed April 23, 2023.

to the lack of car-charging infrastructure and the still lower distance that BEVs can travel than ICEs before they need recharging/refuelling. However, both these problems are being addressed through increased installation of charging facilities and the increased performance of batteries. In Norway, where the purchase of BEVs is heavily subsidised and otherwise promoted (e.g. with BEV-only car lanes) the proportion of new BEV sales in 2021 was 64.5%.[10]

While BEVs are likely to get cheaper over the 2020s, and some projections suggest they will reach cost parity with ICEs before 2030,[11] the real game-changer to their sale and take up in the UK will be the UK Government's ban on the sale of new petrol and diesel vehicles from 2030.[12] There is much to be done in terms of both installation of car-charging facilities and battery improvement before that date.

Energy efficiency policies from the IEA

The International Energy Agency brought together the energy efficiency policies used in its Net Zero Emissions (NZE) scenario and arranged them in a timeline as in Table 3.2. Buildings feature heavily in the years to 2030, with the phase-out of new fossil fuel boilers, constructing and retrofitting buildings to be zero-carbon ready and making heat pumps, LED lighting and high-efficiency appliances the norm, as a way of achieving universal access to modern energy in developing countries.

By 2030 electric vehicles have also become mainstream, and ICE vehicles have been phased out in large cities. By 2035 major industries, including steel, cement and chemicals, are well into the efficiency component of the clean energy transition, and in 2050 all three sectors – industry, buildings and transport – have made the most of energy efficiency opportunities, thereby minimising the quantity of zero-carbon energy that needs to be supplied.

Conclusion

Increasing energy efficiency in buildings, industry and transport will reduce the amount of new low-carbon energy supply that needs to be built, and in many cases will be the most cost-effective way to reduce CO_2 emissions. But in all these sectors it will require strong and consistent government policy to achieve the energy efficiency improvements that are available. Those countries that realise energy efficiency opportunities most effectively will have buildings that are more pleasant and cost-effective to live in, industry that is more competitive and vehicles that cost less to run. And they will be the countries most likely to succeed with achieving deep emissions reductions.

10 https://insideevs.com/news/565428/norway-plugin-car-sales-january2022/. Accessed April 23, 2023.
11 www.theguardian.com/business/2021/may/09/electric-cars-will-be-cheaper-to-produce-than-fossil-fuel-vehicles-by-2027. Accessed April 23, 2023.
12 This date was pushed back to 2035 by the UK Government in September 2023

Table 3.2 Energy efficiency targets in the IEA's NZE scenario

2025	2030	2035	2050
No new sales of fossil fuel boilers	Annual pace of zero-carbon ready building retrofits reaches 2.5% in advanced countries and 2% in emerging countries	New appliances improved to 2020 best available technology	Share of zero-carbon ready buildings in total stock more than 85%
All new buildings zero-carbon ready in advanced economies	Share of zero-carbon ready buildings in total stock: 25%	Motorway speed limit: 100 km/h	Use of scrap for steel production 46%
Total heat pumps installed: 600 million up from 180 million today	All new buildings zero-carbon ready in emerging economies	Public EV chargers installed: 40 million up from 1.3 million today	Share of heat pumps in energy demand for heating: 55%
Share of heat pumps in energy demand for heating: 20%, up from 7% today	New buildings heating and cooling using 50% less energy	Public EV charging capacity 1780 GW	Reduction in use of energy-intensive building materials: 30%
		Share of passenger cars that are EVs in stock: 20% (1% today)	Appliances using 40% less energy
		All industrial electricity motor sales are best in class	Clinker to cement ratio down from 0.71 today to 0.57
		No new ICE passenger cars sold	Share of electricity in total industrial consumption: 46%, up from 21% today
		Process energy intensity of primary chemicals down from 17 GJ today to 16 GJ	Process energy intensity of primary chemicals: 15 GJ per tonne

(Continued)

Table 3.2 (Continued)

2025	2030	2035	2050
Universal access to electricity and clean cooking	LED lighting: 100% sales	New sales of passenger cars that are EVs comprise 64% of total	New buildings heating and cooling: 80% less energy
Excessive hot water temperatures reduced	Set points for heating 19–20°C, cooling 24–25°C	Heavy truck average fuel consumption 19% lower	Heavy truck average fuel consumption: 37% lower
	Appliances using 25% less energy	Share of EV two/three-wheelers in stock: 54%	Business and long-haul leisure air travel does not exceed 2019 levels
	Share of steel production using electric arc furnaces: 37%, up from 24% today	ICEs phased out in large cities	Regional flights are shifted to high-speed rail where feasible
		Clinker to cement ratio down from 0.71 today to 0.65	Global average plastics collection rate: 54%, up from 17% today
		Global average plastics collection rate: 27%, up from 17% today	Public EV chargers installed: 200 million units
		Use of scrap for steel production: 38%, up from 32% today	Private EV chargers in buildings: 3500 million units

Note: Cell shading : no shading is buildings, light grey is industry, darker grey is transport.

Source: International Energy Agency, 2021, p. 15.

References

Climate Change Committee. (2020). *The Sixth Carbon Budget: The UK's path to net zero*. CCC, London.

Dowson, M., Poole, A., Harrison, D., & Susman, G. (2012). Domestic UK retrofit challenge: Barriers, incentives and current performance leading into the Green Deal. *Energy Policy, 50*, 294–305. https://doi.org/10.1016/j.enpol.2012.07.019

Fawcett, T., Rosenow, J., & Bertoldi, P. (2019). Energy efficiency obligation schemes: their future in the EU. *Energy Efficiency, 12*(1), 57–71. https://doi.org/10.1007/s12053-018-9657-1

Grubler, A., Wilson, C., Bento, N., Boza-Kiss, B., Krey, V., McCollum, D. L., Rao, N. D., Riahi, K., Rogelj, J., De Stercke, S., Cullen, J., Frank, S., Fricko, O., Guo, F., Gidden, M., Havlík, P., Huppmann, D., Kiesewetter, G., Rafaj, P., … Valin, H. (2018). A low energy demand scenario for meeting the 1.5 °C target and sustainable development goals without negative emission technologies. *Nature Energy*. https://doi.org/10.1038/s41560-018-0172-6

International Energy Agency. (2021). Energy Efficiency, 2021. IEA, Paris. https://iea.blob.core.windows.net/assets/9c30109f-38a7-4a0b-b159-47f00d65e5be/EnergyEfficiency2021.pdf

Kobayashi, S., Plotkin, S., & Ribeiro, S. K. (2009). Energy efficiency technologies for road vehicles. *Energy Efficiency, 2*(2), 125–137. https://doi.org/10.1007/s12053-008-9037-3

Lees, E., & Eyre, N. (2021). Thirty years of climate mitigation: lessons from the 1989 options appraisal for the UK. *Energy Efficiency, 14*(4), 37. https://doi.org/10.1007/s12053-021-09951-2

Lowe, R., & Oreszczyn, T. (2020). Building decarbonisation transition pathways. CREDS Briefing, Figure 2, p. 3. www.creds.ac.uk/publications/building-decarbonisation-transition-pathways/. Accessed April 23, 2023.

Saunders, H. D., Roy, J., Azevedo, I. M. L., Chakravarty, D., Dasgupta, S., de la Rue du Can, S., Druckman, A., Fouquet, R., Grubb, M., Lin, B., Lowe, R., Madlener, R., McCoy, D. M., Mundaca, L., Oreszczyn, T., Sorrell, S., Stern, D., Tanaka, K., & Wei, T. (2021). Energy efficiency: What has research delivered in the last 40 years? *Annual Review of Environment and Resources, 46*(1), 135–165. https://doi.org/10.1146/annurev-environ-012320-084937

UKERC (UK Energy Research Centre). (2019). Are SUVs sabotaging the green transport revolution? https://ukerc.ac.uk/news/suvs-sabotage-green-revolution/. Accessed August 9, 2023.

Wolske, K. S., Gillingham, K. T., & Schultz, P. W. (2020). Peer influence on household energy behaviours. *Nature Energy, 5*(3), 202–212. https://doi.org/10.1038/s41560-019-0541-9

4 Kicking the addiction to fossil fuels

Summary

This chapter explores the issue of fossil fuels and the companies that supply it. Fossil fuels still supply 80% of the world's primary energy demand. These will need to be phased out for climate stability, but countries' and companies' plans to produce fossil fuels are wildly inconsistent with the temperature targets of the Paris Agreement. Fossil fuel businesses, and their investments, remain overwhelmingly focused on fossil fuels and their public utterances about their activities are at wide variance from the overall profile of their activities. Public pressure against them is growing, but so far neither these nor government policy have proved decisive in reducing fossil fuel supply.

Introduction

This chapter on energy supply opens with a global overview of the world's energy system, which is still dominated by fossil fuels. Figure 4.1 shows that in 2018 fossil fuels still supplied 80% of global primary energy, and over 60% of electricity and final consumption. Coal is the single largest source of electricity, and the use of oil in final consumption is primarily for transport. CO_2 emissions from the energy system essentially all come from fossil fuels. Later chapters will explore the greatly increased contributions to the energy system that will need to be made by renewables, nuclear, bioenergy and hydrogen if the energy system is to be decarbonised. But any chapter on energy supply needs to start with the still dominant role of fossil fuels in energy supply.

Fossil fuels

Fossil fuels derive from dead organisms (mainly plant matter), which, over billions of years, have been compressed by geological processes to produce coal, oil and fossil methane gas (this is sometimes called 'natural gas', presumably to distinguish it from the town gas that was produced from coal and

DOI: 10.4324/9781003438007-5

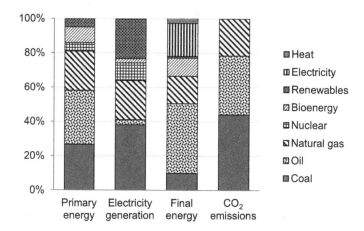

Figure 4.1 Global overview of the energy system, 2018.

Source: International Energy Agency, 2020, p. 20.

that it displaced from cooking, heating and lighting in cities). The plants were composed primarily of carbon and hydrogen, which is why fossil fuels are also called 'hydrocarbons'. Over these billions of years the plants took carbon out of the air, reducing its concentration in the atmosphere, and breathed out oxygen, such that Earth's atmosphere cooled, creating the climatic conditions in which humans have prospered. The burning of these fossil fuels on a colossal scale, thereby putting the carbon back into the atmosphere, is the principal cause of the global warming and climate change discussed in Chapter 1.

Coal, oil and fossil methane gas have different concentrations of carbon per unit of energy, which is released as CO_2 when they are burned. Table 4.1 shows the CO_2 emission factors per unit of primary energy in kWh, when they are burned, from which it can be seen that wood (not sustainable, i.e. when trees are not replanted) and peat, both forms of biomass (see Chapter 6), are the most carbon-intensive fuels when burned, followed by different types of coal. Fossil methane (natural) gas is the least carbon-intensive when burned, but its greenhouse gas (GHG) emissions are larger when the emissions associated with leaks from fracking wells or pipelines (called 'fugitive' emissions) are taken into account, because methane is a potent GHG in its own right as seen in Chapter 1. Schwietzke et al. (2016) estimated that these fugitive methane emissions in the early 2010s were some 2% of fossil methane production.

Figure 4.2 shows the huge growth of fossil fuel consumption since 1850, before which the main energy form was 'traditional biomass' – the burning of wood, crops and animal dung, or human and animal muscles, the energy from which was derived from the food they ate. In 2020 fossil fuels still provided over 80% of global primary energy (oil 31.2%, coal 27.2%, gas 27.4%) with

Table 4.1 CO$_2$ emission factors (grams CO$_2$) per unit of produced energy (kWh$_{PE}$)

Fuel type	gCO_2/kWh_{PE}
Wood not sustainable	395
Peat	382
Lignite	364
Hard coal	354
Fuel oil	279
Diesel	267
Crude oil	264
Kerosene	257
Gasoline	250
LPG	227
Natural gas	201
Wood sustainable	0

Source: www.volker-quaschning.de/datserv/CO$_2$-spez/index_e.php, (Accessed April 27, 2023), derived from figures produced by Umwelt Bundesamt (the German Environment Agency).

hydropower, other renewables and nuclear providing 6.9%, 5.7% and 4.3% respectively.[1] These numbers provide a stark illustration of the decarbonisation challenge.

Fossil fuel use and the Paris temperature targets

Until quite recently there were fears, given the huge increase in the use of fossil fuels, a non-renewable resource, that oil in particular would 'run out', which economists would interpret as becoming unaffordably expensive as depletion occurred. There were estimates as to when 'peak oil' would occur because of the lack of oil resources. Such speculations proved to be unfounded and it is now clear that there are far more fossil fuels underground than humans can afford to burn if they want a stable climate, and that 'peak oil', if it happens, will be a result of climate policy reducing the *demand* for fossil fuels, rather than lack of oil resources.[2]

Papers in the journal *Nature* estimated the proportion of fossil fuel reserves that would need to remain unextracted and unburned (without carbon capture and storage, see Chapter 7) if the carbon budgets for 2°C and 1.5°C global warming by 2100, set out in Chapter 2, were not to be exceeded.

---

1 www.bp.com/en/global/corporate/energy-economics/statistical-review-of-world-energy/primary-energy.html. Accessed April 27, 2023.
2 See for a refutation of peak oil supply any time soon www.forbes.com/sites/michaellynch/2018/06/29/what-ever-happened-to-peak-oil/?sh=6c8149b3731a. Accessed April 27, 2023, and www.forbes.com/sites/jamesconca/2017/03/02/no-peak-oil-for-america-or-the-world/. Accessed April 27, 2023.

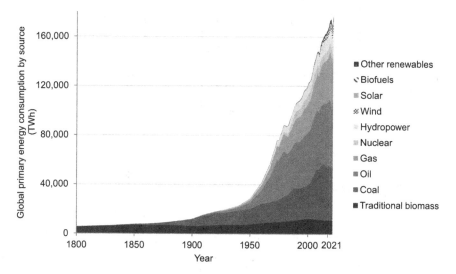

Figure 4.2 The growth of fossil fuel use, 1800–2021.

Source: Our World in Data, https://ourworldindata.org/energy-mix. Accessed April 27, 2023.

Table 4.2a shows the results by region. It can be seen from Table 4.2b that remaining ultimately recoverable resources (RURR), conventional and unconventional, exceed reserves by a large margin for all fossil fuels, and these, of course, will also have to remain unburned to achieve the Paris temperature targets.

Fossil fuels and net-zero emission pathways

To have any chance of staying within the temperature limit of the Paris Agreement, the consumption of fossil fuels will have to fall dramatically before 2050. Figure 4.3 shows the different sources of energy of three of the energy scenarios that were discussed in Chapter 2, that achieve net-zero emissions by 2050 or soon after, and, for the scenarios going through to 2100, that achieve the 1.5°C temperature target in 2100.

It can be seen that in all the scenarios there is a fast reduction in fossil fuel use, especially coal. There is also a huge expansion of wind and solar energy, and an expansion of modern bioenergy. Nuclear energy remains in all the scenarios, but at a level far below the renewables.

The main difference between the scenarios is in their primary energy requirement in 2050: around 550 EJ for the IEA Net Zero Emissions (NZE) scenario, around 530 EJ for Sitra, but only 290 EJ for the LED scenario, which implements far more energy efficiency and energy-saving technologies and

Table 4.2a Regional distribution of unextractable reserves to stay below 1.5°C

Region	Oil				Fossil methane gas				Coal			
	2050		2100		2050		2100		2050		2100	
	(%)	(Gb)	(%)	(Gb)	(%)	(Tcm)	(%)	(Tcm)	(%)	(Gt)	(%)	(Gt)
Africa (AFR)	51	53	44	46	49	6	43	6	86	27	85	26
Australia and other OECD Pacific (AUS)	40	2	40	2	29	0.7	25	0.6	95	80	95	80
Canada (CAN)	83	43	83	43	56	1.1	56	1.1	83	4	83	4
China and India (CHI + IND)	47	17	36	13	29	1.3	24	1.1	76	182	73	177
Russia and former Soviet states (FSU)	38	57	29	44	63	30	55	26	97	205	97	205
Central and South America (CSA)	73	98	62	84	67	4	65	4	84	11	82	11
Europe (EUR)	72	12	72	12	43	2	40	1	90	69	90	69
Middle East (MEA)	62	409	38	253	64	36	49	28	100	5	100	5
Other Developing Asia (ODA)	36	8	31	7	32	2	25	2	42	10	39	9
USA	26	18	20	14	24	2.8	24	2.8	97	233	97	232
Global	**58**	**740**	**42**	**541**	**56**	**87**	**47**	**73**	**89**	**826**	**88**	**818**

Source: Welsby et al., 2021, Table 1, p. 233.

Table 4.2b Best estimates of remaining reserves and ultimately recoverable resources from 2010

Oil (Gb)			Gas (Tcm)			Hard coal (Gt)		Lignite (Gt)	
Res	Con RURR	Uncon RURR	Res	Con RURR	Uncon RURR	Res	RURR	Res	RURR
1,294	2,615	2,455	192	375	300	728	2,565	276	1,520

Source: McGlade & Ekins, 2015, Extended Data Table 1.

Note: 'Reserves' (Res) are the subset of 'resources' that are considered economically recoverable with current technologies. As technologies improve, therefore, resources become reserves. RURR are 'remaining ultimately recoverable resources' divided for oil and gas between conventional and unconventional (e.g. shale oil and shale gas) resources.

Units: Gb are gigabarrels (of oil); Tcm are tera cubic metres (of gas); Gt are gigatonnes (of coal). See the Preface for the meaning of 'giga' and 'tera'.

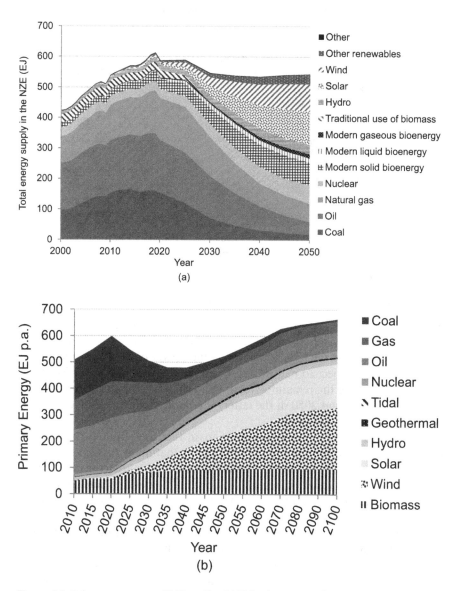

Figure 4.3 Primary energy to 2050 and/or 2100 in three scenarios.
a) The IEA Net Zero Emissions scenario (International Energy Agency, 2021, Figure 2.5, p. 57).
b) The Sitra 1.5°C scenario (Drummond et al., 2021, Figure 16, p. 63).
c) The Low Energy Demand scenario (Grubler et al., 2018, Figure 3c, p. 521).

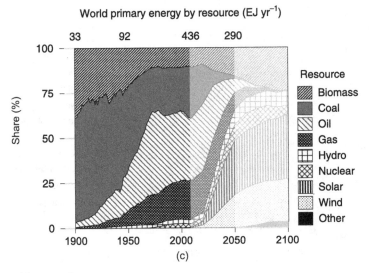

Figure 4.3 (Continued)

behaviours (as discussed in Chapter 3), and thereby manages to do without the carbon capture and storage (CCS) and Negative Emission Technologies (NETs) (discussed in Chapter 7), of which the operation and implementation at scale are still subject to major uncertainty. If they fail to materialise at the required scale, then the remaining fossil fuel use in 2050 in the IEA and Sitra scenarios (and beyond 2050 in the latter) is completely incompatible with the 2100 1.5°C temperature target.

The dramatic and rapid reduction in fossil fuel use required to meet the Paris temperature targets means, as seen above, that many fossil fuel reserves will need to remain unextracted/unburned if these targets are to be met. This in turn means that new fossil fuel reserves discovered by countries' and companies' exploration activities have a high risk of being 'stranded' (i.e. worthless) if the world's economy decarbonises at the required rate. Semieniuk et al. (2022) estimated the value of these potentially 'stranded assets' at more than US$1 trillion. On top of this lost value from fossil fuels having to remain unburned, there are also stranded assets from the investments to extract and refine them. However, this does not prevent the global fossil fuel industry continuing to engage in intensive exploration activity, or the finance sector from continuing to finance such activity, as will be seen below, essentially betting against climate policy being stringent enough to decarbonise to the extent and at the rate required by the Paris targets.

Without new investment, oil and gas supply from existing fields falls at around 8% per annum. The International Energy Agency (IEA)'s Sustainable Development Scenario (SDS, consistent with a 2°C temperature target rather than the 1.5°C target met by the IEA's NZE scenario) estimated that new investment in both existing and new fields would be required to meet energy demand. Even with investment in existing fields, investment in new fields is required to close the gap between demand and supply (International Energy Agency, 2020, p. 11).

However, in the IEA's 2021 more stringent NZE scenario, energy demand declines more quickly to meet the net-zero constraint (required to meet the 1.5°C Paris target), such that the IEA states clearly: "there are no new oil and gas fields approved for development in our pathway, and no new coal mines or mine extensions are required" (International Energy Agency, 2021, p. 21). This assertion – that new coal mines and oil and gas fields are essentially incompatible with the Paris targets – sits very uncomfortably with the fossil fuel plans of the governments which have signed up to them.

Countries' current and projected fossil fuel production

Global coal consumption in 2021 was around 7.9 billion metric tonnes.[3] The top ten (hard) coal producing countries in 2018, with the quantity in million metric tonnes in brackets, were: China (3,530), India (730), USA (634), Indonesia (498), Australia (453), Russia (353), South Africa (253), Kazakhstan (112), Colombia (84), and Poland (64).[4]

Total world oil production in 2020 was around 94 million barrels per day.[5] The top ten countries for oil production, with the numbers of barrels of oil in millions they produced per day in 2019 in brackets, are: USA (12), Russia (11.2), Saudi Arabia (11.1), Iraq (4.5), Iran (4.0), China (4.0), Canada (3.7), United Arab Emirates (UAE) (3.1), Kuwait (2.9), and Brazil (2.5).[6]

Global gas consumption in 2019 was around 3.9 trillion cubic metres.[7] The top ten natural gas-producing countries, with production in billion cubic metres in brackets, are: Russia (47.8), Iran (33.7), Qatar (24.1), USA (15.5), Saudi Arabia (9.2), Turkmenistan (7.5), UAE (6.1), Venezuela (5.7), Nigeria (5.5), and China (5.4).[8]

The Production Gap report published by the United Nations Environment Programme (UNEP) shows that the fossil fuel production plans of countries that produce fossil fuels are completely incompatible with the Paris temperature targets and scenarios in the previous section and, very often, with their own stated 'net-zero' climate commitments.

---

3 https://iea.blob.core.windows.net/assets/f1d724d4-a753-4336-9f6e-64679fa23bbf/Coal2021.pdf. Accessed April 28, 2023.

4 www.statista.com/statistics/264775/top-10-countries-based-on-hard-coal-production/.   Accessed April 28, 2023.

5 www.eia.gov/outlooks/steo/report/global_oil.php. Accessed April 28, 2023.

6 www.offshore-technology.com/features/oil-production-by-country/. Accessed April 28, 2023.

7 www.statista.com/statistics/282717/global-natural-gas-consumption/. Accessed April 28, 2023.

8 https://worldpopulationreview.com/country-rankings/natural-gas-by-country. Accessed April 28, 2023.

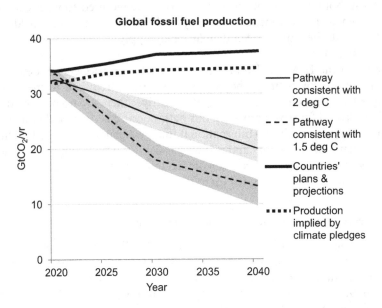

Figure 4.4a The production gap between countries' planned fossil fuel production to 2040 and their climate pledges and production that is consistent with the Paris temperature targets.

Source: SEI et al., 2021, Figure 2.1, p. 15. https://productiongap.org/wp-content/uploads/2021/11/PGR2021_web_rev.pdf. Accessed April 27, 2023.

Figure 4.4 (a,b) shows the extent of the divergence between what the Paris temperature targets require, and fossil fuel producing countries' current production plans. For all the fossil fuels, their production is planned to stay broadly constant (coal) or increase (oil and gas), in contrast to the dramatic reductions required by the Paris temperature targets.

Table 4.3 indicates which countries are responsible for most of the production gap. *All* the major country producers of fossil fuels are planning to expand their oil and gas production, while India and Russia are planning to expand their production of coal, the most carbon-intensive fuel, as well. Unless these plans are radically changed so that coal, oil and gas production all *fall* in the next few years, the temperature targets in the Paris Agreement will be revealed to have been a fairy story.

Given the production gap it is quite extraordinary that the UK, which likes to portray itself as a climate leader, and which both has a national 2050 net-zero target and was the COP26 President (from COP26 in November 2021 until COP27 in November 2022) under the UN Climate Change Convention, charged with encouraging other countries to reduce their emissions, has licensed extraction from new oil and gas fields inhas licensed extraction from new oil and gas fields in the UK Continental Shelf'.[9] The only way production from new UK oil and gas fields would be compatible with global 'net zero' by 2050 is if producers from existing reserves elsewhere agreed to cut their production by the amount of the new UK production. Such an agreement is nowhere in prospect.

The UK with this initiative seems prepared to scupper global net zero even though it has a national net zero by 2050 in its own laws. This is climate vandalism, not climate leadership. Exactly the same arguments apply to the new coal mine in Cumbria to which the UK Government gave its approval in December 2022, a decision condemned by some business groups and by its own climate advisers, the Climate Change Committee.[10] While this mine, if opened, will produce coal for steel-making rather than power generation, its coal will not be used in the UK, but exported to other European countries or further afield.[11] Willis (2023) gives an admirable account of the governance processes and use of evidence in policy that led up to this climate-destructive decision, which seriously damages the UK's claims to be a leader in respect of climate policy.

Fossil fuel companies

The great majority of the world's fossil fuels are produced by large companies, some of which are shareholder-listed multi-national companies, and some of which are state-owned.

Coal companies

Wikipedia lists the top ten companies that mine coal, in order of total (not necessarily coal-mining) revenue, with principal host country in brackets, as: BHP (Australia), Rio Tinto (Australia), China Shenhua Energy (China), Anglo American (UK), Coal India, NTPC (India), Shaanxi Coal and Chemical Industry (China), Sasol (South Africa), Teck Resources (Canada), and CEZ Group (Czechia).[12]

As an example, BHP's coal revenue in 2021 was US$5.2 billion, declining from US$9.1 billion in 2019,[13] out of total revenue in 2021 of US$61 billion.[14] Anglo American is also reducing its holdings of thermal coal (used for power

9 www.nstauthority.co.uk/news-publications/news/2022/nsta-launches-33rd-offshore-oil-and-gas-licensing-round/. Accessed April 28, 2023.

10 www.edie.net/ccc-condemns-cumbria-coal-mine-decision-as-business-groups-continue-voicing-disappointment/. Accessed April 28, 2023.

11 www.theguardian.com/environment/2022/dec/11/new-cumbria-coalmine-backlash-grows-as-steel-industry-plays-down-demand. Accessed April 28, 2023.

12 https://en.wikipedia.org/wiki/List_of_largest_coal_mining_companies. Accessed April 28, 2023.

13 www.statista.com/statistics/1173166/bhp-coal-segment-revenue/. Accessed April 28, 2023.

14 www.bhp.com/-/media/documents/investors/annual-reports/2021/210914_bhpannualreport2021.pdf#page=15. Accessed April 28, 2023.

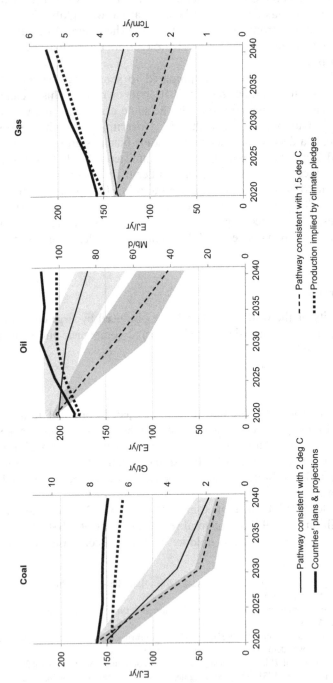

Figure 4.4b The production gap between countries' planned fossil fuel production to 2040 for coal, oil and gas, and their climate pledges and production that is consistent with the Paris temperature targets.
Source: SEI et al., 2021,Figure 2.2, p. 16. https://productiongap.org/wpcontent/uploads/2021/11/PGR2021_web_rev.pdf. Accessed April 27, 2023.

Table 4.3 Projected/planned change in fossil fuel production by 2030 relative to 2019, selected countries

Country	Net-zero commitment, target year	Planned/projected change in national fossil fuel production for 2030 relative to 2019 (EJ)		
		Coal	Oil	Gas
Australia	No commitment	–	+ 0.2	+ 0.6
Brazil	Pledge, 2050	–	+ 5.3	+ 1.3
Canada	Law, 2050	−0.5	+ 1.4	+ 0.3
China	Pledge, 2060	−9.2	+ 0.6	+ 3.8
India	Pledge 2070	+ 6.1	+ 2.5	+ 0.8
Mexico	No commitment	–	+ 0.4	+ 0.5
Russia	No commitment	+ 3.6	–	+ 4.3
Saudi Arabia	No commitment	–	+ 7.1	+ 4.7
UAE	No commitment	–	+ 1.9	na
USA	Policy, 2050	−4.3	+ 5.2	+ 3.8

UAE: United Arab Emirates; – signifies either that annual production in 2019 is less than 0.5 EJ, or change in production by 2030 stays within 5% of 2019 production in energy terms; na, not available. Source: SEI et al., 2021, extracted from Figure 3.1, p.26. https://productiongap.org/wp-content/uploads/2021/11/PGR2021_web_rev.pdf. Accessed April 27, 2023.

generation, as opposed to metallurgical coal, which is used for steel-making), spinning off its South African holdings into a separate company in 2021.[15] With coal being the dirtiest fossil fuel in terms of both CO_2 emissions and local air pollution (see Chapter 12), companies that mine and trade coal, and power utilities that burn coal, are a major target of divestment and other climate-related shareholder campaigns (as discussed later in this chapter). It is likely that companies that are sensitive to such campaigns (mainly stock exchange-listed multi-nationals) will continue to spin off their thermal coal holdings, although it is far from clear that this will lead to reduced coal production.

Oil and gas companies

The top ten oil-producing companies in 2020 were China Petroleum and Chemical Corporation/Sinopec (China), Petrochina (China), Saudi Aramco (Saudi Arabia), Shell (UK), BP (UK), Exxon Mobil (USA), Total (France), Chevron (USA), Marathon (USA) and Lukoil (Russia).[16] The top ten gas-producing companies in 2020 were Petrochina, Saudi Aramco, BP, Exxon, Shell, Total, Chevron, Rosneft (Russia), Lukoil (Russia) and Gazprom (Russia).[17] The revenues of these companies are many times those of the largest

15 www.angloamerican.com/media/press-releases/2021/08-04-2021. Accessed April 28, 2023.
16 www.investopedia.com/articles/personal-finance/010715/worlds-top-10-oil-companies.asp
17 www.investopedia.com/articles/markets/030116/worlds-top-10-natural-gas-companies-xom-ogzpy.asp

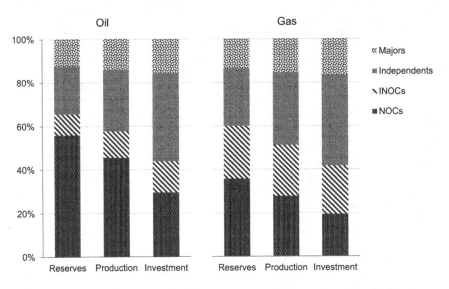

Figure 4.5 Ownership of reserves, production and investment by type of oil and gas company.

Source: International Energy Agency, 2020, p. 19.

coal-mining companies. For example, Saudi Aramco in 2020 had revenues of US$287 billion.[18]

Oil and gas companies are diverse in terms of both size and ownership, and may be placed for analysis in one of four broad groups: National Oil Companies (NOCs), owned predominantly or wholly by their home state; International NOCs (INOCs), NOCs which have considerable investments in other countries; the Majors (large privately owned multi-national companies); and Independents (smaller private oil and gas companies), of which there are many. The majors are BP, Chevron, ExxonMobil, Shell, Total, ConocoPhillips and Eni. Examples of NOCs are Rosneft (Russia), Petrobras (Brazil), China National Petroleum Corporation (CNPC) and Petronas (Malaysia). Examples of INOCs are Gazprom (Russia), Sinopec (China) and India's Oil and Natural Gas Corporation. Figure 4.5 shows that NOCs and INOCs own over 60% of global reserves of oil and gas, and produce over 50%, with the majors owning and producing well under 20%.

When companies measure their GHG emissions, they divide them into Scope 1, Scope 2 and Scope 3 emissions. As shown in Figure 4.6, Scope 1 emissions come directly from the companies' own activities (use of coal, oil

18 https://en.wikipedia.org/wiki/List_of_largest_coal_mining_companies. Accessed April 28, 2023.

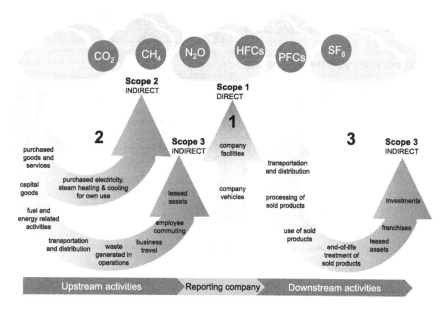

Figure 4.6 Scope 1, 2 and 3 emissions from company activities and products.

Source: Watterson, 2014, p. 13.

and gas in their processes, for heating, or for travel and transport). Scope 2 emissions come from their use of electricity, and depend on their source of power (carbon intensity of grid electricity, or as otherwise purchased). Scope 3 emissions are those associated with the companies' upstream supply chains (all the goods and services they buy to enable them to produce their output), plus any emissions that arise directly from the use of their products (downstream).

For fossil fuel companies, the great majority of their emissions come from their product, when it is burned (part of their Scope 3 emissions). Figure 4.7 breaks down global GHG emissions into multiple parts. The second bar down shows that the great majority of global GHGs comes from fossil fuels, and the third bar down shows that the majority of these emissions come from their consumption (combustion). The fourth bar breaks out emissions from coal from those of oil and gas, showing that they are broadly the same. The fifth bar shows that, as with fossil fuels as a whole, the great majority of oil and gas emissions come from their combustion. The sixth bar shows that only about half of production emissions come from the oil and gas companies' own use of energy (Scope 1 and 2), the other half (Scope 3) coming from the supply chain to build their infrastructure (e.g. refineries, oil rigs).

All fossil fuels are carbon-intensive (Scope 3 emissions when they are burned), and all operations to extract and refine them release carbon emissions (Scope 1 and 2 emissions when they are produced), but their carbon intensity

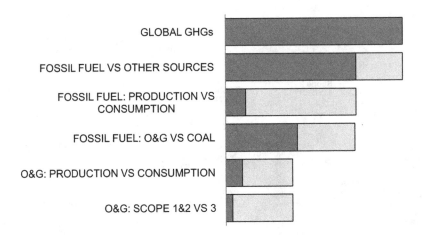

Figure 4.7 Illustrative breakdown of global GHG emissions from fossil fuels.

Source: Al-Kuwari et al., 2021, Figure 2, p. 8.

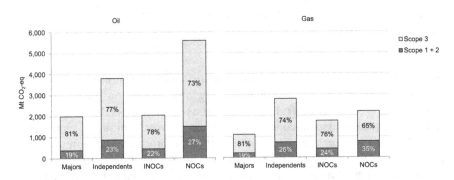

Figure 4.8 Scope 1, 2 and 3 emissions in 2018 for different categories of oil and gas companies.

Source: International Energy Agency, 2020, p. 31.

differs widely not just between different fossil fuels (Table 4.1) but also between different sources of the same fossil fuel, and between the companies that extract them. While Figure 4.7 gives an illustrative average global breakdown between Scope 1, 2 and 3 emissions, Figure 4.8 shows it for the different categories of oil and gas companies. For oil the proportion of production (Scope 1 and 2) emissions, and the level of emissions overall, are largest for the NOCs, with this group of companies producing over 5.5 GtCO$_2$e once their oil is burned. For gas, it is the INOCs who have the largest percentage of Scope 1 and 2 emissions, but the independents' emissions are highest in absolute terms.

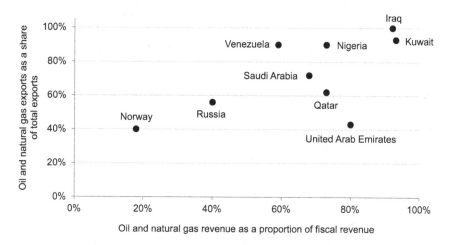

Figure 4.9 Oil and gas as a share of total exports and as a share of total fiscal revenue in selected countries, 2017.

Source: International Energy Agency, 2020, p. 104.

The carbon intensity of oil extraction varies widely between different countries, as well as between companies, depending on the quality of the crude oil and the nature of the various extraction processes used. The emissions from extraction come from the energy required for extraction and refining, the flaring of gas that is sometimes produced with the oil where there is no pipeline to feed the gas into, methane emissions from extraction, and the transport of the fuel. International Energy Agency (2020, p. 109) shows the different contributions to emissions from the various production processes, with Saudi Arabia and UAE having the lowest production emissions, and Venezuela and Iraq the highest. With the combustion of one barrel of oil equivalent (boe) producing around 405 $kgCO_2e$, Venezuela, with its very heavy oil sands, emits over half that in order to produce one barrel of its crude oil.

Many of the major oil-producing countries are very dependent on revenues from fossil fuels, in terms of their share in both exports and fiscal revenues. As shown in Figure 4.9, for Iraq and Kuwait the shares are very high in both categories. Other Middle East countries, Nigeria and Venezuela also have high fiscal and export dependency. For these countries decarbonisation is both a huge industrial challenge and an economic threat. The same goes for the oil and gas companies.

The dismal record of the fossil fuel industry and climate change

Fossil fuel companies have known for nearly half a century, if not more, well before most politicians and the public, that their product would destabilise the climate with potentially catastrophic effects. A 2015 article in *Scientific American* makes clear that Exxon (now ExxonMobil) knew about climate change by 1977, if not before, and that in subsequent decades it publicly refused to acknowledge the issue and actively promoted climate misinformation. Its tactics have been compared to those of the tobacco industry, which spread lies about the health impacts of smoking.[19]

Not content with muddying the scientific water, and sowing doubt and confusion about the threat of climate change in the minds of politicians and citizens, oil and gas companies have tried to prevent or delay climate legislation through engagement in the political system, contributing US$84 million to the campaigns of candidates for the US Congress in 2018 alone. Goldberg et al. (2020) show that they have invested in the political campaigns of politicians with a proven anti-environmental record. In 2019 Forbes reported:

> Every year, the world's five largest publicly owned oil and gas companies spend approximately $200 million on lobbying designed to control, delay or block binding climate-motivated policy. … BP has the highest annual expenditure on climate lobbying at $53 million, followed by Shell with $49 million and ExxonMobil with $41 million. Chevron and Total each spend around $29 million every year.[20]

Now that these tactics, in the face of undeniable evidence of dangerous climate change, have failed to stop countries seeking to reduce emissions, oil and gas companies have changed the tack of their lobbying, claiming that fossil fuels can be 'green' through the implementation of CCS, or by making hydrogen, or that they can contribute to the energy transition by producing more gas, or that they are not really oil and gas companies, but energy companies, and that they are turning to renewables or other energy activities. None of these claims is completely false. CCS does produce less emissions than unabated fossil fuel use, as will be seen in Chapter 7, and it can be used to produce low-carbon hydrogen, as will be seen in Chapter 6.

In addition, some of the majors are indeed investing in renewables, but on a still insignificant scale compared to their oil and gas operations. For the 10 companies it analysed,[21] International Energy Agency (2020, p. 47) found that "aggregate annual capital expenditures for projects outside core oil and gas supply [since 2015] averaged … less than 1% of the total capital expenditures by these companies". Of the four companies analysed by Li et al. (2022, Figure 7a, p.18), the most ambitious, BP, invested just 2.3% of its capital expenditure

---

19 www.scientificamerican.com/article/exxon-knew-about-climate-change-almost-40-years-ago/. Accessed April 28, 7023.

20 www.forbes.com/sites/niallmccarthy/2019/03/25/oil-and-gas-giants-spend-millions-lobbying-to-block-climate-change-policies-infographic/?sh=17a1969a7c4f. Accessed April 28, 2023.

21 BP, Chevron, Eni, ExxonMobil, Shell, Total, CNPC, Equinor (Norway), Petrobras, Repsol (Spain).

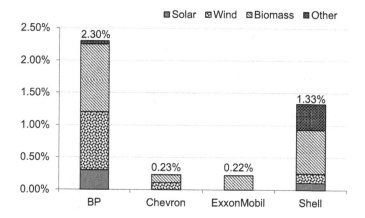

Figure 4.10 Disclosed investment in low-carbon energy production and development, as a proportion of total capital expenditure, 2010–2018.

Note: 'Other' indicates hydropower, CCUS, and smart technologies.

Source: Li et al., 2022, Figure 7a, p. 18.

in renewables over 2010–2018. The figure for ExxonMobil is 0.22%, invested exclusively in biomass (Figure 4.10).

In contrast to this very small share of investment in renewables, Romanello et al. (2022) show that "the current strategies of 15 of the largest oil and gas companies[22] would lead to production exceeding their share of levels consistent with limiting the global average surface temperature rise to 1.5°C by 37% in 2030, and 103% in 2040" (Romanello et al., 2022, Figure 16, p. 1643). For the International Oil Companies (IOCs) the projected excess production in 2040 is 87%, while for the NOCs it is even higher at 111%.

However, the trend of investment in renewables, from the majors at least, may be on an upward path. Table 4.4 shows the change in investments in clean energy and traditional oil and gas activities from different categories of oil and gas companies over 2019–2022. For the European majors, investments in clean energy technologies rose by around US$14 billion, while investments in traditional activities fell by US$15 billion. Only the Middle East NOCs increased their net investment in traditional activities. Given the huge profits of these companies in 2022 because of the high oil and gas prices, it is profoundly to be hoped that all of them will greatly increase their investments on the clean energy side.

22 International Oil Companies (IOCs): BP, Chevron, Eni, Exxon Mobil, Shell, Total, Lukoil, Eni, ConocoPhillips; National Oil Companies (NOCs): Saudi Aramco (Saudi Arabia), Gazprom (Russia), NIOC (Iran), PetroChina, Rosneft (Russia), KPC (Kuwait), ADNOC (Abu Dhabi).

Table 4.4 Change over 2019–2022 (2022 estimated) in investments by different categories of oil and gas companies in clean energy technologies and traditional activities

	Change in investment in traditional oil and gas activities, 2019–2022 (billion US$ 2021)	*Change in investment in clean technologies, 2019–2022 (billion US$ 2021)*	*Net change, 2019–2022 (billion US$ 2021)*
Middle East NOCs	5	0	5
European majors	−15	14	−2
Russian companies	−3	0	−3
Chinese NOCs	−8	0	−7
Other NOCs	−10	3	−7
US majors	−15	1	−14
Independents	−16	0	−16

Note: NOC stands for National Oil Company.

Source: International Energy Agency, 2022, p. 17.

As well as the investments ("production, expenditures and earnings for fossil fuels as well as investments in clean energy") of these four companies (BP, Chevron, ExxonMobil, Shell), Li et al. (2022, p. 3) examined two other aspects of their engagement with the energy transition: their discourse ("frequency of climate- and clean-energy-related keyword use in annual reports"); and their "strategies, pledges and actions related to decarbonization and clean energy".

Their excoriating conclusion is worth quoting at length:

The discourse analysis revealed a distinct increase in keywords in annual reports related to climate change and clean energy, particularly by the European majors, BP and Shell. ... Over the study period, the European majors have more consistently acknowledged climate science, participated earlier in industry climate-change frameworks, adopted internal carbon pricing, spent and pledged more on clean energy, and recently set net-zero transition and energy product decarbonisation goals. Trailing far behind, the American majors [Chevron and ExxonMobil] continuously exhibit defensive attitudes to renewables investment and the need to shift from fossil fuels, explicitly stating ambitions to grow rather than reduce hydrocarbon production. For all majors, however, we caution that most strategy scores have come from "low-hanging fruit" in the form of pledges and disclosure. ... Moreover, we found that some actions contradict pledges. This especially concerns intentions to curb the production of fossil fuels as well as reduce exploration and new developments.

This worrying trend of acting contrary to pledges and public statements has also been highlighted by other sources. This includes reports that all four majors continue to lobby governments to hamper or weaken carbon pricing policies [citations given], to secure favorable fiscal support, and to weaken environmental regulations [citations given]. Also in the goal of obstructing the progress of decarbonization, they continue to redirect the responsibility for reducing GHG emissions to consumers [citations given] while diffusing misleading advertisements that fossil fuels (especially gas) are green [citations given] and exaggerating the scale of clean energy investments [citation given].

> The analysis of financial behavior ... failed to show any major comprehensively transitioning its core business model away from fossil fuels. ... Given the mismatch between discourse, pledges, actions and investments, aligning with recent studies [citations given], we conclude that no major is currently on the way to a clean energy transition. ... Until the three areas of discourse, actions and investment behaviour are brought into alignment, we conclude that accusations of greenwashing by oil majors are well-founded [citations given].

(Each [citation] entry gives one or more references to the scientific literature that backs up the point made.)

Such an analysis is further justified by the recent news that, in response to high global energy prices, in 2023 BP scaled back its emission reduction commitments,[23] while Shell dropped its target to reduce its production of conventional fossil fuels by 55% by 2030.[24]

Future generations suffering under the full impact of climate change will read such words with justified anger. If they are aware of it, they will endorse the characterisation of the fossil fuel industry as an out-of-control supercomputer, pursuing its own objectives to the great detriment of humanity at large.[25]

Climate action against the fossil fuel industry and vice versa

Faced with this seeming determination by the fossil fuel industry to wreck the climate, which policymakers seem unwilling or unable to counter, civil society organisations and citizens have come up with a range of strategies to counter them. These may be broadly characterised as *protest, litigation, shareholder action* and *targeting investors* in this industry.

Protest of course takes many forms, from letter-writing to protests and demonstrations and direct action. In the UK action against fossil fuel companies has come from Extinction Rebellion (XR), 350.org, Fridays for Future,

23 www.theguardian.com/business/2023/feb/07/bp-profits-windfall-tax-gas-prices-ukraine-war. Accessed June 18, 2023.
24 Shell Confirms Change in Strategy to Maintain Fossil Fuel Production Through to 2030. www.edie.net/new-shell-ceo-drops-targets-to-reduce-oil-production/. Accessed June 18, 2023.
25 Professor Stuart Russell, in his 2021 Reith Lectures. 35.30 in www.bbc.co.uk/sounds/play/m0012q21, and personal communication, February 28, 2022.

Table 4.5 Number of climate change litigation cases in different continents

Continent	Number of litigation cases
Africa	13
Asia Pacific	28
Australia	124
European Union & UK	143
North America	1426
Latin America & Caribbean	47

Source: Data from Setzer and Higham, 2022. www.lse.ac.uk/granthaminstitute/wp-content/uploads/2022/08/Global-trends-in-climate-change-litigation-2022-snapshot.pdf, Accessed May 5, 2023.

Greenpeace and others. Another example is an offshoot of XR, called Just Stop Oil, which through 2022 targeted energy infrastructure, blocked roads, and engaged in other disruptive activity to protest against the UK Government's intention to grant new North Sea oil and gas licences.[26] There are such actions and organisations in many other countries.

Litigation against fossil fuel companies has increased dramatically in recent years. Table 4.5 shows, as at April 2023, the number of climate litigation cases in different continents. The USA has easily the largest number (1,419) of cases.

Perhaps the best-known case of climate change litigation is the ruling against Shell in May 2021 in the Netherlands that it should reduce its net emissions (Scope 1, 2 and 3) by 45% (compared to 2019 emissions) by 2030.[27] Shell appealed against this decision in March 2022.[28]

However, civil society groups and citizens are not the only ones engaging in litigation. Fossil fuel companies are engaging in their own actions, suing governments for enacting climate policies that either prevent them from investing in new fossil fuel projects or that reduce the value of past investments. Sky News reported in September 2021 that such lawsuits to the value of US$18 billion had been filed.[29] Four European governments (Netherlands, Slovenia, Italy and Poland) are being sued for €4 billion by five energy groups (including Germany's RWE and Uniper and the UK's Rockhopper) under the Energy Charter Treaty,[30] signed in the early 1990s, with the intention of protecting international energy investments. Many more lawsuits are likely if European

26 www.bbc.co.uk/news/uk-63543307. Accessed April 28, 2023.

27 http://climatecasechart.com/non-us-case/milieudefensie-et-al-v-royal-dutch-shell-plc/. Accessed April 28, 2023.

28 www.reuters.com/business/sustainable-business/shell-filed-appeal-against-landmark-dutch-climate-ruling-2022-03-29/. Accessed April 28, 2023.

29 https://news.sky.com/story/fossil-fuel-companies-are-suing-governments-across-the-world-for-more-than-18bn-12409573. Accessed April 28, 2023.

30 www.ft.com/content/b02ae9da-feae-4120-9db9-fa6341f661ab. Accessed April 28, 2023.

governments push on with their stated climate policies, with the legal arbitration process coming under criticism for bias in favour of investors.[31]

Tienhaara et al. (2022) explore in detail the topic of investor-state investment treaties, of which the Energy Charter Treaty (ECT) is the most important in respect of fossil fuels, but there are 81 other foreign investment treaties that could protect fossil fuel projects, under which there is a potential liability for climate action of up to US$340 billion (Tienhaara et al., 2022, p. 703). The possibility of having to pay compensation to fossil fuel companies on this scale is certain to exert a chilling effect on the ambition of decarbonisation plans, just when it is most urgent that ambition needs to be enhanced.

NGOs including the European Climate Foundation are now engaged in campaigning against the ECT, and by November 2022 Poland, Spain, Germany, France, The Netherlands, Slovenia and Luxembourg had all announced that they intended to leave it. However, the ECT has a 'sunset' clause meaning that even after a country withdraws, it remains subject to litigation for 20 years.[32] This and other issues are being examined under a process of 'modernisation' of the ECT which is being pursued by the European Commission, although few details have so far been released as to what this will entail, despite it being discussed at the ECT meeting in Mongolia in November 2022.[33]

Shareholder action entails investors putting forward motions at company annual general meetings that the companies should accelerate climate action, or even seek to replace Board members with others who will be more active on climate issues. In the same week as the Shell ruling mentioned above, shareholders of ExxonMobil voted two members onto Exxon's 12-person Board, in protest at Exxon's lack of action on climate issues,[34] and Chevron shareholders voted by 61% that Chevron should reduce its Scope 3 emissions.[35] It may be that such shareholder concern is not motivated purely by concern about the climate. There have been voices in the financial sector warning for some years that the climate risk to fossil fuel companies was not only because of its impacts, but also that, if the world's policymakers got serious about reducing emissions, significant oil and gas reserves on companies' balance sheets, and some of their production infrastructure, would lose their value, a phenomenon noted above and known as 'stranded assets'.

31 www.theguardian.com/business/2022/nov/14/revealed-secret-courts-that-allow-energy-firms-to-sue-for-billions-accused-of-bias-as-governments-exit?ref=upstract.com.   Accessed   April 28, 2023.

32 https://europeanclimate.org/stories/lifting-a-climate-block-towards-the-end-of-the-energy-char ter-treaty/). Accessed April 28, 2023.

33 www.energycharter.org/media/news/article/the-33rd-meeting-of-the-energy-charter-conference-held-under-the-chairmanship-of-mongolia/. Accessed April 28, 2023.

34 www.forbes.com/sites/christopherhelman/2021/05/27/shareholders-rebuke-exxonmobil-on-clim ate-in-a-wake-up-call-for-big-oil/?sh=4b02cb4151e4. Accessed April 28, 2023.

35 www.reuters.com/business/energy/chevron-shareholders-approve-proposal-cut-customer-emissi ons-2021-05-26/. Accessed April 28, 2023.

Finally, campaigners against the fossil fuel industry have targeted investors. The world's 60 largest private banks are said to have invested more than US$3.8 trillion into the coal, oil and gas sectors since the 2015 Paris Agreement.[36] The campaigns are seeking to persuade banks and other investors not to invest in fossil fuel companies and, through so-called divestment, to dispose of any such investments which they may have. They have not been wholly unsuccessful: between 2008 and the end of 2020, nearly 1400 organisations, with assets worth over US$14 trillion, committed to divestment (Romanello et al., 2021, p. 1647). Again, there may have been financial considerations, as well as climate concern, behind some divestment. *Forbes* reported in February 2021: "Since its inception in 2012, the S&P 500's Fossil Fuel Free Total Return Index has consistently outperformed the S&P 500 overall."[37] This may well not have been true in 2022 when oil and gas prices increased enormously as a result of the Ukraine war, delivering unprecedented profits to oil and gas companies.

The problem with the shareholder action and divestment campaigns is that they tend to target, and to be more effective with, the majors, who are investor-owned and are much more exposed to public opinion than the NOCs and INOCs. And yet, as seen in Figure 4.5, it is the NOCs and INOCs who both own and produce the majority of the world's oil and gas, but about the extent of whose fossil fuel reserves and resources there is still considerable uncertainty.

In 2019 a campaign was started by civil society organisations to enact a global Fossil Fuel Non-Proliferation Treaty "to phase out fossil fuels and build a globally just transition for every worker, community and country".[38] It calls on governments to end the expansion of fossil fuels, phase out existing fossil fuel and pursue plans to ensure 100% access to renewable energy globally. The Treaty has been endorsed by thousands of academics and scientists, over 100 Nobel Laureates, and cities including Amsterdam, Los Angeles, Sydney, Barcelona, Vancouver and Toronto.

An initiative of the Fossil Fuel Non-Proliferation Treaty is the Global Registry of Fossil Fuels, which has compiled

> the only public domain database of fossil fuel emissions at source. Production and reserves figures from 139 countries give universal coverage at country level and are also broken down into 14,188 projects covering about 70 percent of production at project level.[39]

Such publicly available information will be essential if the world's governments are ever going to call time on the age of fossil fuels.

36 www.ft.com/content/73615213-08ce-4786-9b8c-773029552bbc. Accessed April 28, 2023.
37 www.forbes.com/sites/davidcarlin/2021/02/20/the-case-for-fossil-fuel-divestment/?sh=1dd6d2c17 6d2. Accessed April 28, 2023.
38 https://fossilfueltreaty.org/history. Accessed August 9, 2023.
39 https://fossilfuelregistry.org/. Accessed August 9, 2023.

Other significant initiatives working to bring an end to fossil fuel production are the Beyond Oil and Gas Alliance (BOGA) and the Powering Past Coal Alliance (PPCA). BOGA is "an international alliance of governments and stakeholders working together to facilitate the managed phase-out of oil and gas production".[40] It was launched at COP26 by the governments of Denmark and Costa Rica, and other Core Members include Ireland, France, Portugal and Sweden; it aims "to elevate the issue of oil and gas production phase-out in international climate dialogues, mobilize action and commitments, and create an international community of practice on this issue".

PPCA was founded by the governments of Canada and UK at COP23 in 2017, and by December 2022 had a membership of 48 national governments, 48 sub-national governments and 77 global organisations. It works "to advance the transition from unabated coal power generation to clean energy", with the objective "to phase out coal by 2030 in the OECD and the EU, and by no later than 2040 in the rest of the world".[41]

A characteristic of the membership of both BOGA and PPCA is that no member country in either alliance is a major producer of the fossil fuels that they seek to end the use of. Even so, neither alliance is likely to find a smooth path to its objective. In November 2022 it was reported that Costa Rica was "backing away" from its leadership of BOGA,[42] and the approval of a new coal mine in the UK will do nothing to help the UK Government persuade other countries that its co-founding of PPCA five years ago was little more than a gesture at a time when, without a coal industry and very little coal-fired power generation, it had nothing to lose.

Conclusion

The most important priority if humanity is to get a grip on climate change is to phase out fossil fuels as quickly as possible. This is being strongly resisted by the fossil fuel industry, which, rather than using its great financial wealth to provide other, low-carbon, energy sources, seems determined to extract as much remaining value as it can from its fossil fuel reserves, and explore for more, even at the expense of catastrophic climate change. To this end it has confused the public by denying for a long time the science of climate change, and has done its best to stop or delay climate policy by lobbying governments against such policy and paying large sums of money to the political campaigns of politicians who are opposed to climate policy. While there are now some governments that are prepared to stand up to this lobbying, all too often governments that have fossil fuel resources in their countries are determined to produce them irrespective of the climate consequences. Until climate policy can

40 https://beyondoilandgasalliance.com/. Accessed April 28, 2023.
41 https://poweringpastcoal.org/our-work/. Accessed April 28, 2023.
42 www.climatechangenews.com/2022/11/03/costa-rica-cop27-oil-gas-phase-out-coalition/. Accessed April 28, 2023.

get the fossil fuel industry under control, there is very little chance of humanity preventing the worst effects of climate change. It would be the ultimate irony if the man appointed to be President of COP28, Sultan Al Jaber, who is also the Chief Executive of the Abu Dhabi National Oil Company (ADNOC), one of the largest oil producers in the world, with production plans to greatly exceed its share of the carbon budget for 1.5°C, as seen both above and in recent news reports,[43] turned out to be the man who would perform this historic task.

References

Al-Kuwari, O., Welsby, D., Solano, B., Pye, S., & Ekins, P. (2021). Carbon intensity of oil and gas production. *Researchgate*. https://doi.org/10.21203/rs.3.rs-637584/v1

Drummond, P., Scamman, D., Ekins, P., Paroussos, L., & Keppo, I. (2021). Growth-positive zero-emission pathways to 2050. *Sitra, Helsinki*. www.sitra.fi/en/publicati ons/growth-positive-zero-emission-pathways-to-2050/

Goldberg, M. H., Marlon, J. R., Wang, X., van der Linden, S., & Leiserowitz, A. (2020). Oil and gas companies invest in legislators that vote against the environment. *Proceedings of the National Academy of Sciences, 117*(10), 5111–5112. https://doi. org/10.1073/pnas.1922175117

Grubler, A., Wilson, C., Bento, N., Boza-Kiss, B., Krey, V., McCollum, D. L., Rao, N. D., Riahi, K., Rogelj, J., De Stercke, S., Cullen, J., Frank, S., Fricko, O., Guo, F., Gidden, M., Havlík, P., Huppmann, D., Kiesewetter, G., Rafaj, P., ... Valin, H. (2018). A low energy demand scenario for meeting the 1.5 °C target and sustainable development goals without negative emission technologies. *Nature Energy*. https:// doi.org/10.1038/s41560-018-0172-6

International Energy Agency. (2020). *The Oil and Gas Industry in Energy Transitions*. IEA, Paris. www.iea.org/reports/the-oil-and-gas-industry-in-energy-transitions

International Energy Agency. (2021). *Net Zero by 2050: A Roadmap for the Global Energy Sector*. IEA, Paris.

International Energy Agency. (2022). *World Energy Investment 2022*. IEA, Paris.

Li, M., Trencher, G., & Asuka, J. (2022). The clean energy claims of BP, Chevron, ExxonMobil and Shell: A mismatch between discourse, actions and investments. *PLOS ONE, 17*(2), e0263596. https://doi.org/10.1371/journal.pone.0263596

McGlade, C., & Ekins, P. (2015). The geographical distribution of fossil fuels unused when limiting global warming to 2°C. *Nature, 517*(7533). https://doi.org/10.1038/ nature14016

Romanello, M., Di Napoli, C., Drummond, P., Green, C., Kennard, H., Lampard, P., Scamman, D., Arnell, N., Ayeb-Karlsson, S., Ford, L. B., Belesova, K., Bowen, K., Cai, W., Callaghan, M., Campbell-Lendrum, D., Chambers, J., van Daalen, K. R., Dalin, C., Dasandi, N., ... Costello, A. (2022). The 2022 report of the Lancet Countdown on health and climate change: health at the mercy of fossil fuels. *The Lancet, 400*(10363), 1619–1654. https://doi.org/10.1016/S0140-6736(22)01540-9

Romanello, M., McGushin, A., Di Napoli, C., Drummond, P., Hughes, N., Jamart, L., Kennard, H., Lampard, P., Solano Rodriguez, B., Arnell, N., Ayeb-Karlsson,

---

43 www.theguardian.com/environment/2023/apr/04/revealed-uae-plans-huge-oil-and-gas-expans ion-as-it-hosts-un-climate-summit?CMP=share_btn_tw. Accessed April 8, 2023.

S., Belesova, K., Cai, W., Campbell-Lendrum, D., Capstick, S., Chambers, J., Chu, L., Ciampi, L., Dalin, C., ... Hamilton, I. (2021). The 2021 report of the Lancet Countdown on health and climate change: code red for a healthy future. *The Lancet, 398*(10311), 1619–1662. https://doi.org/10.1016/S0140-6736(21)01787-6

Schwietzke, S., Sherwood, O. A., Bruhwiler, L. M. P., Miller, J. B., Etiope, G., Dlugokencky, E. J., Michel, S. E., Arling, V. A., Vaughn, B. H., White, J. W. C., & Tans, P. P. (2016). Upward revision of global fossil fuel methane emissions based on isotope database. *Nature, 538*(7623), 88–91. https://doi.org/10.1038/nature19797

SEI, IISD, ODI, E3G, & UNEP. (2021). *The Production Gap: 2021 Report.* http:// productiongap.org/2021report

Semieniuk, G., Holden, P. B., Mercure, J.-F., Salas, P., Pollitt, H., Jobson, K., Vercoulen, P., Chewpreecha, U., Edwards, N. R., & Viñuales, J. E. (2022). Stranded fossil-fuel assets translate to major losses for investors in advanced economies. *Nature Climate Change 2022*, 1–7. https://doi.org/10.1038/s41558-022-01356-y

Setzer, J. and Higham, C. (2022) Policy Report, June, LSE Grantham Research Institute, London. www.lse.ac.uk/granthaminstitute/wp-content/uploads/2022/08/Global-tre nds-in-climate-change-litigation-2022-snapshot.pdf. Accessed May 5, 2022.

Tienhaara, K., Thrasher, R., Simmons, B. A., & Gallagher, K. P. (2022). Investor-state disputes threaten the global green energy transition. *Science, 376*(6594), 701–703. https://doi.org/10.1126/science.abo4637

Watterson, J. (2014). *Calculation of mitigation potential / carbon footprint / Life Cycle Assessment (LCA), including application of 2006 IPCC Guidelines.* www.ipcc-nggip. iges.or.jp/public/mtdocs/pdfiles/1407_Sofia/23_Appli2006IPCC_GLs_OtherAreas_ JohnWatterson.pdf

Welsby, D., Price, J., Pye, S., & Ekins, P. (2021). Unextractable fossil fuels in a 1.5 °C world. *Nature, 597.* https://doi.org/10.1038/s41586-021-03821-8

Willis, R. (2023) Use of evidence and expertise in UK climate governance: The case of the Cumbrian Coal Mine. *UCL Open: Environment Preprint.* https://ucl.scienceopen. com/hosted-document?doi=10.14324/111.444/000204.v1

5   The future is electric

Summary

All scenarios of deep decarbonisation see an important role for electricity provided by low- or zero-carbon sources (nuclear or renewables). Most such scenarios see a very great expansion of renewables (mainly solar and wind) and, to a lesser extent, nuclear. These sources of electricity are much less flexible than the fossil fuel generation that they are replacing. Renewables are dependent on time and place – they are not always available and different places have different amounts of sunlight and wind. This has important implications for many aspects of electricity provision, which are explored in this chapter. Energy storage becomes critical for those times when renewables are not available and for balancing the electricity system when there are fluctuations in supply or demand. Interconnection between the grids of different countries can help smooth out the different availabilities of renewables in different countries, and ambitious schemes have been planned to bring solar energy from North Africa to Europe at scale. Electricity networks may need extending to bring distant renewables to centres of demand, or to facilitate the local generation and consumption of renewable electricity. And electricity markets will need adjusting to ensure that consumers are able to benefit from cheap renewables, while still paying enough for their power to enable renewable generators to be profitable.

Introduction

Electricity for human use is not a naturally occurring energy source. It has to be created by conversion of other 'primary' energy sources. At the point of use, electricity produces zero emissions (of both greenhouse gases (GHGs) and local air pollutants). However, the generation of electricity may produce both these kinds of emissions at the power station where the conversion of the fuel to electricity takes place.

DOI: 10.4324/9781003438007-6

Figure 4.1 showed that in 2018 more than 60% of electricity was generated from fossil fuels. These are mainly coal and fossil methane gas, with coal producing more CO_2 and local emissions per unit of electricity generated at the power station, while the climate effects of burning gas for electricity are augmented by the 'fugitive' emissions of gas during its extraction and pipeline transport, and by the energy required to liquefy, transport and re-gasify it when it is converted to liquid natural gas (LNG).

Other significant sources of electricity in Figure 4.1 are nuclear power and renewables (wind, solar, hydro and bioenergy). In addition, electricity can be used to generate hydrogen (through electrolysis of water) and to make liquid synthetic fuels (synfuels, sometimes called e-fuels) through the synthesis of CO_2 and hydrogen. Such fuels can be used, as an alternative to electric vehicles, in internal combustion engines instead of petrol and diesel, or in aeroplane engines, instead of kerosene derived from crude oil.

The production of electricity from renewables and nuclear, the various challenges to which this gives rise, and how much low-carbon electricity is likely to be required by the required shift away from fossil fuels, are the subject of this chapter.

Future low-carbon electricity demand

There is pretty well universal agreement in scenario exercises such as have been reviewed in this book that the low-carbon future is largely electric. This is clearly shown in the three scenarios that have been repeatedly cited in this book.

In the IEA's Net Zero Emissions scenario (NZE) (International Energy Agency, 2021a), electricity demand more than doubles between 2020 and 2050, to comprise, at the end of this period, 66% of energy demand in buildings, 45% in transport and about half in industry (Figure 5.1a) (International Energy Agency, 2021a, pp. 60–61).

In the Low Energy Demand scenario (LED, Figure 5.1b), electricity in 2050 is about 60% of final energy use. In the Sitra 1.5°C scenario (Figure 5.1c) electricity use rises from less than 100 EJ per year to nearly 300 EJ per year by 2050, and to 450 EJ per year by 2100. The great majority of this expansion is provided by wind (onshore and offshore) and solar photovoltaics (PV).

Nuclear power

Nuclear electricity is currently generated by the fission of, usually, uranium. The description of how this is brought about is outside the scope of this book.[1] It may be that electricity one day may also be generated by nuclear fusion,[2]

---

1 For a fairly simple explanation see www.eia.gov/energyexplained/nuclear/. Accessed April 29, 2023.
2 For a fairly simple explanation see https://world-nuclear.org/information-library/current-and-future-generation/nuclear-fusion-power.aspx. Accessed April 29, 2023.

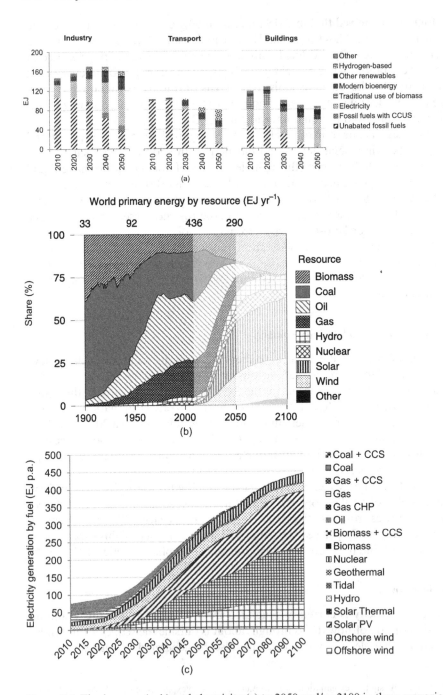

Figure 5.1 Final energy (a, b) and electricity (c) to 2050 and/or 2100 in three scenarios.
a) The IEA Net Zero Emissions scenario (International Energy Agency, 2021a, Figure 2.10, p. 62).
b) The Low Energy Demand scenario (Grubler et al., 2018, Figure 3b, p. 521).
c) The Sitra 1.5°C scenario (Drummond et al., 2021, Figure 17, p. 64).

the process in the sun that produces its energy, but the commercial application of this is likely still to be some decades away, and it will therefore not be in a position to make a significant contribution to decarbonisation for the next 20 years. All discussion of nuclear power in this book therefore refers to nuclear fission reactors.

The world's first nuclear reactor connected to an electricity grid was in Russia (then the Soviet Union) in 1954. The first commercial nuclear reactor was at Calder Hall in the UK and was opened in 1956. Capable of generating around 200 megawatts (MW) of electricity, it produced power until 2003. It is now being decommissioned, a process that, according to the BBC in September 2019, will cost £70 billion, and will not be completed before 2120.[3]

The number of reactors globally grew quickly in the following decades, reaching 400 in 1987. It has remained at around 440 since 1996, and in 2020 was operational in 32 countries and produced around 25,000 TWh, 10% of global, and 28% of low-carbon, electricity.[4] The global nuclear power capacity in 2020 was 393 GW.[5]

Most nuclear power is generated in West and Central Europe and North America. Of the 13 countries in 2020 that generated at least one-quarter of their electricity from nuclear, in France it accounted for three-quarters of its electricity consumption, in Slovakia and Ukraine for more than half, and in Belgium, Slovenia, Bulgaria, Finland and Czechia one-third or more. In South Korea the figure was around 30%.[6]

Nuclear power has for a long time been a controversial technology in some countries. While a detailed exploration of the reasons for this are beyond the scope of this book, they include the possibility of catastrophic accidents (following the accidents at the Chernobyl and Fukushima nuclear plants in 1986 and 2011 respectively); the lack of a long-term solution for the safe storage of nuclear waste; the historic and current links with military nuclear programmes; and the cost of construction, decommissioning and waste storage. The three most recent nuclear power stations to be built in Europe (Olkiluoto in Finland, Flamanville in France and Hinkley Point C in UK) have all run substantively over projected construction times and budgets, while decommissioning old nuclear power stations in the UK has a current estimated cost of £132 billion, to be paid for by current and future taxpayers, and will take 120 years.[7]

3 www.bbc.co.uk/news/uk-england-cumbria-49583192. Accessed April 29, 2023.

4 https://world-nuclear.org/information-library/current-and-future-generation/nuclear-power-in-the-world-today.aspx. Accessed April 29, 2023.

5 www.statista.com/statistics/263947/capacity-of-nuclear-power-plants-worldwide/.   Accessed April 29, 2023.

6 https://world-nuclear.org/information-library/current-and-future-generation/nuclear-power-in-the-world-today.aspx. Accessed April 29, 2023.

7 https://committees.parliament.uk/committee/127/public-accounts-committee/news/136734/sorry-saga-of-disused-nuclear-sites-will-cost-generations-of-uk-taxpayer/. Accessed April 29, 2023.

Most energy scenarios to 2050 have some electricity generation from nuclear power, but generally it is limited. In the IEA's NZE scenario in 2050 "almost 90% of electricity generation comes from renewable sources, with wind and solar PV together accounting for nearly 70%. Most of the remainder comes from nuclear" (International Energy Agency, 2021a, p. 19). In the LED scenario in 2050 nuclear plants generate less than each of solar PV, wind, biomass and hydro (Grubler et al., 2018, p. 520). The Sitra scenario also sees nuclear generating about 10% of electricity in 2050, and both a lower proportion and less in absolute terms in 2100 (Figure 5.1c).

The reason that all these models have such a limited role for nuclear power in 2050 is that they choose the least-cost generating options to meet the scenarios' energy demands, and nuclear is already more expensive than the cheapest renewables in many countries (see the next section) and remains so over the time horizon of the scenarios. A reason that is sometimes given for this is that the model calculations of the cost of renewables omit to include the costs of storage and grid balancing that are incurred by the fact that the major new renewables, wind and solar, are variable and intermittent. This is certainly not the case in respect of modern energy system models. This is an issue which will be explored further in the next section.

Of course, there is always some uncertainty about future costs. For example, a Korean consortium is completing four nuclear power units in the United Arab Emirates (three units of which are already generating[8] with the rest due to come online in 2022/2023) with a total capacity of 5.6 GW at a cost of between US$20 billion[9] and US$32 billion[10] (US$5.7 billion/GW at the high end of this estimate). This compares with the £23 billion estimated cost[11] of the 3.26 GW (£7 billion/GW) Hinkley Point C power station being built in the UK. If this lower cost can be generalised across the industry, and further cost reductions achieved as more units of the Korean type are built, then it may be that nuclear power would be better able to compete with renewables in the future.

One further point worthy of mention at this stage is the notion of 'capacity factor'. When power stations are built they have a rating that indicates the maximum amount of power that they can produce at any one time (e.g. the 3.26 GW for Hinkley Point C mentioned in the previous paragraph). With outages for maintenance and things going wrong, a nuclear power station may be expected over some period of time to produce 80% or more of this power capacity, and this is called its 'capacity factor'. Because the wind doesn't always

---

8 www.world-nuclear-news.org/Articles/Third-Barakah-unit-begins-commercial-operation. Accessed April 29, 2023.

9 www.power-technology.com/analysis/divided-opinion-inside-the-uaes-barakah-nuclear-plant/. Accessed April 29, 2023.

10 www.power-technology.com/projects/barakah-nuclear-power-plant-abu-dhabi/. Accessed April 29, 2023.

11 www.reuters.com/business/energy/edf-announce-new-cost-increase-delay-hinkley-point-nuclear-plant-2022-03-28/. Accessed April 29, 2023.

blow and the sun doesn't always shine, the 'capacity factors' for wind and solar are much lower. In 2018, according to the IEA,[12] the capacity factors for off-shore wind, onshore wind and solar were around 30–50%, 22–44% and 10–20% respectively. This means that much more renewable capacity (in GW) has to be built in order to generate the same amount of electricity (GWh). There are other problems caused by the variable nature of renewable energy which will be further discussed in the sections on renewables, and electricity networks, storage and markets below.

Renewables

The term 'renewables' covers a wide range of energy sources, as shown in Figure 5.1c for electricity (offshore and onshore wind, solar PV, solar thermal [not to be confused with solar thermal panels, which use sunlight to heat water for domestic use], also called concentrated solar power, hydro, tidal, geo-thermal, biomass). In addition, biomass can be used to create liquid fuels (bio-ethanol, bio-diesel) for transport applications, or biogas, as a substitute for fossil methane gas, or used directly for heating purposes by combustion. The term 'bioenergy' covers all these applications, as well as the 'traditional use of biomass' for heating and cooking referred to in Figure 5.1a, and is the subject of the next chapter.

Of the other renewables, offshore and onshore wind are now very familiar to most people, with their high towers and three-pronged blades turning slowly and extracting energy from the wind. An important innovation in offshore wind is the floating turbine, fixed by cables to the seafloor, rather than having a fixed foundation, which enables the turbine to be installed in deeper water. Similarly, solar photovoltaic (PV) panels are very familiar, whether they are in small-scale arrays, for example fixed to roofs (distributed PV), or whether they are in large PV farms (utility-scale PV). Although the latter can cover large areas, the land on which they stand can sometimes also accommodate some agricultural activities, such as sheep grazing. Concentrated solar power (CSP) consists of large arrays of mirrors that focus the sun onto a heat-transfer fluid used to raise steam to drive a turbine. They require large areas of land and very sunny conditions. In Europe they are concentrated in Spain, which has over 2 GW of CSP, and plans to double this by 2025, but Morocco has the largest CSP plant in the world (510 MW, spread over 2,500 hectares), though it will shortly be overtaken by one being built in Dubai.[13]

Large-scale hydropower is produced by large dams on rivers, which create deep reservoirs behind them, which drive turbines as water flows through them. The areas of land submerged, which may have had large populations living there, who have to be resettled, and the associated loss of agricultural

12 www.iea.org/data-and-statistics/charts/average-annual-capacity-factors-by-technology-2018. Accessed April 29, 2023.
13 www.brunel.net/en/blog/renewable-energy/concentrated-solar-power. Accessed April 29, 2023.

land, villages and forests, mean that the construction of large dams has nearly always been controversial and contested. The Three Gorges Dam in China, completed in 2015 and, with a capacity of 22.5 GW, the largest hydropower facility in the world, displaced 1.3 million people, and its reservoir of 600 km length destroyed "natural features and countless rare architectural and archaeological sites".[14] The total number of dams with a reservoir area exceeding 0.01 ha has been estimated at 16.7 million, and while dams big and small are still being built, there is also a movement in Europe and the USA (13 European countries have 230,000 dams, and the USA has 91,000 over 7.2 m in height) to remove dams, for reasons of safety or to restore rivers (Habel et al., 2020). The droughts that are predicted to be more extreme as a result of climate change pose a severe risk to hydropower in some countries in the future, and have already affected power generation in the USA, China and Brazil.[15]

Tidal power arises from the movements of the tides under the gravitational pull of the moon. Electricity may be generated from the movement of tidal currents (also called tidal streams) under the surface of the sea, or by constructing a barrage (dam) across an estuary through which water flows as the tide goes in and out. The first large-scale barrage was constructed at La Rance across the Rance river estuary in Brittany, France, and started generating power in 1966, but its 240 MW size was surpassed by the 254 MW Lake Sihwa Tidal Power Station in South Korea, which opened in 2011.[16] In the UK plans for barrages across the Severn and Mersey estuaries have been discussed for many years (the first proposal for a Severn barrage was put forward in 1849), but have so far been rejected on the grounds of cost. Schemes of various sizes in various places on the Severn estuary could cost up to £34 billion for 8.6 GW peak, or 2.0 GW average, power.[17] Friends of the Earth in Wales opposed the barrage in 2013 on the grounds of both cost (saying that wind and solar were cheaper renewables) and environmental impact.[18]

In comparison with tidal barrages, tidal stream power is at a very early stage of development. The largest proposed power plant is MeyGen, in the Pentland Firth off the north coast of Scotland. Phase 1 of MeyGen comprises four turbines with a combined power of 6 MW, mounted on foundations on the seabed. The turbines are operational and in 2020 generated 37 GWh of electricity. The proposed eventual size of MeyGen is 398 MW.[19]

Geothermal power plants use the heat (or hot water) in Earth's crust to generate steam and drive turbines. Geothermal capacity in 2019 was 15.4 GW, with

14 www.britannica.com/topic/Three-Gorges-Dam. Accessed April 29, 2023.

15 www.reuters.com/business/sustainable-business/inconvenient-truth-droughts-shrink-hydropower-pose-risk-global-push-clean-energy-2021-08-13/. Accessed April 29, 2023.

16 www.nsenergybusiness.com/features/worlds-biggest-tidal-power-plants/#. Accessed April 29, 2023.

17 https://en.wikipedia.org/wiki/Severn_Barrage. Accessed April 29, 2023.

18 www.bbc.co.uk/news/uk-wales-20955412. Accessed April 29, 2023.

19 https://tethys.pnnl.gov/project-sites/meygen-tidal-energy-project-phase-i. Accessed April 29, 2023.

USA having the most capacity with 3.7 GW, Indonesia, Philippines, Turkey and New Zealand having more than 1 GW, and Mexico, Italy, Kenya, Iceland and Japan having more than 500 MW. Philippines, Kenya and El Salvador generate more than 20% of their electricity from geothermal power. Estimates of geothermal power potential range from 35 GW to 2 TW.[20]

Electricity generation from renewables roughly doubled from 2000 to 2008, but fell slightly as a share of total global generation, doubled again to 2020, when its share of global generation rose to 25%, and, in the IEA's main case forecast to 2026, will increase its share to more than 35% by 2026. Hydropower was the largest single source of renewable electricity until 2020, but since then the rate of growth of onshore wind and solar PV have been such that by 2026 its contribution will be surpassed by that of other renewables (International Energy Agency, 2021b Figure 1.9, p. 31).

There are other potential sources of renewable energy which are either at an earlier stage of development, or which may have important applications in some places, but not be significant globally.

Electricity/energy storage

It has been seen that all low-carbon scenarios project a major contribution from renewables, particularly solar PV and wind. An obvious characteristic of these energy sources is that they are variable and intermittent – the wind doesn't blow, and the sun doesn't shine, all the time, and when they do, they do so with varying intensity. This fact is reflected in the concept of 'capacity factor' discussed above, whereby the average power (and energy) outputs of wind and solar plants are below their peak output. But there is also the question of when power is available.

A characteristic of electricity is that it cannot be stored as such. When supply exceeds demand, as it often will in an electricity system mainly powered by renewables, the only two options are to disconnect the supply (called 'curtailing'), and waste the renewable energy, or convert the electricity into another form for energy storage, as is discussed below. Moreover, energy demands have to be met instantaneously by new supply coming on stream immediately if the electricity system is to remain stable. A thought experiment helps to illustrate the issue. Imagine a television showing a crucial football match, Olympic race, or jubilee celebration that had 25 million UK households (about 90% of the total) watching. The watchers all get up at a break in the event for a cup of tea and put on a 3 kW electric kettle. Immediately 75 GW of power demand surges through the network. The overall UK system only has 76 GW in total. The system would be simply overwhelmed in the absence of adequate storage, back-up or interconnection across the channel to other electricity systems with spare capacity. And if this happened at a time when

20 https://en.wikipedia.org/wiki/Geothermal_power#cite_note-IPCC-3. Accessed April 29, 2023.

there was no or little wind or sun, the challenges would be even greater. Some of these issues will be discussed further in the following section on electricity networks. The focus here is on storage.

Balancing the supply and demand of electricity is one of the major problems of electricity grid operators, even when fossil fuels are the principal source of supply, but these problems are much exacerbated, as has been seen, when their role is taken by intermittent and variable renewable energy sources. The scale of the issues that arise can be shown by the typical pattern of energy demand over a year in a country like the UK, and of electricity demand over an average winter's day.

Figure 5.2 shows a number of important characteristics of an energy system like the UK's. Obviously other countries will show similar or different characteristics depending on their fuel sources, level of industrial development and latitude, but many of the points made below will apply generally even if there are major specific differences.

The most obvious conclusion from Figure 5.2 is that winter gas use is far higher than winter electricity use, in terms of GWh used. This has huge implications for decarbonisation if it is envisaged, as it often is, that electricity will replace gas for household heating. The first implication is that if electricity were to replace gas 1:1 in terms of energy used (e.g. through the use of electric resistance heating), the electricity grid would need to have two to three times the capacity of the current UK grid. Given the much lower capacity factor of intermittent wind and solar than the generation it would be replacing, this would increase the required rated capacity of the generating plant to be four to

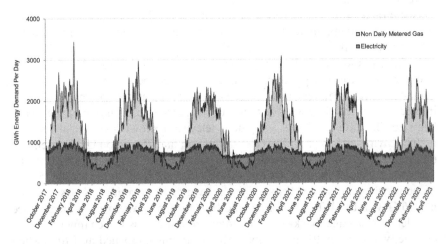

Figure 5.2 UK electricity demand and household use of gas, 2017–2023.

Source: Data taken from the National Grid website (www.nationalgrid.com/uk/) and supplied by Grant Wilson, University of Birmingham.

five times that of the current grid. This emphasises the importance of increasing the energy efficiency of the UK building stock. If UK household gas use could be cut by half in the winter through increased building energy efficiency, then much less new renewable electricity capacity would be needed, greatly reducing the investment required for decarbonisation. If their heat demand could be met by heat pumps (as discussed in a later section), which can generate three units of heat for each unit of electricity, rather than electric resistance heating, which only delivers one unit of heat per unit of electricity input, the required size of the electricity grid would be reduced still further.

The second obvious point arising out of Figure 5.2 is the difference between summer and winter gas use, meaning that electricity generating plant sufficient for winter demand, if electricity substituted for gas, would lie idle through the summer, thereby reducing the return on the investment needed to put it into place. However, if in the summer the generating capacity could use its generation to put energy into storage for the following winter, that would ease the winter burden. The most likely candidate for this storage at present is hydrogen, which is the subject of the next chapter.

The third issue arising from Figure 5.2 is the variation in power generation during the day and between the seasons. These ups and downs in the electricity line are less obvious than the previously mentioned characteristics, but they amount to daily variations in power demand (not visible in Figure 5.2) of 10–20 GW depending on the season. This is shown more clearly in the two weekly power demand charts, in high summer for June 20–27, 2022 (Figure 5.3a) and the peak demand in mid-winter, January 3–10, 2022 (Figure 5.3b).

Both weeks clearly show the difference between the daytime and night-time power demands, with daytime power demands typically being 15 GW more than during the night. Peak weekend power (June 25/26, January 8/9, the last two days in the series) is about 5 GW lower than during the week. The week in summer (Figure 5.3a) shows relatively high solar power, which matches the daytime peaks. For four days of the week wind power is very low (less than 5 GW), but then picks up reaching 15 GW by the end of the week. Gas (combined cycle gas turbine or CCGT) generation is the reverse of the wind, and follows demand, being high when wind is low, and vice versa, with a little coal generation sometimes required to meet the demand peaks. Nuclear provides a steady 5 GW. The grey line shows trading of electricity through interconnectors to the electricity system in mainland Europe, with exports to the continent below the 0 GW line, and imports from it above the line. Power exports tend to be higher overnight. This trade is driven by relative prices between the two systems – the UK exports when power prices are higher on the continent, and vice versa. Interconnection is discussed further below.

Winter peak demand (Figure 5.3b) in the week shown is 10 GW higher than the peak in June. There is much less solar power, but much more wind, only briefly falling below 5 GW on two occasions, and mainly remaining between 10 GW and 15 GW. High wind and low demand can cause gas (CCGT) power

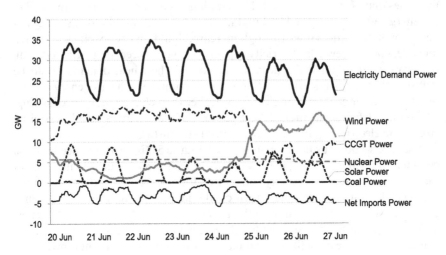

Figure 5.3a UK electrical power demand over the week June 20–27, 2022.

Source: Data taken from the National Grid website (www.nationalgrid.com/uk/) and supplied by Grant Wilson, University of Birmingham.

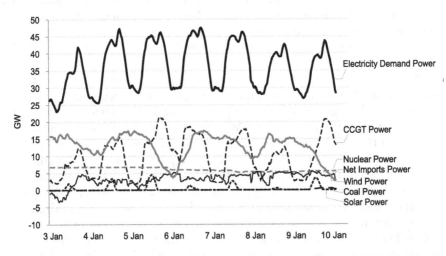

Figure 5.3b UK electrical power demand over the week January 3–10, 2022.

Source: Data taken from the National Grid website (www.nationalgrid.com/uk/) and supplied by Grant Wilson, University of Birmingham.

to fall below 5 GW. In contrast to the week in summer, power is now imported into the UK through the interconnectors.

The power demand across the full month of January (3–31) 2022 (not shown) also illustrates the great variability in the various sources of power in the UK. Peak wind is nearly 20 GW at the end of the month, but has fallen nearly to 0 on several occasions during the month. Gas (CCGT) follows the demand at times of low wind, and often meets 50% of the peak power demand. Nuclear stays steady at 5 GW. Most of the month the UK is importing power from the continent.

It is the Electricity System Operator (currently National Grid in the UK, though that may change soon) that has the challenging task of 'balancing' the grid, to ensure that electricity consumers, who do not want to consume electricity only when the wind is blowing or the sun is shining, can have it always instantly available at the flick of a switch, whatever the weather. With fossil fuels as the major energy source for electricity, this was not so hard to achieve – they are more or less instantly available when required (called 'dispatchable'). Fossil fuels (assuming that supply chains between producer and consumer countries are operating normally) also act as huge stores of always available energy, the most obvious signs of which in the past were the great heaps of coal standing beside coal-fired power stations. These could ensure that the power stations could be flexibly turned up or down to meet demand.

Wind and solar energy require quite different storage arrangements. It is not just that energy needs to be stored so that it can be used when there is no wind or sun. It is also that, as the amount of renewable capacity increases, there will be many times when its available wind and solar electricity exceeds electricity demands, and without storage this renewable electricity will be wasted. Already, when wind and solar provide only around 30% of the UK's electricity on average, one report estimated that wind was curtailed in Great Britain on 75% of days in 2020, with over 3.6 TWh of wind, enough to power a million homes for a whole year, wasted because of the lack of electricity storage.[21]

With a couple of minor exceptions noted below, electrical energy cannot be stored as electricity, but has to be converted to another energy form for storage, and then converted back again to electricity when the power is needed. The conversion process is always less than 100% efficient, so that some energy is lost (usually as heat) in the process. A battery is a means of storing energy as chemical energy, which is converted into a flow of electricity when an electrical circuit is created. Rechargeable batteries enable electricity to be converted back into chemical energy when the battery is charged, so that the battery can be reused many times over.

There is a wide range of means of converting electricity to other energy forms, and then reconverting it back into electricity when electricity is required. The key issues to take into account when considering the effectiveness of an energy storage facility are "how quickly it can react to changes in demand, the rate of energy lost in the storage process, its overall energy storage capacity, and

21 https://renews.biz/65677/storage-could-slash-curtailment-of-british-wind/. Accessed April 29, 2023.

how quickly it can be recharged" (Environmental and Energy Study Institute, 2019). In the IEA's Net Zero Emissions by 2050 scenario,

> installed grid-scale battery storage capacity expands 44-fold between 2021 and 2030 to 680 GW. Nearly 140 GW of capacity is added in 2030 alone, up from 6 GW in 2021. To get on track with the Net Zero Scenario, annual additions have to pick up significantly, to an average of over 80 GW per year over the 2022–2030 period.[22]

This section looks first at batteries, and then at other energy storage technologies that convert electricity to another energy form, before converting it back to electricity as needed. The need for energy storage for a net-zero electricity grid is substantial. In its July 2020 Future Energy Scenarios for the UK,[23] the National Grid estimates that 20–40 GW of electricity storage will be required for net zero by 2050. For comparison, the current UK grid has a total capacity of around 76 GW.

Batteries that are used to support an electricity grid (as opposed to power devices) are called 'utility-scale' batteries. They provide a range of services to an electricity grid, categorised by the International Renewable Energy Agency (2019, Figure 4, p. 9) as System Operation, Investment Deferral, Solar PV and Wind Generators, and Mini-Grids.

The System Operation services include *frequency regulation* (restoring frequency changes due to momentary supply/demand imbalances), *flexible ramping* (enabling a fast increase in supply to match rapid increases in demand), and *black start services* (when the grid requires power to start up again following grid failure – in this case the batteries would substitute for diesel generators).

Investment Deferral services refer to the fact that batteries that store energy at times of low demand can be used to release it at times of peak demand, thereby reducing the need to build new 'peaking plant' to meet such demand. The batteries can also reduce the required investment in transmission and distribution networks at times of peak demand by being located at appropriate places in the network, as discussed further in the next section.

For Solar PV and Wind Generators, batteries allow their energy to be stored when it exceeds current demand, to be used later, thus reducing the amount of energy that is wasted through curtailment, and they enable such generators to offer 'firm' power to the grid, with the batteries able to substitute for these sources' variability and intermittency.

Finally batteries can replace the many diesel generators that are currently in service in places, such as hospitals, that require back-up power, or in other places that operate their own 'mini-grids'.

22 Grid-Scale Storage – Analysis – IEA. www.iea.org/energy-system/electricity/grid-scale-storage. Accessed August 17, 2023.
23 www.nationalgrideso.com/document/173821/download. Accessed April 29, 2023.

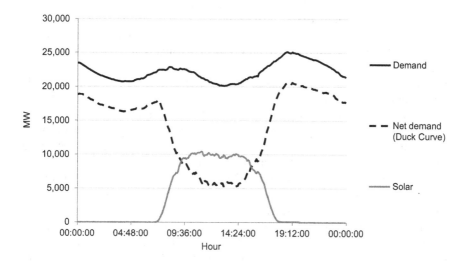

Figure 5.4 The California duck curve, October 22, 2022.

Source: Data from the website of the California Independent System Operator (CAISO), www.caiso.com/TodaysOutlook/Pages/supply.aspx. Accessed April 25, 2023.

The flexible ramping services referred to above are particularly important as the proportion of variable renewables on the grid increases. This is clearly shown in the case of California, which now has over 12 GW of solar PV, about a third on roofs and two-thirds as solar farms.[24] When the sun comes up each day, the amount of solar power rises sharply, and the balancing fossil generation decreases as shown in Figure 5.4, leading to a shape known as the 'duck curve'.

The top line in Figure 5.4 shows total power demand in California on the date shown. This was met by dispatchable sources (the dashed line) plus renewables, mainly solar (the bottom line), which reached a peak of around 10 GW in the middle of the day, before falling back to zero at 19.00. During the increase of solar power in the morning the dispatchable power has to be reduced and if it cannot be reduced fast enough then solar power has to be 'curtailed' and therefore wasted. Then, in the evening as power demand rises towards its peak, the dispatchable power has to be ramped back up again, as the solar power drops off rapidly, at rates that may be difficult to achieve.

With the increased installation of solar power in California from 2010, this 'duck' shape has become more acute. In 2013 the downward-upward 'duck curve' was only just beginning to become apparent, but by 2020, as shown in

---

24 www.greenbiz.com/article/californias-grid-geeks-flattening-duck-curve. Accessed April 29, 2023.

Figure 5.4, it had become deep and steep, with a ramp up of about 15 GW required between about 16.00 and 19.00.

The California power market is reacting strongly to the combined need to continue to reduce fossil fuel generation by increasing renewables, and address the problems related to the duck curve. Utility-scale battery storage is providing a major part of the answer, with battery facilities being increasingly built alongside PV solar farms, so that the solar energy can be stored rather than curtailed when it is surplus to immediate requirements, and then used to reduce the required ramp up rates from fossil fuel plants when the sun goes down, or to help meet peak demand in the evening. In 2020 and 2021 installed utility-scale batteries in California increased to 2.1 GW, an eight-fold increase over the pre-2020 level. As of March 2022, some 256 solar + storage projects, with capacities of 72 GW of solar and 64 GW of batteries, were in the queue for connection to the California power system.[25]

The UK as of 2021 had about 1.3 GW of operational battery storage, but the full pipeline of future projects at that time was calculated to be 16.5 GW on 686 sites, with 10.6 GW of that having reached the stage of having submitted a full planning application, and 9.2 GW of that having had the application approved.[26] Of course, not all the 16.5 GW will be built, but no doubt other projects will come forward. Once planning permission has been received, battery storage plants can be built in a matter of months, rather than years, so that there is every prospect that by 2030 there will be at least 10 GW of utility-scale battery storage in the UK, half the minimum amount of storage estimated by National Grid as being required for 2050.

Although the recent focus has been on the installation of utility-scale battery storage, and this is increasing rapidly, hydropower through 'pumped storage' is still the most widespread of the energy storage technologies that convert electricity into another form – in this case potential energy – for use later. With 'pumped storage' water is pumped up into high-altitude reservoirs at times of low electricity demand (and available electricity supply), to be released through turbines under the force of gravity to provide power at times of high demand or constrained supply. In 2020, China was the country with the highest capacity of pumped storage hydropower, with 30.3 GW. Then came Japan and United States, with 21.9 GW and 19.2 GW, respectively.[27] In the UK there are just four pumped storage facilities, at Dinorwig and Ffestiniog in Wales, and Cruachan and Foyers in Scotland, with a combined power output potential of 2.8 GW. The first was built at Ffestiniog in 1963, and has a power rating of 360 MW. The largest, Dinorwig, opened in 1984 and has a power rating of 1.73

25 https://pv-magazine-usa.com/2022/03/11/californias-solar-market-is-now-a-battery-market/. Accessed April 29, 2023.

26 www.energy-storage.news/large-scale-battery-storage-in-the-uk-analysing-the-16gw-of-proje cts-in-development/. Accessed April 29, 2023.

27 www.statista.com/statistics/689667/pumped-storage-hydropower-capacity-worldwide-by-coun try/. Accessed April 29, 2023.

GW. A number of new projects are planned for Scotland, including SSE's 1.5 GW project at Coire Glas and the more than doubling to more than 1 GW of the capacity at Cruachan, by its operator Drax.[28] The turbine halls deep in the mountain are huge, engineered caverns, so large that Ben Cruachan is already called the 'Hollow Mountain', and the enlargement would take another million tonnes of rock out of it.

The next largest form of energy storage is from concentrated solar power (CSP), which, as described earlier in this chapter, focuses sunlight through a large array of mirrors onto a heat-transfer fluid used to raise steam to drive a turbine. With molten salt as the fluid, the CSP plant can store heat at high temperatures so that the plant can continue to generate electricity when the sun is not shining. The 510 MW plant in Morocco mentioned earlier can store over 3 GWh of energy, enough to keep generating for over 7 hours, i.e. for much of the night.[29]

There are many other forms of energy storage. World Energy Council (2020, Figure 1, p. 9) categorises them as electrical (supercapacitors and superconducting magnetic energy storage, which store electrical energy directly), mechanical (pumped hydro, compressed air, flywheels), electromechanical (sodium sulphur, lithium-ion and Redox flow batteries, with lithium-ion batteries currently dominant, comprising over 90% of the global grid battery storage market), chemical (hydrogen) and thermal (molten salt, used to transfer heat in CSP power stations). All these forms of storage have different maturity, efficiency, response time, lifetime, charge time, discharge time and environmental impact, which makes them differentially suitable for the various storage services described above. These different suitabilities are tabulated and described in detail in World Energy Council (2020, Figure 1, p. 9), but are beyond the scope of this book, except for hydrogen, which is one of the subjects of the next chapter.

Electricity networks

Electricity networks comprise the wires that transport the electricity from where it is generated to where it is used. In the UK (and widely elsewhere) electricity is generated at 25 kV (25,000 volts) and 'stepped up' by transformers (to reduce power losses) to 275 or 400 kV for transmission over long distances, normally (in the UK) by means of aerial cables strung between pylons. Losses from high-voltage transmission, even when the power is transmitted over long distances, are relatively low (around 2%) in the US and UK. Before use the electricity is 'stepped down' to different voltages for different uses: to 33 kV for large manufacturing industry; to 25–33 kV for trains, to 11 kV or 415 V for small industry, schools and hospitals, and 240 V for houses, offices and

---

28 BBC, June 24, 2021, www.bbc.co.uk/news/uk-scotland-highlands-islands-57510870. Accessed April 29, 2023.
29 https://en.wikipedia.org/wiki/Ouarzazate_Solar_Power_Station. Accessed April 29, 2023.

shops.[30] Distribution losses at lower voltages (and distances) are higher than transmission losses (typically 4–5% in the US and UK, though some US states lose more than 10% in transmission and distribution[31]).

When coal dominated UK power generation, power stations tended to be built either in cities, near where the bulk of the power was consumed, to reduce distribution losses, or near coal mines. Gas-fired power stations obviously had to be built where they could access the gas transmission and distribution network. Solar and wind farms, on the other hand, must be built in relatively windy or sunny places, and the power then transmitted to where it is needed. In the UK it has required considerable investment in the transmission network to get the power from the windiest places (offshore in the North and Irish seas, and onshore in Scotland) to the cities where it is predominantly used.

Because the availability of renewables varies in different places at different times, joining up different electricity systems across large distances can enable renewable electricity in one place, where it may not be needed, to be transmitted to another where there is demand, but renewable power might not be available. This is the thinking behind the European Energy Union, which is working towards a single power system across the European Union, making the wind and hydropower in Scandinavia and Northern Europe, and the solar energy in the Mediterranean countries, available across the continent. Although no longer a member of the European Union, the UK still participates in its electricity network through 'interconnectors' running under the sea to the mainland and Ireland. It currently has 6 GW of interconnection, 3 GW to France, 1 GW to each of Belgium and the Netherlands and 500 MW to each of Northern Ireland and the Irish Republic.[32] Many more interconnectors are planned to be built over the next few years.

There have been and are ambitious plans to bring renewable electricity from even further afield. The large amounts of solar energy falling on the very large areas of the Saharan desert, and more widely across the Middle East and North African (MENA) region have long given rise to speculation that it could produce electricity that could then be transmitted to Europe or elsewhere, with a German physicist calculating in 1986 that Earth's deserts received as much energy from the sun every six days as humans used globally in a year.[33] A German diploma thesis from 2005 (May, 2005) estimated that CSP plants covering

254 kilometres x 254 kilometres (the biggest box on the image [Figure 5.5]) would be enough to meet the total electricity demand of the world. The

30 'High-Voltage Networks (national grid)', www.ukfrs.com/promos/17142. Accessed April 29, 2023.

31 http://insideenergy.org/2015/11/06/lost-in-transmission-how-much-electricity-disappears-between-a-power-plant-and-your-plug. Accessed April 29, 2023.

32 www.ofgem.gov.uk/energy-policy-and-regulation/policy-and-regulatory-programmes/interconnectors. Accessed April 29, 2023.

33 www.theguardian.com/environment/2011/dec/11/sahara-solar-panels-green-electricity. Accessed April 29, 2023.

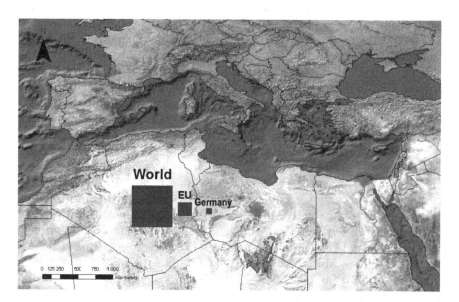

Figure 5.5 Visualisation of the area in the Sahara desert receiving enough solar energy
to provide the electricity of the regions shown.

Source: Adapted from May, 2005, Figure 12, p. 12. www.dlr.de/tt/Portaldata/41/Resources/dokume
nte/institut/system/projects/Ecobalance_of_a_Solar_Electricity_Transmission.pdf. Accessed April
29, 2023.

amount of electricity needed by the EU-25 states could be produced on
an area of 110 kilometres x 110 kilometres (assuming solar collectors that
could capture 100 per cent of the energy).[34]

The smaller boxes show the area required for the power demand of the EU or
Germany.

Clearly the actual numbers in these calculations of solar energy possibilities
are the result of numerous assumptions which can be challenged. But the basic
idea that solar energy in the MENA region could make a significant contribution to meeting Europe's energy demands, as well as its own, received serious
industrial interest in the proposed Desertec project.

It is important to distinguish between the Desertec Foundation,[35] which
promotes the overall idea of the generation of electricity from desert energy
(mainly solar, but also wind), and the Desertec Industrial Initiative (DII), an
idea which was generated by the Desertec Foundation in 2009, and won considerable industrial support for a few years, but later separated from it. DII

34 www.ecomena.org/desertec/. Accessed April 29, 2023.
35 www.desertec.org/. Accessed April 29, 2023.

had plans to bring 15% of Europe's electricity from North Africa through large solar plants connected to Europe through long high-voltage direct-current (HVDC) cables (which lose less energy than AC (alternating current) cables) (Schmitt, 2018).

A 2011 article in *The Guardian* provided a statement of the energy security arguments behind the Desertec project, which was intended to avoid situations "when an energy-rich country holds its neighbours to ransom by restricting or denying supply. Think Russia and its gas",[36] a prescient comment from the perspective of the 2022 war between Ukraine and Russia, when Europe scrambled to wean itself off Russian gas, a dependency that had intensified in the years since 2009 (although it is not clear that swapping energy dependency on Russia for electricity dependency on North Africa would enhance long-term European energy security). Although the original DII consortium fell apart in 2014 for a complex set of reasons including competing visions and accusations of eco-colonialism (Schmitt, 2018), the basic idea lives on through the Desertec Foundation and a number of individual projects. One of these is the Xlinks project, which plans to link 10.5 GW of Moroccan renewable electricity to the UK through 3,800 km HVDC cables, coming ashore at two 1.8 GW connections in North Devon.[37] It is likely that many more such projects will be entered into on an ad hoc basis, rather than as part of a grand plan.

The issue of grid connection is critical, wherever the renewable electricity is coming from. The UK appears to be doing particularly badly in this respect. In February 2023 the *Financial Times* reported that there was a 13-year wait for new renewables generators to be connected to the UK grid,[38] with 600 projects with a combined capacity of 176 GW (compared to a current renewables capacity of 64 GW) in the queue. In addition, the lack of transmission capacity between Scotland (where most of the UK's wind energy is generated) and England means that, according to Carbon Tracker, in 2022 3.4 TWh (equivalent to the average electricity consumption of 1 million homes) of wind energy was lost ('curtailed', see below for further discussion of this issue).[39] This meant that fossil fuel generation in England had to be used instead, resulting in the emission of 1.3 $MtCO_2e$. It seems extraordinary that, having had ambitious decarbonisation plans since 2008, the UK has allowed this congestion to develop. Without urgent investment in transmission capacity the problem will get considerably worse in the future as renewables capacity increases. There are few more urgent priorities than this investment if the UK is to meet its target of decarbonising the UK power grid by 2035.

36 www.theguardian.com/environment/2011/dec/11/sahara-solar-panels-green-electricity. Accessed April 29, 2023.
37 https://xlinks.co/morocco-uk-power-project/. Accessed August 10, 2023.
38 *Financial Times*, February 6, 2023 www.ft.com/content/bc200569-cb85-4842-a59a-f04d34280 5fc. Accessed June 18, 2023.
39 Gone with the wind? Carbon Tracker Initiative. https://carbontracker.org/reports/gone-with-the-wind/. Accessed August 17, 2023.

Electricity markets

Electricity markets have the complex task of getting electricity from the generators through the networks and into homes, factories and businesses, at prices that are both efficient and profitable to businesses (when the electricity supply industry is not fully nationalised), and of ensuring that the electricity is available when and in the quantity that consumers desire it.

All real electricity markets are themselves very complex institutions comprising some mix of price incentives, asset ownership and regulation. The mix is different in different countries and it has changed dramatically over time. In fact, it is changing all the time, as governments, electricity system operators, asset owners and consumers adapt all the time to new technologies, government objectives for energy security and emission reduction, and other environmental challenges.

The topic gets a book length treatment in Glachant et al. (2021), which provides huge detail on the evolution of electricity markets in mainly the US and UK through the 20th century and into the 21st. Most of this is beyond the scope of this book, but no treatment of decarbonisation would be remotely complete without some discussion of the implications for electricity markets of largely or completely removing carbon from electricity systems.

The account here draws heavily on both Glachant et al. (2021) and on Newbery et al. (2018), the latter of which explores the major issues required to effectively manage an electricity system with a high level of variable and intermittent renewables, some of which have been discussed above: interconnection and market integration, electricity storage, the design of RES [renewable energy sources] support systems, distributed generation, efficient electricity pricing and long-term contracts. Interconnection and market integration allow renewables to be transferred from places when and where they are abundant to places when and where they are scarce – the Desertec plan would have been the ultimate in interconnection, while significant market integration already exists in Europe; electricity storage allows electricity to be stored in another form when it is not needed, to be converted back into electricity when it is. Both these mechanisms have already been discussed.

RES support mechanisms were essential when renewable electricity was much more expensive than power from fossil fuels, and are still required for some technologies (e.g. wave and tidal power) which still are. Wind and solar PV are now broadly competitive with fossil fuels, but require ongoing support for two reasons: they are very capital-intensive, but once they are built the electricity they produce is effectively free. If they were to sell their power at this very low price they would never recover the capital cost of building the turbines and PV panels, which means that without support they would not be built. Many countries have offered 'feed-in tariffs' (FITs) to renewable generators, which guarantee a certain price per unit (kWh) of electricity they produce. In the UK the preferred means of support is now through auctions that either guarantee a price for the renewable electricity generated or that commission a certain quantity of power, or that contract for some combination of the two. The current UK approach is through FITs with Contracts for Difference (CfD). The price of electricity on the wholesale electricity market (i.e. the price at which generators sell it to suppliers) is very variable. Under a CfD, generators sell their electricity on the wholesale electricity market at that price, and earn the revenue corresponding to the price at that time. But through the CfD this payment is topped up to the 'strike price' agreed in the CfD. For the first offshore wind plants, the strike price was around £150 per MWh, but in the most recent rounds this has fallen to around £40 per MWh. For the Hinkley Point C nuclear power station the strike price was fixed at £93 per MWh in 2012, but index-linked to inflation, so it is now well over £100 per MWh, and will be higher still when Hinkley actually starts generating power.

With a CfD should the wholesale market price go *above* the strike price, then the generator pays back the difference between the market price and the strike price. With electricity prices very high through the last six months of 2021, generators with CfDs for the first time had to pay back money they had received from the market, which was consistently above the strike price (and remains so). The payment in January 2022 for the last quarter in 2021 amounted to more than £39 million, while that for the first quarter in 2022 was £342 million.[40] This is paid to the suppliers who have paid the market price, and should reduce prices to households – an amount that was around £15 per household over the six months. These payments from generators with CfDs should continue while the electricity price stays high.

Two final mechanisms identified by Newbery et al. (2018) can now be briefly covered. Distributed generation refers to the fact that wind and solar power can be installed at a very local level, by communities and households (most people will have seen solar panels on the roofs of houses). This generation typically connects into the distribution (rather than transmission) part of the power grid, and it is therefore the distribution company (in the UK called the Distribution Network Operator, DNO) which has the responsibility of connecting it into their network and maintaining the balance between supply and demand as the wind speeds or solar radiation change in their area. Distributed generation, where the power has less far to travel to its point of consumption, has several potential economic benefits, reducing transmission costs and the need to build generation plant elsewhere, as well as creating awareness among communities and households of the realities of low-carbon electricity and potentially involving them as 'prosumers' in its production as well as its consumption. However, integrating large quantities of variable renewables into a distribution network in a balanced way can be problematic, as seems to be the case in Cornwall, which in 2020 had 776 MW of low-carbon and renewable generation which the DNO, Western Power Distribution (WPD), found was putting the system under strain and causing it to discourage further building and

---

40 www.lowcarboncontracts.uk/moving-the-money. Accessed April 29, 2023.

connection of distributed generation, which goes against the need to decarbonise the power system as quickly as possible (Bray et al., 2020). Resolving such issues requires complex adjustments to financial incentives and network regulations.

Finally there is the issue of 'efficient electricity pricing'. Because electricity markets have to balance supply and demand second-by-second, there are actually a number of markets in the 'electricity market', the costs of which all feed into the price of electricity. There is the basic technology cost of generation, called the 'Levelised Cost of Electricity' (LCOE) to compare different technologies according to a common method, which is discussed further below for a number of technologies. But each technology has a different impact on the overall electricity system, and this is particularly true for variable and intermittent renewables. These impacts represent further costs for the electricity system.

> For example, a plant built a long distance from centres of high demand will increase transmission network costs, while a 'dispatchable' plant (one which can increase or decrease generation rapidly) will reduce the costs associated with grid balancing by providing extra power at times of peak demand.
>
> (Department for Business, Energy and
> Industrial Strategy, 2020, p. 5)

The impacts are of different kinds and are felt at different parts of the electricity system: the wholesale market (the price that generators receive for their energy), according to the time and quantity of energy that it is available (which varies most for variable and intermittent renewables); the capacity market, through which generating plant is procured to ensure that peak electricity demands can be met; the market for balancing and ancillary services, to ensure that supply and demand are always balanced and the electricity system both operates at the correct frequency, and can recover quickly from unexpected shocks (e.g. the unexpected outage of a large generating plant); the transmission and distribution systems, according to the distance over which the electricity needs to be transmitted, or the extent of the distributed energy that needs to be absorbed in the distribution network, as discussed above. All these impacts need to be managed through appropriate and timely investments in the electricity system, and these investments, if they are to be made by private companies, need to be adequately incentivised. The core institution for this purpose, because it is the principal way in which generators and providers of these other services are paid, is the wholesale market.

The wholesale electricity market

In many ways the wholesale electricity market is unique. In order to balance supply and demand in real time the System Operator (SO) procures commitments from generators to supply a certain amount of power at a certain time the following day, in half-hour periods. There are penalties for not meeting these commitments and the SO continuously monitors the system to contract for more power if there is a shortfall, or 'curtail' it if there is an excess.

It has already been seen that power demand varies dramatically at different times of the day and during different seasons, which can be shown on a 'load duration curve'. This depicts in descending order the number of half hours in the year in which a certain level of power demand was exceeded. A load duration curve in Joskow & Leautier (2021, Figure 3.1, p. 38) for France in 2009 showed that peak demand, for one half hour in the year, was 92.4 GW, and this occurred at 7 p.m. on February 8. Power demand exceeded 60 GW for 5,676 half hours in the year. The minimum power demand (i.e. it was exceeded by all 17,227 half hours in the year), was 31.5 GW, at 7 a.m. on August 5. These half hours were not continuous; they will have occurred at different times of day and different times in the year. But the curve tells the SO how much power the system must provide, and how often during the year. The problem for the SO is to decide which technologies to use and when to use them, in order to minimise the cost of supplying the electricity.

Different technologies cost different amounts to build (fixed costs) per unit of power capacity provided over their lifetime (€/MW/year), and per unit of electricity generated (€/MWh), and different amounts to run (variable or operating cost) per unit of electricity generated, as shown for three representative technologies, in Table 5.1. Given these three technologies (nuclear, combined cycle gas turbine [CCGT], and open cycle gas turbine [OCGT]), how much of each should be used, and when, to meet the demand over the year shown in a 'load duration curve'? From inspection of the numbers in Table 5.1, it is clear that, once nuclear plant has been built, it should be run as much as possible, because it is the cheapest to run. But how much should be built, given its expense? At the other end of the spectrum, OCGT is relatively cheap to build but much more expensive than nuclear to run. At some level of use, the total cost of OCGT will exceed that of nuclear.

The answer is given by plotting a 'total cost curve' for each of the different technologies. Each line starts at the fixed cost of the technology, and then the cost increases per hour used according to its variable cost. For the example load curve described above for France in 2009 (Joskow & Leautier, 2021, Figure 3.1, p. 38), it turns out that the OCGT should only be used for the 475 hours of highest demand (it is therefore called 'peaking plant'), the CCGT is needed for 2363 (2838−475) hours of relatively high demand, while the nuclear plant is the cheapest to run for the remainder, 5922 (8760−2838), of the hours in the year, when demand is relatively low.

Given its low running costs, it therefore makes economic sense, once nuclear plant has been built, for it to run the whole time (called 'baseload'). However, it is not economic to build nuclear plant to satisfy the higher demands that occur for 2838 hours in the year, which are cheaper to serve with CCGTs. For the very highest 'peak' demands that occur for 475 hours per year, it is similarly not worth building extra CCGTs, but to use OCGTs.

Table 5.1 Fixed and variable costs for three technologies

	Fixed cost (€/MW/year)	Fixed cost (€/MWh)	Variable cost (€/MWh)
Nuclear	299,000	34	10
Combined cycle (CCGT)	72,000	8	90
Gas turbine (OCGT)	53,000	4	130

Source: Joskow & Leautier, 2021, Table 3.1, p. 57.

Having identified the technologies that should be built, and the demand half hours over which they should run, the next question is: how should the electricity from the technologies be priced?

A basic economic finding is that in a competitive market, economic efficiency requires that the market price is set by the operating cost of providing the last unit that is required to meet demand (called the 'marginal cost'). This means that the market price is set by the operating cost of the most expensive plant that is operating at any one time.

Figure 5.6 shows a real power system, that of the UK in 2022, with the available generation capacity stacked up in ascending order of operating cost – the so-called 'merit order'. In line with principles of marginal cost pricing, the cheaper plant will always be selected to run first, but the market price of the electricity will be that of the most expensive operating plant that is running. This means that, in the UK, at any power demand between 18 GW and 46 GW, the market price of the electricity will be set by CCGT, which explains why electricity has been so expensive since the Russian invasion of Ukraine, which sent gas prices through the roof, despite over half of UK electricity being provided by much lower cost low-carbon energy.

Several further points can be made about this description of how pricing in such an electricity system works. First, costs can be saved if demand can be shifted from periods of relatively high demand, which require more expensive technologies, to periods of lower demand that use cheaper technologies. Such demand shifting may be through contracts between the SO and large users, or through smart meters with time of use tariffs, which enable appliances to be used when electricity is cheap. In the UK the supplier Octopus is now offering such tariffs. Another option is to charge up electricity storage (e.g. pumped hydro, batteries) when demand is low (and electricity cheap), and release it at times of high demand when electricity is expensive (as battery storage becomes cheaper this may become an increasingly economic option, and batteries in electric vehicles may be used for this purpose on days or at times when the vehicles are not being used).

A second point is that each technology with cheap operating costs relies on periods of high prices to recoup its capital costs. Nuclear power has an operating cost of around £10/MWh, but such power stations are hugely expensive

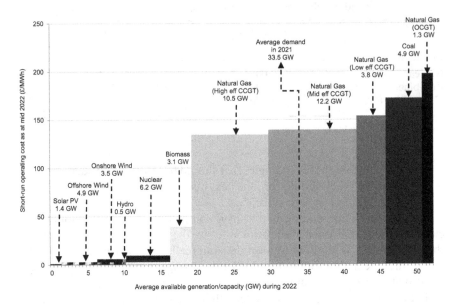

Figure 5.6 The 'merit order' in UK generation, 2022.

Source: Grubb, 2022, Figure 2, p. 12. www.ucl.ac.uk/bartlett/sustainable/sites/bartlett_sustainable/files/ucl_isr_necc_wp3_with_cover_final_070922.pdf. Accessed April 29, 2023.

to build. They only recover these fixed costs when electricity prices are higher, which explains why the CfD strike price discussed above was set in 2012 at £93 per MWh.

A third point visible in Figure 5.6 is that, at a power demand of above 11 GW but below 17 GW, nuclear would be operating below its full capacity. Either the nuclear power plant would need to be 'turned down', which is not easy for this technology, or an alternative load would need to be found. One solution for this in the UK was the Economy 7 tariff, operating at night, which encouraged consumers to install night storage heaters to use available nuclear power at times of low demand at night. Another possible alternative load might be the manufacture of hydrogen, as discussed in the next chapter. This option becomes even more important when the electricity system has a high proportion of renewables on it, which is the situation discussed next.

The exceptionally high global gas prices in 2022, consequent on the Russian invasion of Ukraine, put electricity markets in countries that use gas for power generation, including that in the UK, under exceptional strain, with unprecedentedly high electricity prices. In 2019, gas plants set the electricity price in Great Britain 84% of the time, despite providing just 45% of generation. Much cheaper zero-carbon electricity was supplying over 50% of UK electricity demand but setting the price only 1% of the time (with interconnectors setting the price for the remaining 15% of the time) (Grubb, 2022, p. 14). Suppliers of

gas made huge windfall profits, but so did the nuclear and renewable generating companies, whose costs had not changed but who were getting the very
high price for their power set by the gas price. Those on CfD contracts, of
course, paid the revenue they received from prices above their 'strike price'
back to suppliers, but for the rest it was pure windfall profit.

The UK Government has grappled with the need to reform the electricity
market to take account of the greater proportion of more inflexible low-carbon
energy sources for over ten years (Department of Energy and Climate Change,
2011). But the exacerbation of the problems caused by the huge electricity
price rises in 2021–2022 was the main factor that caused the UK Government
in July 2022 to launch a consultation on a Review of Electricity Market
Arrangements.[41] While the conclusions from this review are still awaited, and
any implementation of what emerges is likely to take a considerable period
of time, it is likely that some suggestions will be made as to how to decouple
the price of renewable and other zero-carbon electricity from the much more
expensive price of gas. One suggestion made independently of government has
been to establish a Green Power Pool, through which vulnerable consumers
can access cheap renewable electricity at a cost much below that generated by
expensive gas.[42]

These explanations permit an assessment of the Climate Change Committee
(CCC)'s Balanced Pathway (BP) decarbonisation projections, which envisage
a considerable expansion of variable renewables capacity and a lesser expansion of new nuclear as shown in Table 5.2: increases in solar PV to 75 GW, in
offshore wind to 85 GW and onshore wind to 25 GW, and nuclear to 10 GW,
all by 2040.

Figure 5.7 illustrates load duration curves, showing the implications of
this capacity expansion. The horizontal axis shows the 8760 hours in the year,
and the graphs show the hours in the year that the power demand exceeds a
given level on the vertical axis. The two upper dashed lines show the electricity demand increasing from 2020 (the lower line) and the BP projections for
2040 (the upper line). The upper dashed line shows that in 2040 the maximum
power demand is around 90 GW (at the left-hand end of the curve), and the
minimum is around 35 GW (at the right-hand end of the curve), i.e. it is above
that level for all the 8760 hours in the year. The dashed line crosses the 60 GW
level at about 5000 hours, meaning that for 5000 hours in the year, demand was
above 60 GW.

The solid lines show the Residual Demand once the contribution of the variable renewables has been accounted for; i.e. the gap between the solid Residual
Demand 2040 line and the top dashed line is the renewable generation. The
minimum renewable power (the smallest gap between the curves, at the far left

41 https://assets.publishing.service.gov.uk/government/uploads/system/uploads/attachment_d
ata/file/1098100/review-electricity-market-arrangements.pdf. Accessed April 29, 2023.
42 www.ucl.ac.uk/bartlett/sustainable/sites/bartlett_sustainable/files/navigating_the_energy-clim
ate_crises_working_paper_4_-_green_power_pool_v2-2_final.pdf. Accessed April 29, 2023.

Table 5.2 UK capacities of generation technologies in 2020, and Balanced Pathway (BP) projections to 2030 and 2040

	GB installed wind & solar capacity (GW)		
	2020	*2030 BP*	*2040 BP*
Solar PV	13.0	28	75
Wind: offshore	10.5	40	85
Wind: onshore	12.7	18	25
Total	36.2	86	185
Peak residual demand exc. nuclear	58.7	54.1	69.7
Min. residual demand exc. nuclear	0.86	−31.7	−85.5
Nuclear capacity	9	7	10.1
Peak residual demand inc. nuclear	50.5	50.0	60.4

Source: Presentation by Keith Bell to UK Energy Research Centre Annual Conference, Manchester, June 13, 2022.

of the curves) is around 20 GW (around 90–69.7). The maximum renewable generation is where the gap is largest, at the right-hand end of the curve, and is around 125 GW, when the renewables meet the full 35 GW of demand and have another 85.5 GW of power available. This additional 85.5 GW of power would be wasted unless there were an opportunity to store it, e.g. through the production of hydrogen. In fact there is surplus power available for hydrogen production for about 3670 hours of the year in 2040 (the Residual Demand curve crosses the 0 line after about 5000 hours of the year, meaning that there is surplus power available for the remaining hours).

There is much less of a gap between the Residual Demand 2020 line and the dashed Demand 2020 line, because there are much less renewables on the system in 2020 than there are projected to be in 2040. In fact, the Residual Demand 2020 line never crosses the 0 Demand level line (its minimum is 0.86 GW, see Table 5.2), so there is never a surplus available for hydrogen production or any other kind of storage. The situation changes gradually to 2040 as more and more renewables are built, as per Table 5.2, and the gap between the Demand curve and the Residual Demand curve increases. In 2040, as has been seen, it is a maximum at the far right of the curves, when the surplus renewable generation is 85.5 GW (more than the current size of the UK grid). The total energy that could be generated by the 3670 hours of surplus power (whenever the Residual Demand curve is negative) is 95 TWh. This could be used to produce hydrogen or stored in some other way, as noted.

With more nuclear capacity, as in Table 5.2, the Residual Demand curves are pushed down further, as shown by the dotted lines Residual Demand inc. Nuc. curves. This means that the total demand is satisfied for more hours in the year by these energy sources (the bottom dotted line crosses the 0 Demand line at around 3500 hours, which means that there is surplus power available

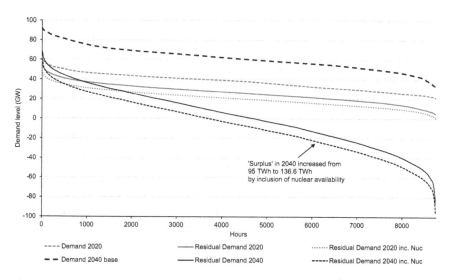

Figure 5.7 Residual Demand curves showing the hours of 'surplus' electricity that would arise from the electricity capacity projections in Table 5.2.

Source: presentation by Keith Bell to UK Energy Research Centre Annual Conference, Manchester, June 13, 2022.

for 5170 hours (8670–3500)). The extra energy that could be used to produce hydrogen in this case rises to 136.6 TWh as shown in Figure 5.7. Hydrogen is discussed in more detail in Chapter 6.

Conclusion

The single most important task in combating climate change is the decarbonisation of the electricity system, so that zero-carbon electricity can increasingly be used beyond lighting and appliances for heating and transport, and for generating hydrogen and syn(e-)fuels for those industrial and transport applications that cannot be electrified. Most zero-carbon electricity, coming from the sun and wind (onshore and offshore), will be variable and intermittent, with fundamental implications for the design of electricity networks. Means of electricity storage will acquire great importance. A number of storage technologies are available to balance the network over relatively short periods, but long-duration, inter-seasonal storage remains a significant challenge in those countries which, like the UK, have great difference in energy demand between summer and winter, and which may experience weeks in winter with little or no wind or solar energy, when demand for heating is highest. To some extent the variability in renewables can also be compensated for through interconnection across significant geographies, in the hope that some countries will experience a surplus of renewable electricity which can be transmitted to others in deficit. Using the abundant solar energy in the Sahara region in this way would certainly be feasible, if not cheap, provided that security issues in North African countries could be resolved.

Another way in which renewable electricity differs from electricity from fossil fuels is that its main cost is upfront capital cost, while its generating cost is very low. This has fundamental issues for the design of electricity markets, which currently price electricity at the cost of the most expensive plant generating at any one time. European countries are currently engaged in the process of redesigning electricity markets so that they become more appropriate for electricity systems with a high proportion of zero-carbon energy.

Provided that these issues with renewable energy can be resolved, then this electricity source offers huge promise for the future, given that, as will be seen in Chapter 12, in many countries it was cheaper than fossil fuel electricity even before the huge increase in the cost of fossil methane gas in 2022. With large solar and wind resources available in most countries, their inhabitants for the first time in their history have the prospect of an abundant, inexhaustible, clean and cheap energy source, for which they are not dependent on the policies and friendship of foreign countries. There is still a long way to go to realise this prospect, but it is eminently do-able, with the investment and political will, by 2050.

References

Bray, R., Woodman, B., & Judson, E. (2020). *Future Prospects for Local Energy Markets: Lessons from the Cornwall LEM*. University of Exeter Energy Policy Group. https://ore.exeter.ac.uk/repository/handle/10871/124214

Department for Business, Energy and Industrial Strategy. (2020). *Electricity Generation Costs 2020*. BEIS, London.

Department of Energy and Climate Change. (2011). *Planning our electric future: a White Paper for secure, affordable and low-carbon electricity*. DECC, London.

Drummond, P., Scamman, D., Ekins, P., Paroussos, L., & Keppo, I. (2021). *Growth-positive zero-emission pathways to 2050*. Sitra, Helsinki. www.sitra.fi/en/publications/growth-positive-zero-emission-pathways-to-2050/

Environmental and Energy Study Institute. (2019). *Fact Sheet. Energy Storage* (2019), White Papers. EESI. EESI, Washington, DC. www.eesi.org/papers/view/energy-storage-2019

Glachant, J.-M., Joskow, P., & Pollitt, M. (Eds.). (2021). *Handbook on Electricity Markets*. Edward Elgar, Cheltenham, UK; Northampton, MA.

Grubb, M. (2022). *Navigating the crises in European energy: Price Inflation, Marginal Cost Pricing, and principles for electricity market redesign in an era of low-carbon transition*. Working Paper #3; Navigating the Energy-Climate Crises. UCL, London. www.ucl.ac.uk/bartlett/sustainable/sites/bartlett_sustainable/files/ucl_isr_necc_wp3_with_cover_final_050922.pdf

Grubler, A., Wilson, C., Bento, N., Boza-Kiss, B., Krey, V., McCollum, D. L., Rao, N. D., Riahi, K., Rogelj, J., De Stercke, S., Cullen, J., Frank, S., Fricko, O., Guo, F., Gidden, M., Havlík, P., Huppmann, D., Kiesewetter, G., Rafaj, P., ... Valin, H.

(2018). A low energy demand scenario for meeting the 1.5 °C target and sustainable development goals without negative emission technologies. *Nature Energy*. https://doi.org/10.1038/s41560-018-0172-6

Habel, M., Mechkin, K., Podgorska, K., Saunes, M., Babiński, Z., Chalov, S., Absalon, D., Podgórski, Z., & Obolewski, K. (2020). Dam and reservoir removal projects: a mix of social-ecological trends and cost-cutting attitudes. *Scientific Reports 2020 10:1*, *10*(1), 1–16. https://doi.org/10.1038/s41598-020-76158-3

International Energy Agency. (2021a). *Net Zero by 2050: A Roadmap for the Global Energy Sector*. IEA, Paris.

International Energy Agency. (2021b). *Renewables 2021: Analysis and forecast to 2026*. IEA, Paris.

International Renewable Energy Agency, T. (2019). *Utility Scale Batteries Innovation Landscape Brief*. IRENA. www.irena.org

Joskow, P., & Leautier, T.-O. (2021). Optimal wholesale pricing and investment in generation: the basics. In Glachant, J.-M., Joskow, P., & Pollitt, M. (Eds.). (2021). *Handbook on Electricity Markets* (pp. 36–72), Edward Elgar.

May, N. (2005) *Eco-balance of a Solar Electricity Transmission from North Africa to Europe*, Diploma Thesis, August, Technical University of Braunschweig www.dlr.de/tt/Portaldata/41/Resources/dokumente/institut/system/projects/Ecobalance_of_a_Solar_Electricity_Transmission.pdf

Newbery, D., Pollitt, M. G., Ritz, R. A., & Strielkowski, W. (2018). Market design for a high-renewables European electricity system. *Renewable and Sustainable Energy Reviews*, *91*, 695–707. https://doi.org/10.1016/J.RSER.2018.04.025

Schmitt, T. M. (2018). (Why) did Desertec fail? An interim analysis of a large-scale renewable energy infrastructure project from a Social Studies of Technology perspective. *Local Environment*, *23*(7), 747–776. https://doi.org/10.1080/13549839.2018.1469119

World Energy Council. (2020). *Innovation Insights Brief – Five Steps to Energy Storage | World Energy Council*. WEC. www.worldenergy.org/publications/entry/innovation-insights-brief-five-steps-to-energy-storage

6 Filling the gaps with bioenergy and hydrogen

Summary

Bioenergy comes from biomass (trees, crops, crop residues, animal manure) that is used to produce power, heat or vehicle fuels ('biofuel'). The main constraint on the quantity of bioenergy that is available is land, given the need also for land to produce food for a growing population and to leave space for forests and wildlife. This chapter explores the availability of biomass under different assumptions and how this matches the extent of bioenergy use in the exemplary pathways that are being cited throughout the book. It also touches on the various controversies around the production of biomass, such as the sustainability or otherwise of the production and use of biomass for power generation or biofuels.

Zero-carbon energy scenarios also usually see a role for hydrogen, the production and potential uses of which are also discussed in this chapter. Elemental hydrogen for energy purposes has to be produced from hydrogen compounds, which is an energy-intensive process, and then more energy is needed to transport it to where it is needed. The most common processes for hydrogen production are steam reformation of fossil methane gas ('grey' hydrogen, when the resulting CO_2 has to be captured and stored for the hydrogen to be 'low-carbon'), or electrolysis of water with renewable electricity, to produce zero-carbon 'green' hydrogen. Hydrogen has a wide range of uses – for heat, power and transport – for which it can be burned, used in fuel cells for power generation, converted into synthetic fuels ('synfuels' or 'e-fuels') or used in the direct reduction of iron for steel-making. The development of hydrogen-using products has advanced markedly in recent years, and there is now considerable investment going into the manufacture of electrolysers to produce 'green' hydrogen, so it may be that after a long gestation period, this energy source will become more widely used over the next 20 years.

DOI: 10.4324/9781003438007-7

Introduction

Before industrial times, bioenergy, generated by burning wood or animal dung, was humanity's principal, even only, energy source, and it remains in use today. However, its requirement for land to grow the biomass, in competition with crops for food and the need to leave some land for ecosystems and other species, means that the amount of biomass available for bioenergy is strictly limited. Once grown, however, biomass can be converted into a number of energy sources, including electricity, liquid fuels and biogas. The carbon emissions associated with bioenergy are contested. Because biomass sucks carbon out of the atmosphere, to release it when it is used, bioenergy can be nearly 'carbon neutral', if the crops are grown in such a way that growing the biomass does not displace forests, deplete soil carbon or produce emissions from fertiliser. It is often accounted as 'carbon neutral' in emission inventories. But this is controversial, given the uncertainties of emissions from soil and land, and often in how the biomass is grown, and it is possible for bioenergy to be more carbon-intensive than the fossil fuels it replaces. These issues are explored in some detail below.

Another fuel with considerable flexibility in its uses is hydrogen. Elemental hydrogen for energy purposes needs to be produced, and its means of production has been characterised by different colours (though these bear no relation to the colour of the hydrogen product, which is a colourless gas): black (from coal gasification), grey (from fossil methane gas), turquoise (from pyrolysis of methane gas), pink (electrolysis using nuclear energy), blue (from fossil methane gas using carbon capture and storage), green (electrolysis using renewable electricity), and yellow (using solar energy).[1] Blue hydrogen is low- (rather than zero-) carbon. Pink, blue, green and yellow hydrogen can be characterised as zero-carbon hydrogen, as can turquoise carbon, because, although it is produced from fossil methane gas, its product apart from hydrogen is carbon, rather than CO_2.[2]

Like bioenergy, hydrogen can be used to substitute for fossil fuels in processes that cannot be electrified, but, again like bioenergy, there are a number of considerations that are likely to limit its uses. These include the amount of energy that is required actually to produce the hydrogen, and the further energy that is required to compress it (sometimes to liquid form at very low temperatures) to transport it. Hydrogen can therefore never be as cheap as the energy source that is used to make it, and this energy source should therefore always be used directly if possible.

Notwithstanding these considerations, both bioenergy and hydrogen have important, if relatively minor, roles to play in decarbonisation, as this chapter sets out.

---

1 www.acciona.com.au/updates/stories/what-are-the-colours-of-hydrogen-and-what-do-they-mean/?_adin=02021864894
2 https://fsr.eui.eu/between-green-and-blue-a-debate-on-turquoise-hydrogen/

Bioenergy

Production and uses of bioenergy

In many pre-industrial societies biomass, in the form of wood or manure ('traditional biomass'), was a common source of fuel for cooking and heating, and Figure 5.1a shows that it is still present in relatively small quantities today, and projected to remain so until 2050.

Today, crops and other forms of vegetation or biomass can be converted into a wide variety of fuels. Trees, usually converted into wood pellets, and energy crops, such as willow coppice or miscanthus, or crop residues, can be used in a power station to generate electricity. The same is true for waste wood, paper and food, which is widely incinerated with 'energy recovery', in the form of electricity or heat, which can be used to substitute for fossil fuels.

For transport applications, some food crops or crop residues (e.g. maize or residues from sugar production), or bio-oils (palm oil, waste food oils), can be converted into liquid fuels (bioethanol, biodiesel), which are mainly mixed in different proportions with gasoline and diesel derived from fossil fuels, respectively. These 'generation 1' biofuels are generally recognised to be limited by the available land, given that their land use competes with land to grow food for humans and their animals, or, in the case of palm oil, with virgin tropical forests – many palm oil plantations in tropical countries have controversially replaced tropical forests, with considerable loss of biodiversity. It is hoped that 'generation 2' biofuels, mainly produced from woody ('cellulosic') biomass rather than food crops, and with far fewer negative environmental impacts, will become widely available and competitive, but progress towards this over the last 15 or so years has been slow and this is not yet the case.

Because the carbon in the biomass has been taken out of the atmosphere when the plant was growing, and returns to the atmosphere when the biomass is burned, it can be, in net terms, a low-carbon (or low-greenhouse gas (GHG)) fuel. How low depends on how much GHG was emitted when the bioenergy crop was grown (e.g. if the crop replaced forests, or from the fertiliser use or disturbance of the soil). If the carbon emissions when the biomass is burned are captured and stored (called 'bioenergy with carbon capture and storage' (BECCS)), this form of electricity generation can be 'net negative', or a 'negative emissions technology' (NET), because it removes more carbon from the atmosphere than it adds to it. The need for such technologies, and the scale required, to reach the Paris temperature targets, is discussed further in the next chapter.

The fact that the production of biofuels will replace other land uses has serious implications for their ability to generate GHG (CO_2e) savings when they substitute for petrol (gasoline), diesel or jet kerosene from fossil fuels. The GHG emissions from biofuel production are made up of two components: those from the cultivation of the crop to be used for biofuel, minus those from the cultivation of any crop (or from the unused land) they are replacing (net emissions from *direct land use change* [dLUC]); plus any net emissions that may arise through the displacement of the replaced crop onto some other land (emissions from *indirect land use change* [iLUC]). Both the dLUC and iLUC emissions could be substantial, for example if the biofuel were to come from palm oil plantations which replaced tropical forest (dLUC), or if the replacement of some crop (e.g. soya) caused the cultivation of soya somewhere else on land that was deforested for the purpose (iLUC).

Njakou Djomo & Ceulemans (2012) reviewed studies on the GHG emissions from producing biofuels taking account of both direct and indirect land use change. They found that the values varied enormously, depending on the period of time considered, and the type of land and feedstock used. Other considerations are the way both the crop for the biofuel and the crop it replaced were cultivated. Table 6.1 shows the results.

A negative number means that, taking account of both the stated kinds of land use change, the crop for the biofuel actually removes more carbon from the atmosphere (or adds less carbon to the atmosphere) than the crop it replaced. A positive number means that it has more GHG emissions than the crop it replaced. It can be seen that the most carbon-intensive biofuel is biodiesel, when palm oil replaces rainforest.

Unless the biofuel has a carbon intensity less than gasoline (replaced by bioethanol) or diesel (replaced by biodiesel), then, far from reducing CO_2

Table 6.1 Ranges of carbon intensity for different biofuels, including land use change

Fuel	Fuel crop	Type of LUC	Crop/use replaced	Net emissions gCO_2/MJ
Bioethanol	Cellulosic (e.g. poplar)	dLUC	Cropland	−52
Bioethanol	Sugarcane (Brazil)	dLUC	Rangeland	34
Biodiesel	Palm oil	dLUC	Degraded	−98
Biodiesel	Palm oil	dLUC	Rainforest	481
Biokerosene	Jatropha	dLUC	Pasture land	−27
Biokerosene	Jatropha	dLUC	Cerrado woodlands	101
Bioethanol	Various	iLUC	Various	0 to 327
Biodiesel	Various	iLUC	Various	0 to 1434
Bioethanol	Various	dLUC + iLUC	Various	−27 to 361
Biodiesel	Various	dLUC + iLUC	Various	−98 to 1677
Petrol (gasoline)	Fossil fuel	na	na	94
Diesel	Fossil fuel	na	na	95[1]

[1] www.eea.europa.eu/ims/greenhouse-gas-emission-intensity-of. Accessed April 29, 2023.

Source: Figures compiled from Njakou Djomo & Ceulemans, 2012, pp. 399–400, except for carbon intensity for diesel, for which the source is as shown.

emissions, the biofuel will increase them. Thus, all bioethanol with a carbon intensity of 94 gCO_2/MJ, or biodiesel with an intensity of 95 gCO_2/MJ, will *add* to global GHG emissions when they replace fossil fuels. The worst bioethanol in GHG terms with a total carbon intensity (dLUC + iLUC) of 361 gCO_2/MJ is nearly four times as carbon-intensive as gasoline, while the worst biodiesel in GHG terms, with a total carbon intensity (dLUC + iLUC) of 1677 gCO_2/MJ is more than 16 times as carbon-intensive as diesel. It is for this reason that biofuels have become a very controversial source of supposed emission reduction, and explains why the European Commission, in its Directives mandating a certain share of biofuel use for transport in the European Union (EU), has stipulated sustainability criteria for the production of biofuels, if they are to be counted towards countries' biofuel targets. This issue is discussed in the transport section of Chapter 9.

A further source of bioenergy is 'biogas', methane gas derived from the anaerobic decomposition in 'anaerobic digesters' of biomass, often manure from large livestock farms, but also waste food and crop residues. The solid residue from the process can be used as a soil conditioner. Biogas can be used as a substitute for fossil methane gas for heating and in (suitably converted) motor vehicles. 'Wood gas' can also be produced from the gasification of biomass, mainly wood, as the name suggests, and this can also be used in motor vehicles instead of fossil methane gas.

World production of bioethanol in 2019 was 115 billion litres in 2019,[3] mainly from the USA (maize, 59.5 billion litres) and Brazil (sugar, 36 billion litres), about 10% of the global gasoline market. World production of biodiesel in 2019 was 48 billion litres (about 3% of the global market), with the biggest producers being the EU (15.7%), USA (8.4%), Brazil (7.2%) and Indonesia (5.9%). While consumption of transport fuels fell in 2020, due to the Covid pandemic, it is expected to bounce back for both gasoline and diesel, and also for biofuels.

Electricity from bioenergy

Power generation from bioenergy is projected in the International Energy Agency (IEA) Net Zero Emissions scenario to increase by 2026, though much less than solar and wind (International Energy Agency, 2021a, Figure 1.9, p. 31). Much power generation with bioenergy by 2050 is projected to use BECCS technology, although by 2026 the amount of this infrastructure that will have been built is very limited.

Biomass power stations can be large. In the UK the Drax power station in Yorkshire was originally a 4 GW coal-fired power station, but over the past ten years Drax has received £4 billion in taxpayer subsidies to convert 3 GW of this to biomass, burning 7 million tonnes of wood pellets each year, sourced

---

3 www.iea.org/reports/renewables-2020/transport-biofuels. Accessed April 30, 2023.

from western Canada, western and south-eastern USA and the Baltic states. It is developing a BECCS facility to store the carbon from burning this biomass, which it claims will enable its electricity production and carbon storage to remove 8 million tonnes CO_2 from the atmosphere each year.[4] Drax claims that its sources of wood are 'sustainable', meaning that new tree growth is removing CO_2 from the atmosphere as the old wood is burned, but this claim is contested by Friends of the Earth UK,[5] who state that Drax is old and inefficient, and should be closed down. The issue of the 'sustainability' of Drax's wood pellets was brought further into question when Drax's own advisers advised Drax against calling its wood fuel 'carbon neutral'.[6]

Availability of bioenergy

As noted above, a major issue surrounding biomass production for energy is the amount of land it might take, where this land will come from, what land use the bioenergy would replace, and what the environmental effects of this land use change (e.g. biodiversity loss, water use, as well as carbon emissions), might be. There have been numerous studies of how much land might be available globally for bioenergy production, on the basis of different assumed constraints on the type of land that could be used for this purpose, given the need also for increased food production, and the conservation of forests and biodiversity.

Haberl et al. (2011) looked in detail at the possible bioenergy/food trade-off. Their baseline projection ('business as usual' [BAU]) of food supply and demand to 2050, involving "improved food supply and rapid agricultural intensification" (Haberl et al., 2011, p. 4764), and associated cropland area, built on work by the UN Food and Agriculture Organization (FAO), such that the 9 billion human population at that time was adequately fed. They then varied different assumptions around: the extent of climate change (negative impacts from climate change outweighed by positive impacts from the 'fertilisation effect' of higher levels of CO_2 in the atmosphere, which stimulates plant growth); yield (higher and lower crop yield assumptions); possible cropland expansion (at the expense of grazing land – cropland is not allowed in the scenario to expand at the expense of forests); and diet (a 'fair and frugal' diet involving less meat consumption). Haberl et al. (2011, Figure 4, p. 4763) shows the results of the scenarios they generated with these different assumptions, giving the results of available bioenergy from the use of crop residues and the conversion of both grazing land and crop land to energy crops.

4 www.drax.com/about-us/. Accessed April 30, 2023.
5 https://policy.friendsoftheearth.uk/insight/future-drax-old-inefficient-damaging-and-expensive
6 Power giant Drax told by own advisers to stop calling biomass 'carbon neutral'. Sky News. https://news.sky.com/story/power-giant-drax-told-by-own-advisers-to-stop-calling-biomass-car bon-neutral-12866031. Accessed August 17, 2023.

The Haberl et al. (2011, Figure 4, p. 4763) scenarios suggest that under BAU 6 Mm^2 can be used to produce over 100 EJ per year of primary bioenergy (i.e. before conversion to other fuels or energy forms). This reduces to less than 4 Mm^2 (around 60 EJ/yr) with lower yields but massive cropland expansion, or increases to 10 Mm^2 (around 160 EJ/yr) with a 'fair and frugal' diet.[7] The scenarios in Figure 5.1 show a range of bioenergy use. The Low Energy Demand (LED) scenario, which excludes BECCS, has the lowest use, at less than 50 EJ/year in 2050; the Sitra scenario has 93 EJ per year in 2050; and the IEA Net Zero Emissions (NZE) scenario has around 100 EJ per year bioenergy use, around 20% of energy needs, in 2050. In the 'lower yield' scenario of Haberl et al. (2011, Figure 4, p. 4763), 100 EJ per year is not available, even with massive cropland expansion, but is feasible in the other scenarios assuming some CO_2 fertilisation effect. However, another assessment (Erb et al., 2012, p. 267) considers that up to 100 EJ/yr might be available from crop residues alone, although using this amount for bioenergy, rather than returning it to the soil, may pose challenges for soil fertility.

Haberl et al. (2011) identify both the climate impacts and the CO_2 fertilisation effect as other areas of key uncertainty, which could cause crop yields either to increase or decrease from their projections, which would allow bioenergy availability also to increase (because the crops would need less land) or decrease. Other issues which may constrain the availability of bioenergy are environmental impacts (soil compaction, fertiliser run-off or pesticide and herbicide pollution) from agricultural intensification to increase yields, or the need to conserve more land to stem biodiversity loss. Also uncertain, obviously, is whether forest land will be maintained, or continue, as at present, to be lost to agricultural conversion, which is a major source of GHG emissions, as discussed in Chapter 10. A further issue identified by Erb et al. (2012) is that producing and getting bioenergy to the major demand areas from the area of most bioenergy supply (Africa) would require substantial foreign investment in and the establishment of secure supply chains, both of which depend on reliable market conditions and sound governance which many African governments struggle with at present. Some of these issues also arise in respect of the major supply areas (China, South East Asia and Central South America) in Butnar et al. (2020).

The final conclusion of Erb et al. (2012, p. 267) on these issues seems a classic academic understatement:

Our assessment indicates that policy strategies are needed that succeed in simultaneously optimizing production and consumption systems and considering the many potential conflicting uses of biomass, land and water. ... Producing some 50–150 EJ/yr of energy crops in 2050, in addition

---

7 1 $Mm^2 = 10^6 m * 10^6 m = 10^{12} m^2$ = 1 million km^2; 6 Mm^2 are approximately 4.6% of Earth's land area excluding Greenland and Antarctica.

to the growth of biomass required to adequately feed a world population of perhaps 9.1 billion, is therefore an endeavor that should not be underestimated.

Hydrogen

It is often said that hydrogen, as one of the two constituents of water (chemical formula H_2O), and as a major component of fossil fuels (often called hydrocarbons), is the commonest element on Earth. This is true, but hydrogen is also a very reactive element, so that elemental hydrogen to be used as a fuel needs to be manufactured, and separated, from one of the compounds of which it is a constituent. Once this has been achieved through one of the processes described below, hydrogen needs to be compressed for transportation and distributed to be available for one of a wide range of possible uses.

Figure 6.1 illustrates the many possible pathways from the initial hydrogen compound through hydrogen production to final end use. Figure 6.1 shows that hydrogen can either be produced centrally (top left quadrant) and then, after possible geologic (underground) storage, distributed through pipelines, tube trailers or tankers to its final end use (top right quadrant); or the electricity or fossil gas can be distributed through an energy network to where the hydrogen is required (bottom left quadrant), and the hydrogen can be produced locally (bottom right quadrant). The final uses (far right) include the generation of electricity, refuelling cars (combustion or fuel cells), and production of heat, and perhaps also power, in industry or buildings. This section will briefly describe all these possibilities.

The US Office of Nuclear Energy gives further detail on the end uses of hydrogen, and also shows that it can be produced from electricity or directly from fossil fuels and renewables, and also by using the heat (as well as the electricity) from nuclear power stations. In addition to the uses mentioned above, hydrogen can be used to create synthetic fuels and ammonia (for fertilisers), and to refine metals.[8]

Hydrogen production

The International Energy Agency (2019, Figure 1, p. 18) shows how the production of hydrogen has grown since 1975. In 2018 around 70 Mt of hydrogen was produced in its 'pure' form (i.e. with only small levels of additives or contaminants), compared with less than 20 Mt in 1975. This is used mainly for oil refining or the manufacture of ammonia. In addition, around 45 Mt of hydrogen in 2018 was used in industry mixed with other gases (compared with

8 US Office of Nuclear Energy, www.energy.gov/ne/articles/could-hydrogen-help-save-nuclear. Accessed April 30, 2023.

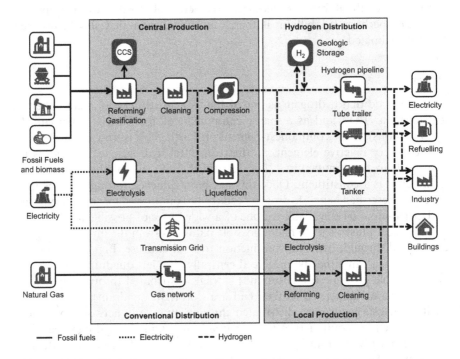

Figure 6.1 Possible hydrogen pathways from production through to end use.

Source: Staffell et al., 2019, Figure 14, p. 476.

less than 10 Mt in 1975), for the production of methanol, the direct reduction of iron in steel-making (DRI) and other uses.

The great majority of hydrogen is currently produced from fossil fuels. Around half is produced by steam reforming fossil methane gas, 30% from partial oxidation of crude oil products, 18% from coal gasification, and 4% from water electrolysis (Staffell et al., 2019, p. 475). The IEA states that CO_2 emissions from current hydrogen production (830 $MtCO_2$ per year) are "equivalent to the CO_2 emissions of Indonesia and the United Kingdom combined" (International Energy Agency, 2019, p. 17).

Emerging methods of hydrogen production routes include high-temperature steam electrolysis, solar thermo-chemical water splitting (artificial photosynthesis) and hydrogen production from biomass gasification. Hydrogen production from fossil fuels can only be characterised as 'low-carbon' if the CO_2 deriving from the process is captured and stored, as described in the next chapter. As noted above, 'green' hydrogen is zero-carbon (e.g. from electrolysis using renewable electricity), 'blue' hydrogen is 'low-carbon' (e.g. steam reforming natural gas with carbon capture and storage [CCS]), while grey or black hydrogen is produced from fossil fuels with no CCS. 'Coal gasification'

Table 6.2 Efficiency and energy consumption of different hydrogen production routes

	Efficiency (LHV) (%)	Energy requirement (kWh per kgH₂)
Methane reforming	72 (65–75)	46 (44–51)
Electrolysis	61 (51–67)	55 (50–65)
Coal gasification	56 (45–65)	59 (51–74)
Biomass gasification	46 (44–48)	72 (69–76)

Source: Staffell et al., 2019, Table 4, p. 477.

was widely used in countries like the UK (before the large-scale extraction of fossil methane gas) to produce 'town gas', a mixture largely of hydrogen and carbon monoxide (CO), distribution of which in urban areas was the basis of the first gas grids. The CO in town gas was what made it poisonous.

The production of hydrogen requires the use of energy. Table 6.2 shows the efficiency and energy consumption of different hydrogen production routes. It can be seen that methane reforming of fossil methane gas is the most efficient. But it still loses a minimum of 25% of its energy (lower heating value, LHV[9]) when it is converted to hydrogen, while electrolysis and coal gasification lose at least a third, and biomass gasification loses over half. One conclusion of such numbers is that electricity from renewables should only be used for climate reasons to produce hydrogen when it cannot substitute directly for carbon-based electricity, or would otherwise be wasted.

Different net-zero scenarios project very different quantities of hydrogen in final energy use in 2050 and 2100. For example, Figure 6.2 shows some of the scenarios in the IPCC database. A small amount of hydrogen is used in 2050 in three of the scenarios (S2, LED, S5), and then more in 2100, with most (around 80 EJ) in S5 (see the dappled block. second from the top, in the S5 2100 bar).

International Energy Agency (2021a, Figure 2.19, p. 75) shows global hydrogen demand by sector in the IEA NZE scenario. The proportion of low-carbon hydrogen increases through to 2050. For comparison, the 500 Mt hydrogen in the NZE in 2050 is around 60 EJ, so the NZE is more ambitious in respect of hydrogen than even the S5 scenario in Figure 6.2.

Hydrogen distribution

When hydrogen is produced centrally, as in the top left quadrant of Figure 6.1, it needs to be distributed to where it will be used. The main means of overland distribution are pipelines (compressed hydrogen gas), tube trailers (compressed hydrogen gas) and tankers (liquefied hydrogen), also as shown in Figure 6.1.

9 For definitions of Lower Heating Value (LHV) and Higher Heating Value (HHV), see https://h2tools.org/hyarc/calculator-tools/lower-and-higher-heating-values-fuels. Accessed April 30, 2023.

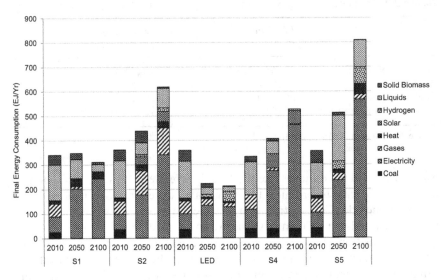

Figure 6.2 Final energy consumption in different scenarios from the IPCC database.

Source: Drummond et al., 2021, Figure 13, p. 54.

In addition, shipping can be used to transport liquid hydrogen overseas, in a similar way to that currently used for liquid natural (fossil methane) gas (LNG).

As with hydrogen production, the compression or liquefaction of hydrogen requires the use of energy. Table 6.3 shows the energy penalty and energy consumption involved. It shows that there is a considerable energy penalty for hydrogen compression (at least 12% and 18% to compress hydrogen to 500 bar and 900 bar[10] respectively) and liquefaction (at least 66% of the energy in the hydrogen). A common pressure used for hydrogen in the tank of a motor vehicle is 700 bar.

These energy losses, and the other economic factors involved in the various distribution options, mean that there is a complex trade-off between scale and distance to determine the cheapest means of hydrogen distribution. Table 6.4 shows the various factors involved in qualitative terms. On-site production (for example, on forecourts for refuelling vehicles) is the cheapest to set up, but becomes expensive for large volumes. Tube trailers, too, are relatively cheap for low-volume distribution over short distances. Tankers with liquefied hydrogen become economic for larger volumes over long distances. Pipelines are the most expensive option to install, but are the cheapest way to distribute large volumes over long distances. Around 3,000 km of pipeline exist in the US and Europe to distribute hydrogen for industrial processes (Staffell et al., 2019, p. 479). The

10 1 bar is approximately atmospheric pressure at sea level. The associated metric unit is the Pascal (Pa), where 1 bar = 100 kPa.

Table 6.3 Energy penalty and electricity requirement for hydrogen compression and liquefaction

	Energy penalty (vs. LHV) (%)	Electricity requirement (kWh per kgH₂)
Compression to 500 bar (including cooling)	15 (12–24)	2.6 (2–4)
Compression to 900 bar (including cooling)	21 (18–30)	3.5 (3–5)
Liquefaction	78 (66–90)	13 (11–15)

Source: Staffell et al., 2019, Table 6, p. 478.

Table 6.4 Factors that affect the economics of hydrogen distribution

Distribution route	Capacity	Transport distance	Energy loss	Fixed costs	Variable costs
On-site production	Low	Zero	Low	Low	High
Gaseous tube trailers	Low	Low	Low	Low	High
Liquefied tankers	Medium	High	High	Medium	Medium
Hydrogen pipelines	High	High	Low	High	Low

Source: Staffell et al., 2019, Table 8, p. 479.

construction of long-distance pipelines, however, will probably need to await greater certainty that they will be justified by the demand.

Many countries already possess a network of gas pipelines, today used predominantly to distribute fossil methane ('natural') gas. The question arises as to whether this network could be used to distribute hydrogen. Like electricity, there is a distinction in gas networks between transmission and distribution. The methane gas distribution network into homes and offices is at relatively low pressure and is increasingly being replaced by plastic pipes, which are compatible with hydrogen.

This has led to thoughts in the UK to convert whole communities to hydrogen in a self-contained kind of way, as explored in the Leeds City Gate H21 project, which produced a substantial report showing how the UK natural (fossil methane) gas distribution network could be incrementally converted to hydrogen,[11] place by place. Initial projects for 'Hydrogen Neighbourhoods' are now being explored in Redcar on Teesside[12] and in Scotland.[13]

11 www.northerngasnetworks.co.uk/wp-content/uploads/2017/04/H21-Report-Interactive-PDF-July-2016.compressed.pdf. Accessed April 30, 2023.
12 https://redcarhydrogencommunity.co.uk/. Accessed April 30, 2023.
13 https://hydrogen-central.com/hydrogen-homes-uk-neighbourhood-green-energy-revolution/. Accessed April 30, 2023.

The pipelines in the gas transmission network, which transport natural gas over longer distances and at higher pressures, are made of steel, and hydrogen would make them brittle, so these would need to be replaced to carry 100% hydrogen. Hydrogen could be added to the current fossil methane gas transmission network, up to about 20%, to reduce its carbon intensity, but of course this would leave 80% of fossil methane gas and so would not be compatible with the 2050 net-zero target.

Hydrogen storage

One of the major potential uses of hydrogen, as discussed in Chapter 5, is as a storage medium for 'surplus' electricity from renewables, the so-called 'power-to-gas' use of electricity. If enough hydrogen could be stored in this way, it could provide an effective answer to the intermittent and variable nature of solar and wind power, particularly in respect of the need for inter-seasonal storage – the storage of electricity in the summer, when demands for electricity and heat are relatively low, for use in winter when demands are much higher, solar energy is less available, and there may be periods of low or no wind. The question then arises as to how this hydrogen from power-to-gas, produced when electricity demands are low and renewable electricity is abundantly available, can itself be stored.

The main means of storing relatively small amounts of hydrogen, for use in vehicles or in vehicle refuelling stations, or for industrial use, is in high-pressure tanks. For large volumes of hydrogen, the most promising option currently seems to be salt domes or caverns, which are already used to store fossil methane gas. Staffell et al. (2019, p. 480) report that there are already operational storage projects of this kind in the UK and US of 24 GWh and 83 GWh respectively. Caglayan et al. (2020) have identified a very large technical potential for hydrogen storage in Europe.

A third means of 'storing' hydrogen is to convert it to a synthetic fuel, which can substitute directly for the whole range of liquid and gaseous fossil fuels, by combining the hydrogen with water and CO_2. If the CO_2 is taken from the atmosphere, to which it returns when the synthetic fuel is burned, then the process is carbon neutral. If the hydrogen comes from electrolysis of water, such fuels are called 'e-fuels', and are generically called 'power-to-X', where the X may be direct substitutes for petrol and diesel for vehicles, kerosene for aviation and fossil methane gas for heating. The advantage of the liquid e-fuels over hydrogen is that they can be stored at atmospheric pressure and have a higher energy density, so they are easier to transport. The disadvantage of such synthetic fuels is the energy used in their production, on top of the energy penalty from producing the hydrogen in the first place. This means that they are still much more expensive than the fossil fuels they would replace.

Hydrogen end uses

One of the great attractions of hydrogen is that, as shown in Figure 6.1 and like fossil fuels, it has many possible end uses: in heating, transport, and industry. These uses are described in more detail in the relevant sections in Chapter 9. Here these uses are introduced in more generic terms.

It has been seen that hydrogen can be converted to synthetic fuels to substitute directly for fuels used in transport and heating. But hydrogen itself can also be used for heating and transport.

For heating it is relatively straightforward to convert boilers and appliances from burning fossil methane gas to burning hydrogen. Indeed this is largely a reverse of the process carried out in the UK in the 1970s when there was wholesale conversion from 'town gas' (a mixture of hydrogen and carbon monoxide) to natural gas from the North Sea. As then, converting the whole country to hydrogen would need to be a huge national campaign, with areas simultaneously converted to supply hydrogen and the appliances adapted to burn it. It would take 10–20 years. It will be interesting to see how the conversion process works out in the Hydrogen Neighbourhoods mentioned above. However, it is very unlikely that it will ever be economically or energetically sensible to seek to replace the current grid for fossil methane gas with a similar one for hydrogen for large-scale home heating.

For transport, hydrogen can be burned directly in combustion engines, which need minimal adaptation to use the new fuel. A number of major carmakers, including Renault and Toyota, have prototype hydrogen combustion engine (HCE) vehicles.[14] However, the combustion process means that HCEs still emit nitrogen oxides, even if they emit no CO_2, so they are not strictly zero-emission vehicles (ZEVs).

Hydrogen can also be used to power vehicles using fuel cells. Fuel cells can convert the chemical energy in a wide range of energy sources directly to electricity. When the energy source is hydrogen, the only emissions are of water vapour. Both Toyota and Hyundai have a commercial fuel cell electric vehicle (FCEV). Although these cars can charge faster than a battery electric vehicle (BEV) and have longer ranges before needing refuelling, they are currently significantly more expensive than a comparable BEV, and the provision of refuelling infrastructure is also both more expensive and well behind that of BEV charging points. It may be that, in transport, hydrogen fuel cells (HFCs) will be more successful than batteries in heavy-duty, longer-distance applications, like buses, trucks, trains and ships. However, given improvements in battery technology, some experts doubt that HFCs will be able to outcompete batteries even in heavy-duty road applications.[15]

14 https://fuelcellsworks.com/news/hydrogen-combustion-engines-are-coming/. Accessed April 30, 2023.
15 https://electrek.co/2022/02/15/study-hydrogen-fuel-cells-cannot-catch-up-battery-electric-vehicles/. Accessed April 30, 2023.

Table 6.5 Uptake of hydrogen and fuel cell applications in 2018

Country	CHP units	Fuel cell vehicles	Refuelling stations	Forklift trucks
Japan	223,000	1800 cars	90	21
Germany	1200	467 cars, 14 buses	33	16
China	1	60 cars, 50 buses	36	N/A
US	225 MW	2750 cars, 33 buses	39 public, 70 total	11,600
South Korea	177 MW	100 cars	11	N/A
UK	10	42 cars, 18 buses	14	2

Source: Staffell et al., 2019, Table 11, p. 482.

Fuel cells can also be used to produce power and hot water in buildings when they are operated to produce 'combined heat and power' (CHP). Japan is the world leader in this application, with its ENE-FARM programme, which by 2020 had installed over 300,000 fuel cell CHP devices in buildings, running off hydrogen produced from fossil methane gas.[16] Life cycle analysis (LCA) of micro-CHP fuel cell systems in Europe, compared against condensing gas boilers and heat pumps using average grid electricity, found that, for the same output, they had lower emissions of GHGs and other air pollutants (Nielsen et al., 2019, p. 343). Provided fossil methane gas and water are available, micro-CHP systems produce electricity independently from the grid and so can act as back-up in case of power cuts. To be zero-carbon, such systems would have to run off hydrogen produced from natural gas with CCS (see next chapter) or from renewables.

Other potential markets for fuel cells include forklift trucks, where they are already well established commercially, because of their clean indoor operation, motorcycles, with Suzuki showing interest, and auxiliary power units (APUs) for overnight accommodation in trucks and caravans (to provide air conditioning, cooking, and heating and cooling), and also for aeroplanes (to reduce ground emissions), unmanned aerial vehicles (drones), tractors and golf carts. Table 6.5 summarises the uptake of hydrogen and fuel cell applications as of 2018. This is a fast moving field, and the numbers of many of these applications are likely to have increased substantially since then.

Investment in electrolysers for hydrogen

2020–2021 saw a very significant increase in the capacity of electrolysers to produce green hydrogen. International Energy Agency (2021b, p. 30) shows hydrogen production growing from about 30 MWe in 2019, to around 60 MWe in 2020, to an estimated 280 MWe in 2021. Around half of the hydrogen was intended to be used in vehicles, with blending in the gas grid the next biggest,

16 www.challenge-zero.jp/en/casestudy/469. Accessed April 30, 2023.

followed by synthetic fuels, electricity storage and industrial applications. After a number of years of false starts and excessive hype, there are now, for the first time perhaps, solid grounds for thinking that the time for hydrogen has come.

Activity related to the development of hydrogen as an energy source is stepping up in many countries. As an example, Scotland, which is a relatively small country, but one well-endowed with renewable energy, has published a roadmap through the 2020s and 2030s detailing its proposed 'hydrogen journey', involving GW-scale production, uses in transport, industry and the manufacture of synthetic fuels, a transmission pipeline and significant exports. Scottish Government (2022, p. 18) lists 60 current and planned hydrogen projects in Scotland that are contributing to this roadmap, across end users, production, storage, transmission, distribution, and multi-vector combinations of these. It is planned that these should be served from 12 'hydrogen hubs' at different locations round the country.

Of course, Scotland and the UK more broadly are not the only countries with hydrogen ambitions and strategies – the US, Germany, Korea and Japan are also mentioned often as countries investing heavily in hydrogen, but also key lower-income countries such as India.[17] In fact, fears surfaced in 2022 that the UK was falling behind other countries in terms of both its ambition and its investment in this area.[18]

Conclusion

The heavy lifting of decarbonisation will be done by electricity, mainly in the form of solar and wind. But not all energy use can be electrified, as will be seen in the following chapters, and for these uses both bioenergy and hydrogen have a not unimportant role to play. Bioenergy comes with substantial caveats concerning both competition with other land uses (food and biodiversity) and the environmental sustainability of its production. Where forests are cleared for the wood to be burned for electricity production, it may be that the carbon intensity of the operation is as high or higher than fossil fuel electricity. Similarly, where biofuels derive from palm oil plantations that have replaced tropical rainforests, it is likely that their carbon intensity is as high or higher than the fossil diesel they are replacing. Any use of bioenergy should therefore be accompanied by maximum transparency as to its source and how it was produced.

The limitations of hydrogen derive from the energy sources from which it is produced. The great majority of current hydrogen production is 'grey' (i.e. from steam methane reforming [SMR] of fossil methane gas) with no CCS. It therefore plays no role in carbon reduction. Globally, only 1% of global

17 https://fsr.eui.eu/the-green-hydrogen-economy-insights-from-india/. Accessed April 30, 2023.
18 See, for example, www.ft.com/content/7d7420bc-079e-4730-b849-5a18ccd7258f. Accessed April 30, 2023.

SMR hydrogen production is 'blue',[19] and in the UK the first CCS plants for hydrogen production are just now being planned. Green hydrogen meanwhile largely remains a gleam in the eye, albeit one that is growing brighter as more renewable electricity capacity is constructed. It remains the case that it is far more energy-efficient to use renewable electricity directly, so that its rational use for hydrogen production is limited to periods when potential electricity supply exceeds the demand for it and it would otherwise be wasted. At present this happens relatively infrequently, but its occurrence will increase as more and more renewable energy capacity is constructed. It is desirable that electrolysers are developed and deployed now to be able to take full advantage of the availability of surplus renewable electricity when it arises.

References

Butnar, I., Broad, O., Solano Rodriguez, B., & Dodds, P. E. (2020). The role of bioenergy for global deep decarbonization: CO_2 removal or low-carbon energy? *GCB Bioenergy*, *12*(3), 198–212. https://doi.org/10.1111/GCBB.12666

Caglayan, D. G., Weber, N., Heinrichs, H. U., Linßen, J., Robinius, M., Kukla, P. A., & Stolten, D. (2020). Technical potential of salt caverns for hydrogen storage in Europe. *International Journal of Hydrogen Energy*, *45*(11), 6793–6805. https://doi.org/10.1016/j.ijhydene.2019.12.161

Drummond, P., Scamman, D., Ekins, P., Paroussos, L., & Keppo, I. (2021). *Growth-positive zero-emission pathways to 2050*. Sitra, Helsinki. www.sitra.fi/en/publications/growth-positive-zero-emission-pathways-to-2050/

Erb, K. H., Haberl, H., & Plutzar, C. (2012). Dependency of global primary bioenergy crop potentials in 2050 on food systems, yields, biodiversity conservation and political stability. *Energy Policy*, *47*, 260–269. https://doi.org/10.1016/J.ENPOL.2012.04.066

Haberl, H., Erb, K. H., Krausmann, F., Bondeau, A., Lauk, C., Müller, C., Plutzar, C., & Steinberger, J. K. (2011). Global bioenergy potentials from agricultural land in 2050: Sensitivity to climate change, diets and yields. *Biomass and Bioenergy*, *35*(12), 4753–4769. https://doi.org/10.1016/J.BIOMBIOE.2011.04.035

International Energy Agency. (2019). *The Future of Hydrogen*. IEA, Paris.

International Energy Agency. (2021a). *Net Zero by 2050*. IEA, Paris.

International Energy Agency. (2021b). *World Energy Investment*. IEA, Paris.

Nielsen, E. R., Prag, C. B., Bachmann, T. M., Carnicelli, F., Boyd, E., Walker, I., Ruf, L., & Stephens, A. (2019). Status on Demonstration of Fuel Cell Based Micro-CHP Units in Europe. *Fuel Cells*, *19*(4), 340–345. https://doi.org/10.1002/fuce.201800189

Njakou Djomo, S., & Ceulemans, R. (2012). A comparative analysis of the carbon intensity of biofuels caused by land use changes. *GCB Bioenergy*, *4*(4), 392–407. https://doi.org/10.1111/j.1757-1707.2012.01176.x

Scottish Government. (2022). *Hydrogen Action Plan*. Scottish Government, Edinburgh. www.gov.scot/publications/hydrogen-action-plan/. Accessed April 29, 2023.

Staffell, I., Scamman, D., Velazquez Abad, A., Balcombe, P., Dodds, P. E., Ekins, P., Shah, N., & Ward, K. R. (2019). The role of hydrogen and fuel cells in the global energy system. *Energy and Environmental Science*, *12*(2). https://doi.org/10.1039/c8ee01157e

---

19 www.globalccsinstitute.com/wp-content/uploads/2021/04/Circular-Carbon-Economy-series-Blue-Hydrogen.pdf. Accessed April 30, 2023.

7 Carbon capture, use, storage and removal, and geoengineering

Summary

It has been seen from previous chapters that most emission scenarios that meet the Paris temperature targets make sometimes very extensive use of technologies that involve carbon capture and storage (CCS), some of which also use carbon (CCUS), or which may also, or separately, involve removing carbon dioxide (carbon dioxide removal, or CDR) from the atmosphere. In addition, there is growing interest, given already apparent effects of climate change, in techniques that involve 'geoengineering' the climate through direct, large-scale interventions in natural processes in the atmosphere, oceans or on land. This chapter discusses these technologies, their current deployment, and possible effectiveness, limitations and dangers. It also explores the use of some of these techniques by companies and, potentially, governments, to 'offset' their continuing emissions. In respect of the different geoengineering technologies, the chapter discusses how the use of these technologies should be governed and what their potential unintended or negative consequences might be.

Introduction

As seen in Chapter 2, the world has left large-scale emission reduction very, very late. There are three fundamental, but very challenging, implications of this.

The first implication is that, given that all countries are still very dependent on fossil fuels for their energy, and cannot cease their use immediately, a technology will have to be implemented that will capture the emissions from burning fossil fuels, or making cement, and store them securely (i.e. for the indefinite future). Such technology is called 'carbon capture and storage' (CCS). To take account of the possibility of capturing and *using*, rather than just storing, the CO_2, the acronym is expanded to CCUS. This is the subject of the first section of this chapter.

DOI: 10.4324/9781003438007-8

The second implication is even more challenging. To stay within the 1.5°C Paris temperature target, it will now not be enough simply to reduce greenhouse gas (GHG) emissions to zero by 2050. CO_2 will actually have to be removed from the atmosphere at a very large scale. Figure 2.13 showed that these 'negative emissions' would need to start soon after 2030, and reach 10 $GtCO_2$ per year by 2100, in order to bring the average level of global warming back down from nearly 1.9°C, which it reaches soon after 2050, to the Paris target of 1.5°C by the end of the century. How this might be achieved is the subject of the second section of this chapter.

The third implication is attracting increasing attention, and involves either enhancing the reflection of solar radiation back into space, or stopping it from reaching Earth's atmosphere in the first place. These techniques are collectively called 'radiation management' and are discussed in the final section of this chapter.

Overview

Figure 7.1 illustrates the various techniques that have been proposed for carbon removal from the atmosphere, and for radiation management, some of which will be discussed later in this chapter. A number of the techniques are also described in lay-persons' terms by the Royal Geographical Society in a page on its website.[1]

In respect of carbon removal from the atmosphere on land, and moving along the horizontal line from the left of the picture, agriculture can grow bioenergy crops, the emissions from which, when burned, can be captured and stored through BECCS, as discussed in the previous chapter, and further below. Alternatively, biomass may be converted to biochar through pyrolysis (combustion largely in the absence of oxygen) and the carbon sequestered in the soil; or trees may be planted, which will remove CO_2 from the atmosphere as they grow. Finally, air capture machines may be built to remove CO_2 from the atmosphere, after which it can be stored underground.

In the oceans, the addition of nutrients (phosphorus, nitrogen, iron) may enhance the removal of CO_2 from the atmosphere by vegetable matter in the ocean, and surface nutrients may also be enriched through enhancing 'upwelling' from the deep ocean. Carbonates may be added to the ocean to reduce its acidity and enable it to absorb more CO_2 from the air. Enhanced 'downwelling' may cause surface carbon to sink to the deep ocean, allowing more to be absorbed at the surface from the atmosphere.

The radiation management techniques in Figure 7.1 are shown as reflecting solar radiation back into space from the inclined line, at different levels in the atmosphere. At ground level, increasing surface albedo (enhancing the levels

---

1 www.rgs.org/CMSPages/GetFile.aspx?nodeguid=f8d0c915-ce64-4a1a-a3bf-eafd2697b c9e&lang=en-GB. Accessed April 30, 2023.

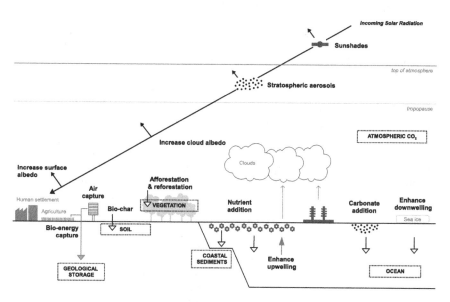

Figure 7.1 Various technologies for carbon removal and radiation management.

Source: Lenton & Vaughan, 2009, Figure 1, p. 5541.

of reflectivity of incoming solar radiation) can be achieved by installing white surfaces on buildings or on deserts, and by planting light crops or other vegetation. Clouds also reflect back solar radiation, and 'marine cloud brightening', increasing the albedo of clouds above the ocean, may be stimulated through the large-scale generation of sea spray. Though not shown in Figure 7.1, the albedo of the surface ocean itself may also be increased by injecting it with micro-bubbles of air in a process that mimics the formation of the white surfaces of breaking waves or turbulent seas.[2] In the upper atmosphere aircraft can be used to inject aerosols (a suspension of fine solid or liquid particles in gas), which serve a similar reflective purpose, in a process which mimics the cooling of the atmosphere that follows major volcanic eruptions. Finally, reflective sunshades may be deployed in space.

Some the techniques or technologies described above, especially those related to radiation management, have been called 'geoengineering', which has been defined as "the deliberate large-scale manipulation of an environmental process that affects Earth's climate, in an attempt to counteract the effects of global warming".[3] The Convention on Biological Diversity (CBD) (Williamson & Bodle, 2016) has produced a useful diagram (see Table 7.1) linking the

2 https://geoengineering.global/ocean-albedo-modification/. Accessed April 30, 2023.
3 www.brookings.edu/research/preparing-the-united-states-for-security-and-governance-in-a-geo engineering-future/. Accessed April 30, 2023.

Table 7.1 Techniques for climate mitigation (emission reduction), CO_2 removal and solar radiation management

Climate mitigation (conventional mitigation)	Negative emission techniques (carbon dioxide removal)	Sunlight reflection methods (solar radiation management)	Other techniques (enhanced heat storage or enhanced heat escape)
Energy efficiency	Reforestation and afforestation (large scale)	Surface albedo: urban	Other ocean circulation changes
Renewable energy: solar, wind, wave & tidal	Biochar (large scale)	Surface albedo: cropland	Cirrus cloud reduction
Nuclear energy	Bioenergy with CCS (BECCS)	Surface albedo: desert & grassland	
Renewable energy: biofuels	Direct air capture of CO_2	Surface albedo: ocean (including sea-ice)	
Carbon capture & storage (CCS) for fossil fuels	Land biomass burial/ sequestration	Marine cloud brightening	
Agricultural techniques to increase soil carbon, e.g. no till	Enhanced weathering (land)	Stratospheric aerosols	
	Enhanced weathering (ocean)/alkalinity Ocean fertilisation Ocean upwelling	Space mirrors	

Source: Adapted from Williamson & Bodle, 2016, Figure 1.1, p. 15.

techniques and technologies discussed in this section with the other methods of reducing GHG emissions that have been described in other chapters.

The main 'climate mitigation' technologies in Table 7.1 have been described in Chapters 3, 5 and 6. Carbon capture and storage (CCS) is discussed in the next section of this chapter, followed by the carbon dioxide removal techniques. Then come those techniques that are marked in Table 7.1 as narrowly defined 'geoengineering', basically comprising solar radiation management (SRM).

The processes through which CDR and SRM occur are complex, as are the calculations as to how much global warming they can prevent. Lenton & Vaughan (2009) provides a detailed description of both. Here the processes and their effectiveness are described only in basic terms, considering only those techniques which according to Lenton & Vaughan (2009) have reasonable potential to slow global warming in the short and medium term.

Carbon capture, use and storage (CCUS)

Figure 7.2 gives a basic diagram of what CCUS entails. The CO_2 is captured from a power station or industrial facility, compressed, and either used or transported for storage in an appropriate geological formation underground, onshore or offshore. For the UK, depleted oil and gas wells in the North Sea are often considered as suitable sites. Where the CO_2 is just stored, rather than used, the acronym CCS will be used.

CCS began in the 1970s when CO_2 was pumped into oil wells to increase the pressure and increase the amount of oil that could be recovered. Two-thirds of stored CO_2 are used for this 'enhanced oil recovery' (EOR) (International Energy Agency, 2020, p. 37), though the carbon stored is thereby offset to some extent by that in the new oil extracted, which will enter the atmosphere when it is burned.

The development of CCS has been bumpy. Operational CO_2 capture has grown only from around 15 $MtCO_2$ to 40 $MtCO_2$ per year in the period 2010–2021. Many of the plants in both early and advanced development early in that

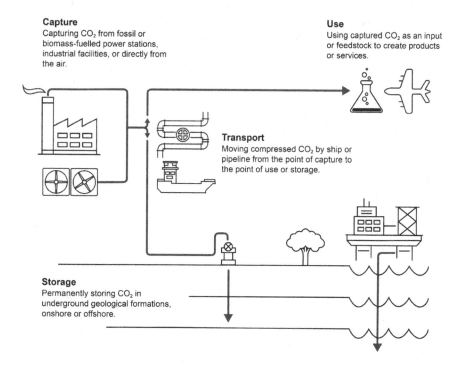

Figure 7.2 Processes of carbon capture, use and storage.

Source: International Energy Agency, 2020, p. 20.

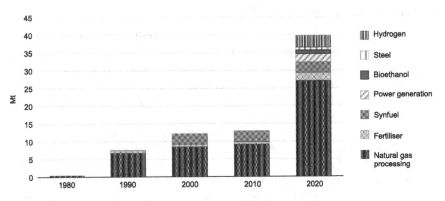

Figure 7.3 Plans for large-scale CCS facilities in different years and sectors.

Source: International Energy Agency, 2020, Figure 1.2, p. 27.

decade did not come to fruition, and some even that were under construction did not become operational. By 2020 construction was very limited. However, in 2022 development plans went well beyond the level of 2010, such that in that year there were plans for CCS facilities that would capture nearly 250 $MtCO_2$ per year (Global Carbon Capture and Storage Institute, 2022, Figure 3, p. 6). Of course, it remains to be seen whether all these planned plants actually become operational.

However, Figure 7.3 shows that current plans for CCS relate to many more sectors than they did 10 or 20 years ago. Then CO_2 capture from fossil methane gas (here called natural gas) processing[4] was absolutely dominant, with very minor amounts from the manufacture of fertiliser and synfuels.[5] Fossil methane gas processing still provides most of the CO_2 for CCS in 2020, but there are contributions now from power generation and the manufacture of bioethanol, steel and hydrogen.

The increase in 2021 of planned growth in CCS is apparent in the number of facilities in development, which in that year was close to 200, compared to around 80 at the previous peak in 2010–2011.[6] Moreover, the International Energy Agency (IEA) believes that there are three reasons why a greater proportion of these plants will be seen through to operation than in the 2010s:

4 When extracted, fossil methane gas often contains CO_2, which has to be removed for the gas to reach the required purity.
5 Synfuels are created from CO_2 and hydrogen from electrolysis of water, and are used predominantly in transport as fossil fuel substitutes.
6 www.iea.org/commentaries/carbon-capture-in-2021-off-and-running-or-another-false-start. Accessed April 30, 2023.

Table 7.2 CCUS deployment in the IEA Net Zero Emissions (NZE) scenario

	2020 (MtCO₂)	2030 (MtCO₂)	2040 (MtCO₂)	2050 (MtCO₂)
Electricity sector				
Coal with CCUS	3	280	918	625
Gas with CCUS (MtCO₂)	0	63	246	237
Bioenergy with CCUS (MtCO₂)	0	87	457	572
Industry				
Industry processes (MtCO₂)	0	160	902	1,627
Industry combustion (MtCO₂)	3	212	842	1,170
Fuel supply				
Other energy (MtCO₂)	30	173	296	412
Biofuel transformation (MtCO₂)	1	149	406	624
Hydrogen transformation (MtCO₂)	3	454	929	1,353
Other				
Direct air capture (MtCO₂)	0	87	623	983

Source: Data from International Energy Agency, 2021b, Figure 2.21. p. 80.

- New business models have emerged, through proposals to develop industrial hubs, generating economies of scale and reduced commercial risks, rather than stand-alone plants.
- The investment environment has improved, with more support now for CCUS, in the US with price support of US$50 per tonne of carbon stored, with proposals in Congress to increase this to US$85, and similar carbon prices in the EU Emissions Trading System (EU ETS).
- The realisation that for net zero, CCUS is a necessity, not an option, as seen in the IEA's Net Zero Emissions (NZE) scenario (see Table 7.2).

By 2050 in the NZE scenario CCS is capturing 7.6 GtCO₂ per year. Table 7.2 shows that the biggest use of CCS by 2050 is in capturing emissions from hydrogen transformation (from fossil methane gas), and industrial processes and combustion, with each capturing over 1 GtCO₂ per year. Direct air capture (DAC) (explained in the next section of this chapter) is removing nearly 1 GtCO₂ per year from the atmosphere, and biofuel transformation and bioenergy with CCUS are both capturing over 0.5 GtCO₂. It may be noted that this contribution to decarbonisation of CCS, BECCS and DAC in 2050 in NZE is very similar to that in the Sitra 1.5°C scenario (see Figure 2.13). The investment requirements of building out CCS and DAC to this extent are huge – by 2050 the annual investment in these technologies exceeds US$160 billion (International Energy Agency, 2021b, p. 82).

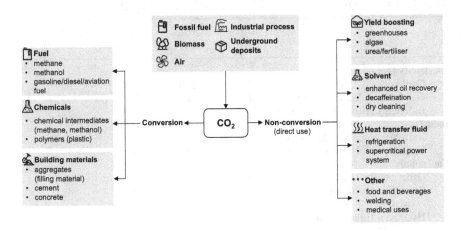

Figure 7.4 Uses of CO_2.

Source: International Energy Agency, 2021a.

Uses of CO_2

It can be seen from Figure 7.4 that there are many uses for CO_2, for chemicals and building materials, making synthetic fuels, enhancing oil recovery (EOR), in refrigeration and food and beverages, and for welding and medical uses. However, these uses are relatively small compared to the scale of capture and storage required for net zero. Over half of 2020's production of around 230 MtCO2 is used for fertilisers, while 70–80 Mt/year is used for EOR (International Energy Agency, 2021a) – both of which uses engender further GHG emissions. The uses that actually lock up CO_2 (for example, in building materials) are still very small.

Carbon dioxide removal

It has already been seen that, in the great majority of scenarios that keep the global average temperature increase to below 1.5°C, and even more so to get back to atmospheric concentrations of GHGs of 350 ppm or below, reducing emissions is not enough. Past CO_2 emissions will need to be removed from the atmosphere on a large scale.

There is obviously some uncertainty as to whether the climate, having been changed by the rise in atmospheric CO_2 concentrations, would in fact return to its previous state if those concentrations were to fall through the removal of CO_2 from the atmosphere. Natural systems do not always behave in this orderly, symmetrical way. And of course it would not happen if climate change itself triggered emission feedbacks, the kinds of 'tipping points' discussed in Chapter 1, that acted to reinforce climatic changes.

On this issue some reassurance may be taken from the Intergovernmental Panel on Climate Change (IPCC)'s statement with high confidence in its most recent assessment report:

> If global net negative CO_2 emissions were to be achieved and be sustained, the global CO_2-induced surface temperature increase would be gradually reversed but other climate changes would continue in their current direction for decades to millennia. For instance, it would take several centuries to millennia for global mean sea level to reverse course even under large net negative CO_2 emissions.
>
> (IPCC, 2021, D.1.6)

This suggests that atmospheric carbon dioxide removal (CDR), while not in any sense being a 'quick fix', may at least be able to prevent further global heating.

There are basically three ways through which CDR can be achieved: through machines that suck the CO_2 from the air, called direct air capture (DAC); enhancing the take up of CO_2 emissions by the oceans; and increasing the take up of CO_2 emissions by vegetation or other land use. Table 7.3 gives examples of a few of these methods, with their potential side effects. Most positive in terms of impact on food supply and biodiversity is to sequester carbon in the soil, with probable positive effects on water quality and quantity as well, but there is an order of magnitude uncertainty (0.3–3 $GtCO_2$ per year) over how much carbon can be sequestered in this way. Next most positive is 'managing forests', which includes looking after current forests (i.e. stopping deforestation), reforestation of previously destroyed forests, and afforestation which involves establishing new forests on previously unforested land. Greatly expanding the area of land under forests could have a negative impact on food supply (as suggested in Table 7.3) if the land concerned is currently being used for food production. Impacts on biodiversity and water will depend on the kinds of forests being established, with forests composed of multiple native species being better for biodiversity, and probably for water, than monoculture plantations.

BECCS, as noted in the previous chapter, stands for 'bioenergy with carbon capture and storage'. The issues related to bioenergy have been discussed in Chapter 6, with the carbon capture and storage simply involving capturing the emissions when the biomass is burned for either electricity or heat production. The scale of BECCS envisaged in net-zero scenarios can be very large indeed. In the IEA's NZE scenario BECCS captures nearly 1.4 $GtCO_2$ per year in 2050 from power, industry and biofuel production (International Energy Agency, 2021b, p. 79). Butnar et al. (2020, p. 199) report that as of 2017 "only one BECCS plant is operational. Located in Decatur, Illinois, United States, the US\$ 208 million plant captures 1 $MtCO_2$/year from corn bioethanol production." The 2050 IEA NZE scenario would thus require 1,400 such plants to be operational. The scenarios reviewed by Fuss et al. (2018) require 0.5–5 $GtCO_2$ per year of BECCS.

Table 7.3 Examples of methods of carbon dioxide removal (CDR) from the air

	Biosphere				Geosphere			
	Storage in plants and soils				Storage in rocks and minerals			
	Managing forests		**Soil sequestration**		**BECCS***		**Air capture**	
	Potential side effects	Evidence base	Potential side effects	Evidence base	Potential side effects	Evidence base	Potential side effects	Evidence base
Impacts on:								
Water quality	Positive	Low	Positive	Low	Negative	Low	Negative	Low
Water quantity	Positive & negative	Low	Positive	Low	Negative	Medium	Positive & negative	Low
Food supply	Negative	Medium	Positive	Medium	Negative	Medium	None	N/A
Biodiversity	Positive & negative	Medium	Positive	Medium	Positive & negative	Low	None	N/A
Sequestration per year	>3 GtCO$_2$		0.3–3 GtCO$_2$		>3 GtCO$_2$		>3 GtCO$_2$	
Measurement and verification	Fair		Poor		Fair		Good	

* BECCS: bioenergy with carbon capture and storage.
Source: Data from Joppa et al., 2021, p. 630.

The final example in Table 7.3 is the capture of carbon by machines directly from the air. DAC technologies are still very much in their infancy. According to the IEA "There are currently 18 DAC plants operating worldwide, capturing almost 0.01 Mt [10,000 t] CO_2/year, and a 1 Mt CO_2/year capture plant is in advanced development in the United States".[7] The IEA also estimates that the cost of DAC CO_2 is US\$135–345 per tCO_2 (International Energy Agency, 2020, p. 68). More discussion of how DAC technologies work can be found in International Energy Agency (2020, pp. 82ff). The rest of this section will focus on enhancing carbon dioxide removal by the oceans, and on land, before discussing the opportunities, and dangers, of seeking to use such removals as 'carbon offsets' for existing emissions.

Ocean carbon dioxide removal

It was seen in Chapter 1 (Figure 1.6) that oceans currently absorb just under a third of human CO_2 emissions, so it is natural, when thinking of ways of removing more carbon dioxide from the atmosphere, that attention should turn to whether the amount being absorbed could be increased.

There are three kinds of processes which could enhance ocean CDR: biological processes, which enhance photosynthesis in the ocean; chemical processes, which remove dissolved CO_2 from the ocean and so allow it to absorb more from the atmosphere; and electrical processes which remove dissolved CO_2 from the ocean by passing electrical current through seawater, again allowing it to absorb more CO_2 from the atmosphere.[8]

A 2018 review analysed 13 ways in which the ocean could reduce the scale and impacts of climate change (Gattuso et al., 2018), and a later paper (Gattuso et al., 2021) focused on the four means whereby the ocean could remove carbon dioxide from the atmosphere: growing marine biomass, called 'macro-algae' (including seaweed and seagrass) at scale; restoring and increasing coastal vegetation (such as mangroves); fertilisation of the ocean through the addition of iron or other nutrients, which enhance the growth of phytoplankton; and adding alkaline rocks to the ocean to remove dissolved CO_2 and therefore increase the ocean's ability to absorb more CO_2 from the atmosphere. These four CDR techniques were assessed against different criteria as shown in Table 7.4.

Growing seaweed and then using it as bioenergy with CCS (Marine BECCS) is assessed as moderately effective in carbon uptake, with high duration of effect and cost effectiveness. However, its feasibility is still very low because it still has to be proved at scale. Increasing coastal vegetation is feasible with high durability, and has co-benefits (e.g. fisheries, storm breaks) but is of limited effectiveness because of the relatively small number of sites where it could be

---

7 www.iea.org/reports/direct-air-capture. Accessed April 30, 2023.
8 See www.wri.org/insights/leveraging-oceans-carbon-removal-potential (Accessed April 30, 2023) for further description of the three processes.

Table 7.4 Assessment against different criteria of four marine CDR technologies

	Marine BECCS	Restoring and increasing coastal vegetation	Enhancing open-ocean productivity	Enhancing weathering and alkalinisation
Effectiveness to increase carbon	Low	Very low	Very low	Very high
Effectiveness to reduce OW/OA/SLR	Low	Very low	Very low	High
Feasibility	Very low	Moderate	Low	Very low
Duration of effect	High	Very high	Moderate	Very high
Cost effectiveness	High	Very low	Very low	Moderate
Global governability	Moderate	Very high	Very low	Very low
Co-benefits	Low	High	Low	Low
Disbenefits	Very low	Low	High	High

Source: Data from Gattuso et al., 2021, Figure 1, p. 4.

Note: OW is ocean warming; OA is ocean acidification; SLR is sea level rise.

implemented at scale. Enhancing ocean productivity would also have a durable effect, but its feasibility is very low because of the amount of material (e.g. iron) required to implement it at scale and the same feasibility considerations arise with alkalinisation, despite its high effectiveness, because of the large quantities of rock required and the associated mining and logistics. In terms of potential quantity of CDR, this is only estimated for increasing coastal vegetation (cumulative 95 $GtCO_2$ until 2100, around 2.5 years of current anthropogenic emissions) and fertilisation (5.5 $GtCO_2$ per year) (Gattuso et al., 2021, pp. 2, 3).

On the basis of these kinds of considerations, which are much expanded upon in the paper, it is clear that none of these ocean CDR techniques offers an easy policy route to mitigating climate change. Increasing coastal vegetation is 'low regrets', in the sense that there are good reasons to do it apart from climate change (as with reducing deforestation on land), but its overall effectiveness as climate action is low, and it is expensive. Gattuso et al. (2021) classify the other three measures as at 'concept stage', meaning that none of them is likely to be implemented at scale in the next ten years, and some may never reach this stage, because of feasibility or public acceptability issues.

Land-based carbon dioxide removal

Options for land-based CDR are sometimes called 'natural climate solutions' (NCS), to distinguish them from engineered CDR, such as through DAC machines. Griscom et al. (2017, p. 1645) define NCS as "options to mitigate climate change by increasing carbon sequestration and reducing emissions of carbon and other greenhouse gases through conservation, restoration, and improved management practices in forest, wetland, and grassland biomes". By including the avoidance of emissions through improved land management, this definition goes beyond strict CDR, but it gives a good idea of the role that land management could play in reducing climate change.

Griscom et al. (2017) quantified this role as follows (where 1 $PgCO_2e$ is the same as 1 $GtCO_2e$):

> The net emission from the land use sector is only 1.5 petagrams of CO_2 equivalent ($PgCO_2e$) per year, but this belies much larger gross emissions and sequestration. Plants and soils in terrestrial ecosystems currently absorb the equivalent of ~20% of anthropogenic greenhouse gas emissions measured in CO_2 equivalents (9.5 $PgCO_2e$ [9.5 $GtCO_2e$]) per year. This sink is offset by emissions from land use change, including forestry (4.9 $PgCO_2e$ per year) and agricultural activities (6.1 $PgCO_2e$ per year), which generate methane (CH_4) and nitrous oxide (N_2O) in addition to CO_2. Thus, ecosystems have the potential for large additional climate mitigation by combining enhanced land sinks with reduced emissions.

Chapter 10 discusses how emissions from agriculture can be reduced. This section explores how NCS can increase CO_2 absorption from the atmosphere.

Twenty NCS options are reviewed by Griscom et al. (2017) and the contributions of these to CDR or emission prevention are shown in Table 7.5. Further detail of precisely what these and other (e.g. enhanced rock weathering) land management options consist of, how they work, and what needs to be done to remove or avoid the emissions may be found in the 2022 Energy Transitions Commission report on the subject (Energy Transitions Commission, 2022). However, it should be apparent that they involve large-scale changes in land use, land management and farming (e.g. stopping deforestation, reforesting already-deforested areas, regenerative agriculture) that will be extremely challenging to achieve on anything like the scale required. There are currently few signs that the necessary transformation has even begun.

In total the calculations suggest that the maximum additional mitigation potential of all the 20 NCS considered, consistent with safeguarding land for food, fibre and biodiversity habitat, is 23.8 $GtCO_2e$/year. Of that, about half is available at a cost of less than US$100 per tCO_2, when this is put into a cost-optimal scenario consistent with a 66% probability of staying within 2°C. Two-thirds of this 'cost-effective' mitigation potential comes from forest NCS. About half comes from nine CDR NCS, and the other half comes from avoided emissions through the other NCS. Of the 20 NCS considered, only coastal restoration was included in the ocean CDR section, so other ocean CDRs could potentially add to the estimates of CDR potential above. One of these, which is said to have the potential to remove 5.5 $GtCO_2e$ from the atmosphere, would

Table 7.5 Potential contribution by 2030 of 20 natural climate solutions to GHG emission prevention and CDR

	Climate mitigation potential in 2030 ($GtCO_2e\ yr^{-1}$)			Other benefits			
	<2°C ambition <10 US$/ tCO₂e/year	<2°C ambition <100 US$/ tCO₂e/year	maximum with safeguards	Soil	Water	Bio-diversity	Air
Forests							
Reforestation	0	3.037	10.124	✓	✓	✓	✓
Avoided forest conv.	1.816	2.897	3.603	✓	✓	✓	✓
Natural forest management	0.441	0.882	1.470	✓	✓	✓	
Improved plantations	0	0.266	0.443			✓	
Avoided woodfuel	0	0.110	0.367	✓	✓	✓	✓
Fire management	0	0.127	0.212	✓	✓	✓	✓
Agriculture & Grasslands							
Biochar	0	0.331	1.102	✓			
Trees in croplands	0	0.439	1.040	✓	✓	✓	✓
Nutrient management	0.635	0.635	0.706	✓	✓	✓	✓
Grazing – feed	0	0.204	0.680				
Conservation agriculture	0.248	0.372	0.413	✓	✓		
Improved rice	0.080	0.159	0.265		✓		
Grazing – animal management	0	0.060	0.200				
Grazing – optimal intensity	0.045	0.089	0.148	✓	✓	✓	
Grazing – legumes	0.088	0.132	0.147	✓		✓	
Avoided grassland conversion	0	0.035	0.116	✓	✓	✓	
Wetlands							
Coastal restoration	0	0.200	0.841	✓	✓	✓	✓
Peat restoration	0.149	0.394	0.815	✓	✓	✓	✓
Avoided peat impacts	0.452	0.678	0.754	✓	✓	✓	✓
Avoided coastal impacts	0.182	0.273	0.304	✓	✓	✓	✓

Source: Data from Griscom et al., 2017, Figure 1, p. 11646.

Note: Climate mitigation 'safeguards' are constraints imposed on the options to protect the production of food and fibre and habitat for biological diversity. The < 2°C <100 US$/tCO₂e/year column represents mitigation levels for a greater than 66% chance of holding global warming to <2°C. Other benefits of the options include greater biodiversity, water filtration and flood control, soil enrichment, and air filtration.

be to allow marine pelagic fish stocks to recover at scale (Schmitz et al., 2023, Table 1, p. 325) – something which, when achieved, would also permit a larger sustainable fish harvest. The 23.8 GtCO$_2$ estimate in Griscom et al. (2017) is around 30% higher than previous estimates in the literature, for reasons that are explained in the paper, but too detailed to go into here.

Figure 7.5 shows how such changes in land use could help reduce net emissions through to 2050.

Several conclusions emerge from Figure 7.5. The first is that NCS can play an important role in reducing emissions in the short term, as some measures can be implemented relatively quickly. However, the second conclusion is that CDR and emission avoidance make a much smaller contribution to a 2°C-compliant scenario than the reduction of emissions from fossil fuel use. It is therefore extremely concerning that, in some quarters, CDR and avoided emissions are being touted as ways to 'offset' carbon emissions rather than reduce them. These 'carbon offsets' are the subject of the next section.

Carbon offsets

GHGs are a global pollutant which have the same effect on the global climate wherever they are emitted. This has given rise to two market-based trading activities, whereby people, companies or economic sectors that find it difficult or expensive to reduce their emissions can pay for other activities to reduce their emissions instead, thereby 'avoiding' emissions, or to remove CO$_2$ from the atmosphere, thereby 'offsetting' their continuing emissions. The 'carbon credits' generated through the emission avoidance or removal can be bought and sold in carbon markets. The buyers may then count these credits as 'theirs',

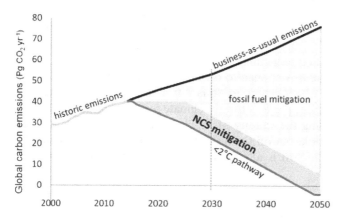

Figure 7.5 Potential contribution of NCS to a 66% 2°C-compliant scenario.

Source: Griscom et al., 2017, Figure 2, p. 11647.

and count or 'offset' them, against any voluntary or regulated emission reduction targets they may have.

In a legislated emission trading system, such as the EU emission trading system (EU ETS), the total quantity of emissions (called emission allowances) from the participants in the scheme is fixed by regulation, and declines each year (see Chapter 12). The companies in the system have to surrender each year the number of allowances corresponding to their calculated emissions, by reducing their emissions and buying allowances on the market. Provided that the scheme is well administered and enforced, such a system guarantees that emissions do indeed fall along the regulated trajectory.

It has been seen in Chapter 2 that, to get anywhere near the 1.5°C or 2°C targets arising from the Paris Agreement, GHG emissions will need to fall to net zero, by 2050 for the 1.5°C target, and fairly soon thereafter for the 2°C target, depending on the trajectory. The national net-zero targets that have been adopted have put pressure on companies to show how their own activities will be compliant with these targets, and many of them have adopted their own net-zero targets, to be achieved at some time in the future.[9] The broader economics of achieving these targets will be discussed in Chapter 11. The focus here is on how much trust should be put in the robustness of the carbon offsets with which some companies are saying that they will meet their targets.

Unlike with regulated emission trading systems, where authorities strictly control the number of emission allowances issued and how they can be traded, voluntary carbon and offset markets have no such statutory regulatory issuance, verification and assurance systems, although there are a number of non-statutory standards, an issue discussed further below. Companies and individuals can buy offsets from providers, who claim that they have avoided or removed emissions, often with very little scrutiny or assurance that the offset really will lead to the emission avoidance or removal that is claimed.

Offset projects may include clean cookstoves, tree planting or forest protection, installing biogas plants or other renewable energy systems. The project developer then calculates the supposed CO_2 emission savings from these projects and 'sells' them to a company or other agent who wants to set them against some emission reduction target, while they actually go on emitting.

One of the standards applied to offsets is called Gold Standard, the most recent version of which claims to "quantify and certify the impacts of climate and development projects toward climate security and sustainable development".[10] Assessing the validity of its claims is both complex and outside the scope of its book, but on the face of it is made more problematic by the fact that Gold Standard is also an offset provider. On the day of my visit

9 For a discussion of these targets see https://capitalmonitor.ai/factor/environmental/which-sectors-are-the-most-ambitious-on-net-zero/. Accessed April 30, 2023.

10 www.goldstandard.org/blog-item/new-standard-launched-accelerate-and-measure-progress-toward-sustainable-development-goals. Accessed April 30, 2023.

to the 'marketplace' section of its website,[11] it was listing 34 projects from many different countries (including Timor Leste, Honduras, Uganda, India, Myanmar, China and others), with carbon offset prices in the range of US$10–45/tCO$_2$. There was a wide range of projects, including community forestry, small-scale hydropower, wind turbines, improved cookstoves, biogas, and clean water access. Each project has a brief write-up about its achievements and supposed carbon savings. For example, a cookstoves project in Mexico and other central American countries claimed to have installed 22,000 improved cookstoves and to have saved over 98,000 tCO$_2$, in addition to achieving other social and economic benefits. For this project, 1167 carbon credits were said to be 'in stock' at a cost of US$30 each.

All the projects listed may have been excellent, worthwhile development projects, deserving of charitable support. But as a robust way of offsetting other people's or companies' emissions they are very questionable. Those paying their US$10–45 will be using these offsets to justify their continuing emissions, which will be emitted now, and, if they are CO$_2$, stay in atmosphere contributing to climate change for perhaps hundreds of years. The forests underlying the offsets may be cut or burned down before they offset the carbon emitted, the hydropower or wind turbines may not work or stop working, the cookstoves may not be used. In each case, assuring that the money actually went to the projects, that the projects delivered the benefits claimed, and that they would not have delivered those benefits had the offset not been purchased (known as the 'additionality' of the offset, see below) would require a monitoring, reporting and verification system of comparable sophistication to that underlying the EU ETS. There is no evidence from the marketplace website that Gold Standard itself operates such a system, in the absence of which very little trust can be placed in the reported carbon 'savings'. Though the 'standards' part of the Gold Standard website claims its standards are robust, the conflict of interest between these claims and its carbon offset provision is apparent.

It is almost certain that some offset providers will not incur the trouble and cost of putting a robust verification system in place and using it conscientiously, and that some projects will not deliver the promised carbon reductions. In such circumstances we could see the carbon offset market booming, as emitters continue emitting while claiming they were meeting their carbon targets (or salving their consciences), with net emissions also continuing their relentless increase.

The potential downsides of a booming market in carbon offsets go well beyond the non-delivery of the carbon savings that is their ostensible purpose. Seeking to manage large areas of land exclusively for carbon absorption could lead to a range of other environmental, social and economic problems. The environmental problems include loss of biodiversity (many environments

---

11 https://marketplace.goldstandard.org/collections/projects. Accessed May 5, 2022.

will exhibit a trade-off between maximum carbon sequestration and biodiversity, similar to the trade-offs between maximising bioenergy and biodiversity, discussed in Chapter 6). The social problems include using large areas of land for carbon sequestration instead of food production, and potentially expropriating this land from indigenous peoples, again similar to the possible conflicts between bioenergy and food production. The economic problems may arise from increases in the price of land caused by the demand for it from the new carbon markets, excluding those whose economic activities are less profitable than 'carbon farming'. Such effects in Scotland[12][13] and Wales[14] have already been reported in the press.

The potential scale of carbon markets, currently running at about £50 million per year, is mind-boggling. Seddon et al. (2021) report that there are already three 'trillion-tree initiatives', one led by a group of conservation charities,[15] another a platform set up by the World Economic Forum,[16] and a third established by an NGO,[17] supported by the UN, which states on its website that "Trees can be planted almost anywhere". In its article of April 2023 on three 'trillion trees campaigns',[18] the *Financial Times* (FT) reported that these aimed, through tree planting and land restoration, to change the land use of 1.2 billion hectares, equivalent to around a quarter of global agricultural land, while "India's climate pledge involves changing the use of nearly two thirds of its land". The FT article expresses scepticism from a number of experts that such land use change could be accomplished without massive displacement of local, including indigenous, people, that it was likely to involve large-scale plantations with small biodiversity benefits, that on current evidence many trees would not survive, and that if they did much of their carbon could be released to the atmosphere when the trees were cut down in a few decades' time.

Much of the money for this planting of dubious value to the climate is scheduled to come from corporate offset programmes reported in Seddon et al. (2021, Table 2, p. 1522) as amounting to several billions of US dollars, with Shell alone then planning to spend £300 million per year over 2019–2021. A report in 2021 claimed that the voluntary carbon market in that year was

12 www.theguardian.com/uk-news/2022/mar/05/tree-planting-drive-scottish-highlands-risks-widening-inequality. Accessed April 30, 2023.

13 *Financial Times* article entitled 'Carbon capture pitches smallholders against big business', March 11, 2022. www.ft.com/content/2ae63752-cefd-45b9-9282-a97584cc2cb2. Accessed April 30, 2023.

14 *Financial Times* article entitled 'Welsh village becomes battleground over ethics of afforestation', November 6, 2022. www.ft.com/content/a0ba8969-9d9b-4f86-ab3a-aa527502bebd. Accessed April 30, 2023.

15 https://trilliontrees.org/. Accessed April 30, 2023.

16 A platform for the trillion trees community. 1t.org. Accessed August 10, 2023.

17 www.trilliontreecampaign.org/. Accessed April 30, 2023.

18 https://ig.ft.com/one-trillion-trees/. Reporting by Alexandra Heal, April 12, 2023. Accessed April 14, 2023.

worth US$1 billion,[19] and the consultancy Deloitte said "the market could reach US$ 50 billion in the near future".[20] The Clean Energy Wire website in October 2021 gave several examples of offset providers falling into disrepute, of companies discontinuing their relationships with offset providers because of doubts about the robustness of the supposed carbon savings, and of companies making claims of 'carbon neutrality' that were later ruled by authorities to be unjustified.[21] Seddon et al. (2021, p. 1518) say:

> There are serious concerns that [tree planting for carbon sequestration] is distracting from the need to rapidly phase out use of fossil fuels and protect existing intact ecosystems. There are also concerns that the expansion of forestry framed as a climate change mitigation solution is coming at the cost of carbon rich and biodiverse native ecosystems and local resource rights.

It is well understood what the major challenges are in constructing a robust system of voluntary carbon markets. The Energy Transitions Commission (2022, p. 63) lists them as: *leakage* – the possibility that avoiding carbon-emitting activities (e.g. deforestation) somewhere will simply shift them to somewhere else; *additionality* – being sure that the carbon removal or avoidance would not have happened in the absence of the carbon-reducing project; *permanence* – carbon emissions can stay in the atmosphere for centuries, so offsets need to sequester carbon for a similar period, which is obviously very difficult to ensure with land-based projects such as tree planting, given that trees are vulnerable to fires and pests; *measurement* – it is very difficult to account accurately for the quantity of carbon removed or avoided through some measures (e.g. carbon sequestration in soils through different agricultural techniques).

Years of experience with the Kyoto Protocol's Clean Development Mechanism (CDM)[22] and, more recently, the global scheme to make payments for projects that Reduce Deforestation and Forest Degradation in developing countries (REDD+)[23] have shown that, even in the context of the UN Framework Convention on Climate Change, these challenges have been very difficult to address effectively. This is likely to be much more the case with a voluntary carbon market in which offset providers are keen to compete on price, and offset purchasers, where individuals or companies are chasing net zero, are keen to buy their offsets as cheaply as possible.

19 www.ecosystemmarketplace.com/articles/voluntary-carbon-markets-top-1-billion-in-2021-with-newly-reported-trades-special-ecosystem-marketplace-cop26-bulletin/. Accessed April 30, 2023. For deeper analysis see Forest Trends Ecosystem Marketplace (2021).
20 www2.deloitte.com/uk/en/focus/climate-change/zero-in-on-carbon-offsetting.html. Accessed April 30, 2023.
21 www.cleanenergywire.org/news/carbon-offset-market-booms-despite-nagging-greenwash-concerns. Accessed April 30, 2023.
22 https://cdm.unfccc.int/about/index.html. Accessed April 30, 2023.
23 https://redd.unfccc.int/. Accessed April 30, 2023.

Academic concerns with this situation led to the formulation in 2020 of the Oxford Principles for Net Zero Aligned Carbon Offsetting,[24] the core principles of which are: prioritise emission reduction *first* (emphasis added); ensure the environmental integrity of offsets; disclose offset use; shift offsetting towards carbon removal from the atmosphere, and its permanent storage; support the development of a robust offset market. A particularly egregious example of using offsets rather than emissions reduction was Bill Gates, in the context of the launch of his book on climate change in 2021, citing offsets in his defence of his use of a private jet,[25] and emitting 1,600 tonnes of CO_2 in a single year.[26] Another example is that of the UK Government Minister who claimed that the new coal mine he was approving would be 'carbon neutral' because of its use of carbon offsets from the Gold Standard standard-setting body. To its credit, Gold Standard dismissed the claim as "greenwashing nonsense – obviously nonsense, morally nonsense and technically insane".[27]

Grappling with these issues of robustness in the voluntary carbon market are two new initiatives, both of which launched consultations on their core principles in 2022. The Voluntary Carbon Markets Integrity initiative (VCMI) was established "to help ensure that voluntary carbon markets make a significant, measurable, and positive contribution to achieving the Paris Agreement goals while also promoting inclusive, sustainable development."[28] VCMI consulted on its Provisional Claims Code of Practice in July 2022. The Claims Code is intended to give guidance on a process for companies to purchase and make claims about having purchased 'high-quality carbon credits', once they have adopted a credible and validated net-zero target and interim targets for some of the years in between. For example, to make a VCMI Gold claim, a company must be on track to meet its next interim target and have purchased 'high-quality carbon credits' that cover 100% of its emissions (Scope 1, 2 and 3) that are in excess of this interim target. VCMI is not itself seeking to define 'high-quality carbon credits'. This is one of the objectives of the second recent initiative in this area, the Integrity Council for the Voluntary Carbon Market (ICVCM).[29]

In pursuit of its theory of change ('build integrity and the scale will follow') ICVCM is seeking to develop Core Carbon Principles (CCP) and an

24 www.ox.ac.uk/news/2020-09-29-oxford-launches-new-principles-credible-carbon-offsetting. Accessed April 30, 2023.

25 www.newsweek.com/bill-gates-defends-using-private-jets-big-homes-climate-change-carbon-emissions-1571005. Accessed April 30, 2023.

26 www.theguardian.com/technology/2021/jan/09/bill-gates-joins-blackstone-in-bid-to-buy-british-private-jet-firm. Accessed April 30, 2023.

27 www.theguardian.com/environment/2022/dec/13/gove-defence-of-uk-coalmine-dismissed-as-greenwashing-nonsense. Accessed April 30, 2023.

28 https://vcmintegrity.org/wp-content/uploads/2022/06/VCMI-Provisional-Claims-Code-of-Practice.pdf. Accessed April 30, 2023.

29 https://icvcm.org/. Accessed April 30, 2023.

Assessment Framework that "will set new threshold standards for high-quality carbon credits, provide guidance on how to apply the CCPs, and define which carbon-crediting programs and methodology types are CCP-eligible".[30] With these Principles it will "provide governance and oversight over standard setting organizations on adherence to CCPs as well as on market infrastructure and participant eligibility".[31] Its consultation on its CPPs closed in September 2022, receiving more than 5,000 responses. It published its 'Core Carbon Principles and Assessment Framework' in March 2023.[32]

The ICVCM's CCPs bear some relation to the Oxford Offsetting Principles and include requirements for additionality; permanence; information on mitigation activity; no double counting; governance; a registry; robust independent third-party validation and verification; robust quantification of emission reductions and removals; sustainable development impacts and safeguards; and a transition towards net-zero emissions.

These are lofty objectives, but their realisation in practice will be very difficult, and perhaps impossible. For example, in respect of the principle of permanence, the CPPs state that any carbon removals for use as offsets "shall be permanent, or if they have a risk of reversal, any reversals shall be fully compensated". Now the tree-planting offset programme of the US State of California, which is a regulated, not voluntary, carbon market, has a provision that any tree planting for offsets should plant 20% more trees than are granted offsets, precisely to guard against the 'risk of reversal' in respect of the offset trees. This may have seemed adequately prudent, but in the California wildfires of 2021 it seems that pretty well the whole of this insurance planting went up in flames (Badgley et al., 2022). The *Financial Times* reported this study with the headline 'Wildfires destroy almost all forest carbon offsets in 100-year reserve'.[33] It is not clear how any 'Principles' can guard against these sorts of events, and one must ask in the light of the recent extent of Californian wildfires how long it will be before the offset trees themselves go up in flames, cancelling out the entire tree-planting offset programme.

Everyone with an interest in climate stability must wish the VCMI and ICVCM initiatives well in their most challenging endeavour, given the importance of carbon removals and avoidance in the achievement of the Paris targets. But troubling questions remain around the whole notion of a voluntary market where it seems that there is so much money to be made, and where the corporate buyers of offsets have strong incentives to buy them as cheaply as possible. Will self-regulation of the offset industry with oversight from an independent body provide the kind of strict market regulation that seems to be

---

30 https://icvcm.org/the-core-carbon-principles/. Accessed April 30, 2023.
31 https://icvcm.org/about-the-integrity-council/. Accessed April 30, 2023.
32 https://icvcm.org/wp-content/uploads/2023/03/CCP-Foreword-FINAL-28Mar23.pdf
33 www.ft.com/content/d54d5526-6f56-4c01-8207-7fa7e532fa09. August 5, 2022. Accessed April 4, 2023.

required to avoid the kinds of problems that have beset similar initiatives such as the Marine Stewardship Council or the Roundtable on Sustainable Palm Oil? Both initiatives have so far signally failed to put their industries on a sustainable basis,[34][35] despite clearly the best of intentions from many participants and stakeholders in these initiatives, and substantial resources devoted to their monitoring and verification.

The Report of the UN's High-Level Expert Group on the Net Zero Emissions Commitments of Non-State Entities[36] (basically companies), published at COP27 in 2022, had some strong words to say about the current state of the voluntary carbon market:

A system to define and ensure standards for both the integrity of the credits themselves and how non-state actors claim them is not yet in place. As a result, too many non-state actors are currently engaging in a voluntary market where low prices and a lack of clear guidelines risk delaying the urgent near-term emission reductions needed to avoid the worst impacts of climate change.

(United Nations' High-Level Expert Group on the Net Zero Emissions Commitments of Non-State Entities, 2022, p. 19)

In her Chair's Note to the report entitled 'It's Time to Draw a Red Line around Greenwashing', Catherine McKenna made clear her views in this area: "Non-state actors cannot buy cheap credits that often lack integrity instead of immediately cutting their own emissions across their value chain." The UN Secretary-General is quoted in the report as saying: "We cannot afford slow movers, fake movers or any form of greenwashing" (United Nations' High-Level Expert Group on the Net Zero Emissions Commitments of Non-State Entities, 2022, p. 7).

It is still very much an open question as to whether a carbon market, through which large companies seek to persuade their customers and governments that they are taking climate change seriously and that they themselves are on a pathway towards 'net-zero' emissions, can remain voluntary, or whether it will require government legislation and enforcement to provide the 'integrity' that the ICVCM is aiming for. In November 2022 the European Commission seemed to be moving in this direction with its proposal for certification of the voluntary carbon market, using the criteria of quantification, additionality, permanence, and sustainability.[37] It remains to be seen whether, and how, such a scheme could work and be enforced at the global level.

---

34 https://ethicalunicorn.com/2019/02/22/is-msc-certified-fish-really-sustainable/. Accessed April 30,2023. www.theguardian.com/environment/2021/jul/26/blue-ticked-off-the-controversy-over-the-msc-fish-ecolabel. Accessed April 30, 2023.

35 www.ethicalconsumer.org/food-drink/rspo-criticisms-investigated. Accessed April 30, 2023.

36 www.un.org/sites/un2.un.org/files/high-level_expert_group_n7b.pdf. Accessed April 30, 2023.

37 https://ec.europa.eu/commission/presscorner/detail/en/ip_22_7156. Accessed April 30, 2023.

There are therefore currently many more questions than answers in relation to offsetting and the voluntary carbon market more generally. It is therefore much to be regretted that humanity has left decarbonisation so long and so late that the Paris Agreement targets, which would have been well within reach had emission reduction started in earnest 30 years ago, now require much more carbon dioxide to be removed from the atmosphere than can be achieved without the most concerted global effort, which as yet has hardly begun.

Solar radiation management

Solar radiation management entails either stopping solar energy reaching Earth, or reflecting it back into space by various means.

Albedo enhancement

Once solar energy (sunlight) has got through to Earth's surface, whether it is reflected back into the atmosphere, and thence into space, or absorbed, depends on the albedo (reflectivity) of the surface it strikes. Broadly, the darker the surface, the greater the proportion that will be absorbed, serving to warm the surface that absorbs it, while the higher the concentration of GHGs in the atmosphere, the less reflected energy gets back into space. Figure 7.6 shows how the albedo varies for different surfaces, from above 75% for snow to nearly zero for the dark ocean.

A whole range of methods for enhancing the albedo of various land surfaces have been proposed, including covering ice sheets or glaciers to stop them melting, painting roofs and pavements white, planting more reflective crops, or putting a layer of white plastic over deserts. Many of these techniques are reviewed, with a description of their possible negative effects, by the Geoengineering Monitor.[38]

Stopping solar energy reaching Earth

Preventing some solar energy from reaching Earth's atmosphere would require large structures in space that would either reflect or scatter solar energy away from Earth, or would effectively cast a shadow over it. It has been estimated that preventing some 1.7–1.8% of solar radiation reaching Earth would be sufficient to counteract a doubling to 560 ppm of the atmospheric concentration of CO_2 from pre-industrial times, when it was around 280 ppm. The current concentration is around 412 ppm, as seen in Chapter 1.

Angel (2006) modelled the impact of a 'space sunshade' in synchronised orbit with Earth, that would deflect enough solar radiation away from Earth

---

38 www.geoengineeringmonitor.org/wp-content/uploads/2021/04/surface-albedo-modification. pdf. Accessed April 30, 2023.

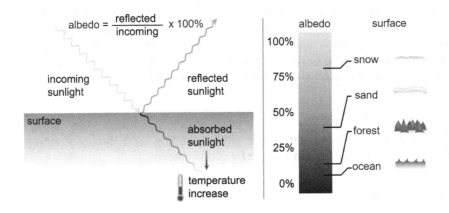

Figure 7.6 The albedo of different Earth surfaces.

Source: Golja, 2017.

to achieve the 1.8% reduction in solar radiation reaching it. The necessary size of the structure would be 4.7 million km^2 and it would weigh some 20 million tonnes. Consisting of a cloud of small spacecraft 100,000 km long, with each spacecraft weighing 1 gram and measuring 1 metre across, it would take 25 years to construct and the spacecraft would be propelled into space 800,000 at a time, using electromagnetic radiation, at a total estimated cost of "a few trillion dollars". The BBC gave a rather sceptical review of this proposal in 2016.[39] While it may well be that any implementation of the idea of 'mirrors in space' would differ markedly from this relatively early design, this description gives some idea of the scale of deployment that would be necessary to achieve the necessary solar reflection or shading to counteract a doubling of atmospheric CO_2 concentrations.

At sea, 'marine cloud brightening' has been proposed. This involves spraying a large number of droplets of sea water from the surface of the ocean into the marine clouds above it, thereby enhancing the reflectivity of the clouds. One study (Wood, 2021) estimated that offsetting the global warming from a doubling of CO_2 in the atmosphere from pre-industrial levels could be achieved by the annual spraying of 50–70 Mt of salt spray into the atmosphere. This would require 10,000–100,000 salt sprayers operating over most of the 54% of Earth's surface that is ocean and remote from land. Another sea-based proposal is to

---

39 www.bbc.com/future/article/20160425-how-a-giant-space-umbrella-could-stop-global-warming

spray or inject micro-bubbles of air into the surface of the ocean to simulate the whiteness of the sea that is formed by breaking waves.[40]

A technique that seems to have received more scientific attention than any of the above is stratospheric aerosol injection, which entails the injection of clouds of particles into the upper atmosphere (stratosphere, 10–50 km from Earth's surface), which would absorb or scatter solar energy, having a cooling effect on Earth. It is known that large numbers of particles in the atmosphere would have this effect because of the eruption of the volcano Mount Pinatubo in 1991. This shot vast quantities of ash and gas, including some 15 Mt of sulphur dioxide (SO_2), into the stratosphere. The SO_2 reacted with water to form droplets of sulphuric acid which spread round the world, cooling Earth over the next 15 months by around 0.6°C.[41] Once the aerosols had been cleared out of the atmosphere by natural processes, the temperature returned to the temperature it would have reached without the eruption, because it had no effect on the GHG concentration in the atmosphere, which is why this technique would have to be carried out on a regular basis if it were to offset global warming.

The aerosol injection would have to be carried out at a very large scale in order to have a significant effect on global warming. One study (Smith et al., 2022) explored the implications of a programme of aerosol injections (using a sulphur dioxide aerosol) that was designed to reduce polar temperatures by 2°C. Such a programme, injecting 13.4 Mt of SO_2 into the stratosphere, 13 km above the poles over eight months in the year (four months at each pole) would require a minimum of a fleet of 125 specially designed aircraft flying 175,000 sorties per year, with the programme estimated to cost around US$11 billion per year. There are still numerous uncertainties around whether such a programme would have the estimated effects on the global temperature, what its other environmental, social and economic effects might be, and how it might be launched and governed. Moreover, once started and in the absence of GHG emission reduction to net zero, such a programme would have to be continued indefinitely. This is because under the aerosol shield GHG concentrations would continue to increase. Therefore any cessation of the programme would cause the global warming that had been suppressed by the aerosols to rebound as the aerosols fell out of the atmosphere. The fact that such a risky programme of symptom alleviation is even receiving serious attention in the scientific literature is a searing indictment of the global community's failure over many years to get to grips with the increase of GHG emissions, which are the source of the problem.

The governance implications of a number of the geoengineering (SRM and some CDR) techniques remain very undeveloped. It would be quite possible for the government of a large country, if suffering from unbearable impacts from climate change (e.g. lengthy searing heat waves or drought) to seek unilaterally

---

40 https://geoengineering.global/ocean-albedo-modification/. Accessed April 30, 2023.
41 https://earthobservatory.nasa.gov/images/1510/global-effects-of-mount-pinatubo.   Accessed April 30, 2023.

to deploy some of the techniques above (e.g. injection of aerosols into the stratosphere), with potentially very serious implications for other countries. The IPCC's Sixth Assessment Report expresses the issue thus:

> Stratospheric aerosol intervention (SAI) – the most researched SRM method – poses significant international governance challenges since it could potentially be deployed uni- or mini-laterally and alter the global mean temperature much faster than any other climate policy measure, at comparatively low direct costs [citations]. While being dependent on the design of deployment systems, both geophysical benefits and adverse effects would potentially be unevenly distributed [citation]. Perceived local harm could exacerbate geopolitical conflicts, not the least depending on which countries are part of a deployment coalition [citations], but also because immediate attribution of climatic impacts to detected SAI deployment would not be possible. Uncoordinated or poorly researched deployment by a limited number of states, triggered by perceived climate emergencies, could create international tensions [citations]. An additional risk is that of rapid temperature rise following an abrupt end of SAI activities [citations].
>
> (IPCC, 2022, pp. 14–63)

And

> Currently, there is no targeted international law relating to SRM, although some multilateral agreements – such as the Convention on Biological Diversity, the UN Convention on the Law of the Sea, the Environmental Modification Convention, or the Vienna Convention on the Protection of the Ozone Layer and its Montreal Protocol – contain provisions applicable to SRM [citations]
>
> (Intergovernmental Panel on Climate Change, 2022, pp. 14–64)

(Note: The [citations] in the above quote refer to references that are given in the original text.)

The Swiss Government put forward a draft resolution on 'Geoengineering and its governance' at the fourth UN Environment Assembly (UNEA-4) in 2019, which recommended that the Executive Director of the UN Environment Programme should assess, for UNEA-5:

> Criteria to define SRM and CDR technologies; The current state of science surrounding such technologies, including as related to risks, benefits and uncertainties; The current state and challenges of governance frameworks; Actors and activities with regard to research and deployment; Possible future global governance frameworks.
>
> (as reported in Jinnah & Nicholson, 2019, p. 878)

The resolution was not passed, apparently due to opposition from the United States and Saudi Arabia, revealing differences in views among governments between those who wanted minimal or no regulation, and those who favoured a precautionary approach (Jinnah & Nicholson, 2019).

For the present the most concrete proposals, also cited by the IPCC, seem to come from Nicholson et al. (2018), who recommend that steps should be taken to: "guard against the risks of uncontrolled SRM development; enable potentially valuable research; build legitimacy for research and any future policy through broad public engagement and ensure that SRM is only considered as one part of a broader mitigation agenda". They further "propose three interventions to work towards those objectives in the near term by: developing a transparency mechanism for research; creating a global forum for public engagement; and including consideration of SRM in the global stocktake under the Paris Agreement" (Nicholson et al., 2018, p. 322). Clearly a robust regulatory regime for these technologies is some way off. Meanwhile the research continues.[42]

Conclusion

CCS, uses of carbon dioxide, and techniques of CDR are still in their relative infancy and will require huge investment to reach anything like the deployment levels that modelling suggests are necessary to keep global warming within 1.5°C or 2°C. The first priority of emission reduction must remain cutting back to zero by mid-century the use of fossil fuels. CCS for fossil fuels may be part of that journey, but by or soon after mid-century this too will need to be replaced by cleaner energy sources. CDR will then be required, at scale, to reduce the atmospheric concentrations of CO_2 down to the 350 ppm that was estimated some years ago to be the highest they should be allowed to go. The use of dubious claims of emission avoidance or removal by companies seeking to meet net-zero targets, or individuals seeking to ease their consciences, so that they can go on emitting, gives an illusion of climate action while emissions keep rising. The UN Secretary-General might have been talking to fossil fuel companies or governments that continue opening new mines or wells, while claiming to be on track for 'net zero', through the use of offsets, when he said at COP27 in 2022:

> We must have zero tolerance for net-zero greenwashing. ... Using bogus 'net-zero' pledges to cover up massive fossil fuel expansion is reprehensible. It is rank deception. This toxic cover-up could push our world over the climate cliff. The sham must end.[43]

---

42 www.science.org/content/article/ocean-geoengineering-scheme-aces-its-first-field-test. Accessed April 30, 2023.
43 https://news.un.org/en/story/2022/11/1130317. Accessed April 30, 2023.

References

Angel, R. (2006). Space sunshade might be feasible in global warming emergency. *Science Daily*, November 5, www.sciencedaily.com/releases/2006/11/061104090 409.htm

Badgley, G., Chay, F., Chegwidden, O. S., Hamman, J. J., Freeman, J., & Cullenward, D. (2022). California's forest carbon offsets buffer pool is severely undercapitalized. *Frontiers in Forests and Global Change, 5*. https://doi.org/10.3389/ffgc.2022.930426

Butnar, I., Broad, O., Solano Rodriguez, B., & Dodds, P. E. (2020). The role of bioenergy for global deep decarbonization: CO_2 removal or low-carbon energy? *GCB Bioenergy, 12*(3), 198–212. https://doi.org/10.1111/GCBB.12666

Energy Transitions Commission. (2022). *Mind the Gap: How carbon dioxide removals must complement deep decarbonisation to keep 1.5°C alive.* www.energy-transitions. org/publications/mind-the-gap-cdr/

Forest Trends Ecosystem Marketplace. (2021). 'Market in Motion', State of Voluntary Carbon Markets 2021, Instalment 1. Washington, DC: Forest Trends Association. www.ecosystemmarketplace.com/articles/voluntary-carbon-markets-top-1-billion-in-2021-with-newly-reported-trades-special-ecosystem-marketplace-cop26-bulletin/

Fuss, S., Lamb, W. F., Callaghan, M. W., Hilaire, J., Creutzig, F., Amann, T., Beringer, T., De Oliveira Garcia, W., Hartmann, J., Khanna, T., Luderer, G., Nemet, G. F., Rogelj, J., Smith, P., Vicente, J. V., Wilcox, J., Del Mar Zamora Dominguez, M., & Minx, J. C. (2018). Negative emissions – Part 2: Costs, potentials and side effects. *Environmental Research Letters, 13*(6), 063002. https://doi.org/10.1088/1748-9326/ AABF9F

Gattuso, J.-P., Magnan, A. K., Bopp, L., Cheung, W. W. L., Duarte, C. M., Hinkel, J., Mcleod, E., Micheli, F., Oschlies, A., Williamson, P., Billé, R., Chalastani, V. I., Gates, R. D., Irisson, J.-O., Middelburg, J. J., Pörtner, H.-O., & Rau, G. H. (2018). Ocean Solutions to Address Climate Change and Its Effects on Marine Ecosystems. *Frontiers in Marine Science, 5*. www.frontiersin.org/article/10.3389/fmars.2018.00337

Gattuso, J.-P., Williamson, P., Duarte, C. M., & Magnan, A. K. (2021). The Potential for Ocean-Based Climate Action: Negative Emissions Technologies and Beyond. *Frontiers in Climate, 2*. www.frontiersin.org/article/10.3389/fclim.2020.575716

Global Carbon Capture and Storage Institute. (2022). *Global Status of CCS 2022.* GCCSI, Melbourne.

Golja, C. (2017). Solar geoengineering: Is controlling our climate even possible? Blog, December 20. Harvard University. https://sitn.hms.harvard.edu/flash/2017/solar-geo engineering-controlling-climate-possible/. Accessed August 17, 2023.

Griscom, B. W., Adams, J., Ellis, P. W., Houghton, R. A., Lomax, G., Miteva, D. A., Schlesinger, W. H., Shoch, D., Siikamäki, J. V., Smith, P., Woodbury, P., Zganjar, C., Blackman, A., Campari, J., Conant, R. T., Delgado, C., Elias, P., Gopalakrishna, T., Hamsik, M. R., ... Fargione, J. (2017). Natural climate solutions. *Proceedings of the National Academy of Sciences, 114*(44), 11645–11650. https://doi.org/10.1073/ pnas.1710465114

IPCC. (2021). *Summary for Policymakers.* In: *Climate Change 2021: The Physical Science Basis. Contribution of Working Group I to the Sixth Assessment Report of the Intergovernmental Panel on Climate Change* [Masson-Delmotte, V., P. Zhai, A. Pirani, S. L. Connors, C. Péan, S. Berger, N. Caud, Y. Chen, L. Goldfarb, M. I. Gomis, M. Huang, K. Leitzell, E. Lonnoy, J. B. R. Matthews, T. K. Maycock, T. Waterfield, O.

Yelekçi, R. Yu, and B. Zhou (eds.)]. Cambridge University Press, Cambridge, United Kingdom and New York, NY, USA, pp. 3–32, doi:10.1017/9781009157896.001.

IPCC. (2022). *Climate Change 2022: Mitigation of Climate Change. Contribution of Working Group III to the Sixth Assessment Report of the Intergovernmental Panel on Climate Change* [P. R. Shukla, J. Skea, R. Slade, A. Al Khourdajie, R. van Diemen, D. McCollum, M. Pathak, S. Some, P. Vyas, R. Fradera, M. Belkacemi, A. Hasija, G. Lisboa, S. Luz, J. Malley, (eds.)]. Cambridge University Press, Cambridge, UK and New York, NY, USA. doi: 10.1017/9781009157926.001.

International Energy Agency. (2020). *Energy Technology Perspectives: CCUS in Clean Energy Transitions.* IEA, Paris. https://iea.blob.core.windows.net/assets/181b48b4-323f-454d-96fb-0bb1889d96a9/CCUS_in_clean_energy_transitions.pdf

International Energy Agency. (2021a). *About CCUS.* www.iea.org/reports/about-ccus

International Energy Agency. (2021b). *Net Zero by 2050: A Roadmap for the Global Energy Sector.* IEA, Paris.

Jinnah, S., & Nicholson, S. (2019). The hidden politics of climate engineering. *Nature Geoscience, 12*(11), 876–879. https://doi.org/10.1038/s41561-019-0483-7

Joppa, L., Luers, A., Willmott, E., Friedmann, S. J., Hamburg, S. P., & Broze, R. (2021). Microsoft's million-tonne CO_2-removal purchase – lessons for net zero. *Nature, 597*(7878), 629–632. https://doi.org/10.1038/d41586-021-02606-3

Lenton, T. M., & Vaughan, N. E. (2009). The radiative forcing potential of different climate geoengineering options. *Atmospheric Chemistry and Physics, 9*(15), 5539–5561. https://doi.org/10.5194/acp-9-5539-2009

Nicholson, S., Jinnah, S., & Gillespie, A. (2018). Solar radiation management: a proposal for immediate polycentric governance. *Climate Policy, 18*(3), 322–334. https://doi.org/10.1080/14693062.2017.1400944

Schmitz, O., Sylvén, M., Atwood, T., Bakker, E., Berzaghi, F., Brodie, J., Cromsigt, J., Davies, A., Leroux, S., Schepers, F., Smith, F., Stark, S., Svenning, J-C., Tilker, A. & Ylänne, H. (2023). Trophic rewilding can expand natural climate solutions. *Nature Climate Change, 13*, 324–333. https://doi.org/10.1038/s41558-023-01631-6

Seddon, N., Smith, A., Smith, P., Key, I., Chausson, A., Girardin, C., House, J., Srivastava, S., & Turner, B. (2021). Getting the message right on nature-based solutions to climate change. *Global Change Biology, 27*(8), 1518–1546. https://doi.org/10.1111/gcb.15513

Smith, W., Bhattarai, U., MacMartin, D. G., Lee, W. R., Visioni, D., Kravitz, B., & Rice, C. V. (2022). A subpolar-focused stratospheric aerosol injection deployment scenario. *Environmental Research Communications, 4*(9), 095009. https://doi.org/10.1088/2515-7620/ac8cd3

United Nations' High-Level Expert Group on the Net Zero Emissions Commitments of Non-State Entities. (2022). *Integrity Matters: Net Zero commitments by businesses, financial institutions, cities and regions.* www.un.org/sites/un2.un.org/files/high-level_expert_group_n7b.pdf

Williamson, P., & Bodle, R. (2016). *Update on Climate Geoengineering in Relation to the Convention on Biological Diversity: Potential Impacts and Regulatory Framework.* Technical Series No.84, Secretariat of the Convention on Biological Diversity, Montreal.

Wood, R. (2021). Assessing the potential efficacy of marine cloud brightening for cooling Earth using a simple heuristic model. *Atmospheric Chemistry and Physics, 21*(19), 14507–14533. https://doi.org/10.5194/acp-21-14507-2021

8 Enabling decarbonisation

Digitalisation, the circular economy and critical minerals for the clean energy transition

Summary

This chapter discusses three key enablers for decarbonisation: the digitalisation of energy use, the circular economy approach to the use of materials, and the metals and minerals required for low-carbon energy technologies.

New digital technologies already play an important role on the supply side of the energy system through, for example, optimisation of energy efficiency and smart grid sensing. Such uses will doubtless be extended but the big new opportunities for digitalisation in energy relate to the demand side – not only by enabling more demand response in electricity, but also through its application to energy use in industry, transport and buildings, in everything from the increasing sophistication of smart meters to autonomous vehicles to the Internet of Things. This chapter discusses the many opportunities in this area and how the barriers to their realisation may be addressed.

Given trends in urbanisation and development, and the consequent massive need for infrastructure in many countries, the extraction and use of materials of all kinds is bound to increase, especially in lower-income countries. The extraction of materials is currently responsible for 50% of global greenhouse gas (GHG) emissions, and both the tonnage extracted and the associated emissions are projected on current trends to increase dramatically by 2060. However, this chapter shows that this is not inevitable, and that a combination of strong climate action, and policies for resource efficiency and a circular economy can reduce both material extraction and GHG emissions.

Changing the energy system to eliminate carbon emissions will also be material-intensive in a different way, with batteries for electric vehicles, and with renewable energy technologies, such as wind turbines and solar panels, using many metals that are currently extracted in small quantities. These quantities will need to be scaled up dramatically if these technologies are to replace fossil fuels over the next three decades. This is a huge

DOI: 10.4324/9781003438007-9

challenge for companies and governments involved in the production of minerals and metals. This chapter explores which materials are required for the energy transition, how much of them might be needed, where they might come from, what the problems might be in supplying them in the required quantity, and how these problems might be overcome.

Introduction

This chapter discusses three different issues, each of which will have an enormous influence on the costs and efficiency with which decarbonisation will be achieved, or whether it will be achieved at all. The three issues are the digitalisation of energy use, the circular economy approach to the use of materials, and the metals and minerals that are critical for the construction of low-carbon energy technologies.

Digitalisation of energy use

The International Energy Agency (IEA) defines digitalisation as "the growing application of information and communications technologies (ICT) across the economy, including energy systems" (International Energy Agency, 2017, p. 21). It is already well advanced in both the supply and demand sides of the energy system. The IEA lists the key drivers of energy digitalisation as advances in ICT that result in falling costs of data generation (e.g. through sensors) and storage, the growing application of artificial intelligence (AI) and machine learning, and increasing connectivity between people, their energy suppliers, and their energy-using appliances at home (the Internet of Things, IoT). According again to the IEA

> investment in digital electricity infrastructure and software has grown by over 20% annually [over 2014–2016], reaching USD 47 billion in 2016. This digital investment in 2016 was almost 40% higher than investment in gas-fired power generation worldwide (USD 34 billion).
> (International Energy Agency, 2017, p. 25)

The number of connected devices in the IoT has been projected to grow from 15 billion in 2015 to 75 billion in 2025 (Rhodes, 2020, p. 23).

Digitalisation of the electricity supply system

The electricity supply systems in industrialised countries already make extensive use of digital technologies, but this will be greatly expanded in the future to facilitate distributed generation (locally based 'microgeneration' of renewables connected directly into the distribution network), and demand response (shifting of electricity demand away from times of peak usage), as well as providing a range of other services, examples of which are shown in Table 8.1.

Table 8.1 Examples of digitalisation of the electricity system

	Generation	Transmission	Distribution	Consumer
Big Data	Optimisation of operation efficiency through analytics	Forecasting of future load and pricing	Analytics for optimising microgeneration and storage in communities	Advice on usage and saving energy
AI/Machine Learning	Optimisation of wind farms through wind speed forecasting	Autonomous agents trading energy	Optimisation of networks against physical faults	Automation of demand response
Internet of Things	Drone inspection of equipment	Smart grid sensing, monitoring and asset management	Enabling local microgrids through embedded control	EVs, in-home/ building sensors
Blockchain	Emissions certificates and certificates of origin	Direct energy trading	Microgrids and local markets	Billing, metering, demand response

Source: Rhodes, 2020, Table 1, p. 8.

Most of the cells in Table 8.1 are self-explanatory, and some on the consumer side are discussed in more detail below. Blockchain, the best-known application of which is its use in cryptocurrencies like Bitcoin, is a technology that stores multiple synchronised and identical copies of data across a network of devices (Rhodes, 2020, p. 28), which permits secure trading of certificates or energy.

Implementation of digital technologies across the generation, transmission, distribution and consumer parts of the electricity system leads to the creation of a 'smart grid', which allows real-time communications and adjustments between these different elements of an electricity system, potentially increasing flexibility and security, and reducing costs, as described further below.

Demand response is likely to be a particularly important use of digital technology. It was seen in Chapter 5 how electricity demand varies greatly during the day, and how the price is set by the most expensive plant that is operating. At times of peak demand, it has to be met by the most expensive generating plant on the system. Zakeri et al. (2022, Figure 5, p. 19) showed the variation in average 'day-ahead' electricity prices (those that are bid into the system by generators and consumers 24 hours in advance) for several European countries from 2015 to 2019. The advance prices vary enormously, and have even been negative in Germany, at times when it is cheaper for some generators to pay some consumers to increase their demand rather than turn down their supply.[1] One of the largest price ranges was in Ireland in 2019, when the range of prices was between 0 and over €130/MWh. The low prices tend to occur when renewables (wind and solar) are abundant compared to demand. The high prices tend to occur when demand is high (e.g. in winter) but renewables are limited. The actual prices in real time can turn out to be even higher (or lower) if market conditions turn out to be different from what was expected.

There are clearly significant savings to be made if energy consumers can be given incentives to shift their energy consumption from peak times with high prices to off-peak times when prices can be much lower. This is known as providing for *demand response*, which the US Department of Energy has defined as:

> Changes in electric usage by end-use customers from their normal consumption patterns in response to changes in the price of electricity over time, or to incentive payments designed to induce lower electricity use at times of high wholesale market prices or when system reliability is jeopardized.
>
> (U.S. Department of Energy, 2006, p. ix)

Demand response already happens to some extent in the UK, in that some major industrial consumers are paid through demand response contracts to reduce their energy consumption when the grid is struggling to cope with high demand. Through these contracts, the companies gain more from the contracts than they lose from the (hopefully) infrequent times when they are asked to consume less than they otherwise would. Other electricity consumers gain because the reduction in energy consumption by the large companies reduces electricity prices below where they would otherwise be.

In the past this kind of demand response has tended to be used only at 'critical times', "typically only a few hours per year, when wholesale electricity market prices are at their highest or when reserve margins are low due to contingencies such as generator outages, downed transmission lines, or severe weather conditions" (U.S. Department of Energy, 2006, p. 6). Digitalisation allows an extension of demand response measures across a much larger number and range of consumers than large energy-intensive companies.

Demand response may be effected through one of two mechanisms: price-based or incentive-based demand response. Incentive-based demand response works through the kinds of contracts discussed in the previous paragraph. Digitalisation enables the aggregation of larger numbers of small consumers so that they can benefit from lower prices if they are prepared to reduce their consumption at peak times. Price-based demand response works through time-of-use tariffs, or real-time pricing, whereby customers pay the actual price of

---

1 For an explanation of the negative electricity price phenomenon, see www.cleanenergywire.org/factsheets/why-power-prices-turn-negative. Accessed April 30, 2023.

Table 8.2 Benefits from demand response

Benefit
Benefits from relative and absolute reductions in electricity demand
Benefits resulting from short run marginal cost savings from using demand response to shift peak demand
Benefits in terms of displacing new plant investment from using demand response to shift peak demand
Benefits of using demand response as 'stand-by' reserve for emergencies/unforeseen events
Benefits of demand response in providing stand-by reserve and balancing for wind
Benefits of demand response to distributed power systems
Benefits in terms of reduced transmission network investment by reducing congestion of the network and avoiding transmission network re-enforcement
Benefits from using demand response to improve distribution network investment efficiency and reduce losses

Source: Bradley et al., 2011, Table 1, p. 11.

the electricity at the time they are consuming it, rather than an average price. This means they will get a cheaper price if they shift their consumption away from peak periods. Digitalisation permits these price-based demand responses through smart meters that can show consumers the different prices of electricity at different times of day. The Internet of Things also makes it possible for appliances (e.g. washing machines, freezers) to interact with electricity prices, switching themselves on when it is cheap, and off (in the case of freezers for short periods only if the food is not to be spoiled) when it is expensive, or as programmed by the consumer. Some UK electricity suppliers are already giving their consumers a financial incentive to shift their consumption of electricity away from peak times.[2] While time-of-use tariffs are not yet widely available in the UK, a study for UK Citizens Advice found that they could shave 5–10% off peak demand and, in a high-renewables scenario, save consumers £20–27 million a year.[3] The biggest savings accrued to consumers who had electric vehicles (EV) and electric heating. Consumers with EVs may also be able to earn money by making their batteries available to feed electricity into the grid at times of peak demand and prices.

The benefits from demand response go beyond the reduction in peak prices. Table 8.2 lists the full range of benefits that can derive from demand response.

As noted in Table 8.2, demand response can reduce the power generation capacity that needs to be built. A US study in 2015 found that "typically 10 percent of our electric system capacity is built to meet demand in just 1 percent

2 www.whatissmartenergy.org/featured-article/timing-is-everything-3-tips-for-shifting-energy-usage. Accessed April 30, 2023.
3 www.citizensadvice.org.uk/Global/CitizensAdvice/Energy/Citizens%20Advice%20summ ary%20of%20the%20value%20of%20time%20of%20use%20tariffs.pdf. Accessed April 30, 2023.

Table 8.3 Sources of flexibility in electricity systems in the IEA Net Zero Emissions (NZE) scenario, 2020 and 2050

	Advanced economies (%)		Emerging market and developing economies (%)	
	2020	2050	2020	2050
Coal	20	0	18	1
Natural gas	36	4	11	3
Oil	6.34	0.14	4.88	0.34
Hydrogen-based	0.00	13.98	0.00	17.53
Nuclear	3.57	5.56	1.09	2.70
Hydro	26.68	16.14	15.99	23.39
Other renewables	3.14	4.88	0.99	7.32
Batteries	0.83	28.28	0.11	27.89
Demand response	3.50	26.90	0.99	16.73

Source: International Energy Agency, 2021b, Figure 4.18, p. 177.

of hours during the year",[4] meaning that if this peak demand can be reduced in these hours, this 10% of capacity need not be built. Reducing peak demand also allows for the deferral of investments in electricity transmission and distribution (T&D) systems. The same US study estimated that, over 2016–2020, savings in T&D in Illinois for a low and high demand response simulation were US$362 million and US$488 million respectively. For Massachusetts the numbers were US$82 million and US$150 million (Navigant Consulting, 2015, pp. 26–27). Overall, the study estimated that in a high demand response scenario, every US$1 invested in demand response could save US$4.07 in Massachusetts, and US$2.73 in Illinois (Navigant Consulting, 2015, p. 35).

Demand response becomes particularly important in low-carbon scenarios as a provider of flexibility, as intermittent and variable renewables increase their share of generation. Table 8.3 shows that demand response increases from low levels (3.5% and 0.99% in advanced and emerging market or developing economies respectively) in 2020, when fossil fuels predominate (62.34%) in the provision of system flexibility, to 26.9% in advanced economies, and 16.73% in other economies, by 2050.

Digitalisation of buildings, transport and industry

Apart from enabling widespread demand response, digitalisation may be expected to bring about widespread changes in all the main demand sectors: transport, buildings and industry. Figure 8.1 shows the potential

4 https://blog.advancedenergyunited.org/articles/new-report-reducing-peak-demand-saves-money-for-electricity-customers. Accessed April 30, 2023.

change to energy demand (vertical axis) that may be caused by digitalisation, and the barriers which it may face to its introduction (horizontal axis). The biggest changes may come in transport, through autonomous vehicles and shifts to Mobility-as-a-Service (MaaS) companies, which offer multi-modal, interconnected travel options on their platforms. In respect of buildings, user programming of Building Information Management (BIM) systems and smart thermostats offer the greatest potential to change energy demand. As is often the case, the applications with the highest potential to change energy demand also face the highest barriers to their widespread introduction.

Not all the digital innovations in Figure 8.1 will necessarily reduce energy demand. For example, autonomous cars may increase rather than reduce travel. A transport system that is automated, connected, electric and shared (ACES) may not necessarily be low-energy. In addition, technologies that increase energy efficiency (e.g. LED lighting technology) may be energy-saving in themselves but may lead to an increase in consumption because they reduce the costs of the desired energy service (a phenomenon known as the rebound effect).

In respect of freight transport, however, digitalisation is likely to reduce energy demand and CO_2 emissions. International Energy Agency (2017, Figure 2.3, p. 38) shows how digitalisation could reduce CO_2 by an extra 20% through systemic measures (e.g. coordination of freight movements), vehicle efficiency and fuel switching.

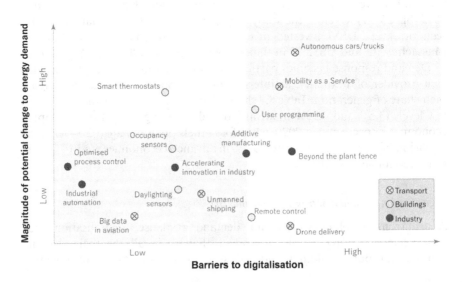

Figure 8.1 Potential impact of digitalisation on transport, buildings and industry.

Source: International Energy Agency, 2017, Figure 2.1, p. 30.

Digitalisation could also save significant energy from both commercial and residential buildings (Figure 8.2). In commercial buildings the main savings are likely to come from more intelligent management of the building, using BIM systems to learn when and how a building is being used, to optimise temperature control, and ensure through sensors that energy use matches occupation. The largest potential saving is in space heating, with space cooling, water heating, lighting (through widespread installation of LEDs) and appliances also making a contribution to savings of over 60,000 TWh (60 PWh).

However, in buildings too digitalisation could also lead to an increase in energy demand, enabling new applications (e.g. interactive doorbells and security cameras) or expanding the use of existing technologies (e.g. LEDs). All devices or appliances with connectivity need to be on stand-by, and this requires power. The amount per device or appliance is small, but given the huge projected growth in connected devices, the total power consumption is non-negligible and reduces the potential energy savings from digitalisation.

Industry already uses many digital technologies across a range of applications, both inside factories and 'beyond the plant fence' (Table 8.4). Of course, different industrial sectors can take advantage of the different opportunities for digitalisation to a different extent, but overall the energy and CO_2 savings could be significant. In the US investments of US\$235 million in improved process controls in small and medium manufacturing businesses over

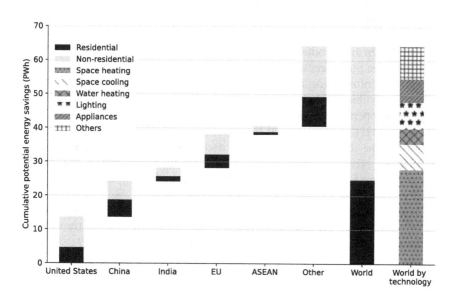

Figure 8.2 Cumulative potential energy savings from digitalisation of buildings.

Source: International Energy Agency, 2017, Figure 2.4, p. 42.

Table 8.4 Various applications of digital technologies in industry

	Readiness level for wide adoption		
	Low	Medium	High
Beyond the plant fence			
Connectivity	Cloud-connected workers, remote-controlled operations	Connected industrial equipment	Connected supply value chains
Inside the plant			
Analytics and enabled workforce	Advanced process control systems	Digital twin, decision support from advanced data analytics	Artificial intelligence
Industrial equipment	Smart sensors	3-D printing, industrial robotics	

Source: International Energy Agency, 2017, Figure 2.8, p. 50.

1987–2015 saved US$330 million, with average payback periods of typically less than a year. In Australia such measures enabled a range of sectors (food product manufacture; minerals and metal products; beverage, tobacco and textiles; other manufacturing) to save over 4% of their energy in less than two years (International Energy Agency, 2017, pp. 51–52). 3D printing has considerable potential to reduce scrap and waste in a number of sectors. Digital twins which model industrial processes or buildings can save resources in both product design and product testing.

Beyond the plant fence, digital communications between firms allow them to exchange information about surplus raw materials, by-products or wastes (industrial symbiosis)[5] between sites of the same firm to allow them to benchmark their operations, identify bottlenecks and spread best practice, and between workers, facilitating real-time learning and reducing the need to travel.

Because the digital revolution, and its application to the energy system, are relatively recent, many more applications in many different fields may be expected. Table 8.5, from the IEA, gives some further examples as to how digitalisation can contribute to increased energy efficiency.

These opportunities and implications of digitalisation, which will be pervasive throughout society in coming years, should be borne in mind throughout the discussion of the major individual demand sectors in Chapter 9.

5 For example, Materials Marketplace in the US. https://go.materialsmarketplace.org/. Accessed April 30, 2023.

Table 8.5 Strategies for digital energy efficiency deployment

Strategy
Institutional arrangements and platforms for data sharing and data management
Cyber security frameworks and guidelines; data protection frameworks
Stakeholder awareness raising, removal of interoperability barriers
Methodologies for valorising energy efficiency and flexibility
Programmes focusing on digital solutions to benefit disadvantaged groups and communities
Re-skilling and up-skilling programmes, inclusion of digital skills in training, education and academic curricula
Measures to improve the efficiency of digital solutions, promotion of circular economy for ICT products
Finance for pilot and demonstration projects, funding for start-ups, removal of barriers for new market entrants

Source: International Energy Agency, 2021a, p. 96.

Material efficiency and the circular economy

The great majority of economic activity involves the use of materials, the extraction of which has grown fast over the last ten years, and is projected to grow further in the future. Both the quantity of materials, and the waste involved in processing and using them, have been increased by the dominant 'take – make – use – throw away' model of economic production and consumption that has evolved in richer countries since the Second World War. Given the quantity of energy and GHG emissions, and other environmental impacts, involved in producing materials for the economy, it is imperative that this model is replaced by greatly increased material efficiency and a more circular use of resources. This section explores the extent to which the adoption of such a model could both reduce material extraction and the GHG emissions and other environmental impacts, that are involved in the whole lifecycle of material production and use.

Growth of extraction of materials

The use of materials, driven by population and economic growth, has grown dramatically over the past 50 years. The International Resource Panel (2019, Figure 2.7, p. 43) computed the growth from 1970 to 2017 in global material use in four categories: biomass (including food), fossil fuels, metal ores and non-metallic minerals (e.g. sand and gravel used for construction). Material use more than trebled over that time, from 27 Gt in 1970 to 92 Gt in 2017. All four categories of materials showed growth, with non-metallic minerals growing the most, as emerging economies, especially China, created infrastructure in the 1990s and 2000s.

While Asia and the Pacific extracted about 60% of the global total of materials in 2017, on a consumption basis (i.e. adding the materials in imports, and subtracting those in exports) the high-income countries used 27 tonnes of materials per person in that year, which is 60% more than the emerging economies and 13 times the material use per person in low-income countries.

The extraction and processing of materials on this scale are responsible for enormous environmental impacts: around 50% of global GHG emissions, and some 90% of water stress and impacts on biodiversity (International Resource Panel, 2019). Continuation of these trends will, of course, make these matters worse, as is seen in the next section.

Projection of materials extraction to 2060

The Organisation for Economic Co-operation and Development (OECD) estimated that on current trends global material use would more than double from 79 Gt in 2011 to 167 Gt in 2060 (Organisation for Economic Cooperation and Development, 2018). The OECD (2018, Figure 10, p. 15) disaggregated the four categories above into 24 more specific materials. There is huge projected growth in construction materials, with sand, gravel and crushed rock increasing from nearly 25 Gt in 2011 to 55 Gt in 2060, and limestone (for cement) more than doubling its use to nearly 15 Gt by 2060, as more and more countries put infrastructure in place. Other materials exhibiting strong growth are bituminous coal for power generation and steel-making, iron ores to produce steel, and copper ores, the latter a crucial input for the energy transition as discussed below.

On these trends, it is not only material extraction and use that will double over 2011–2060. Associated GHG emissions increase dramatically too, rising from 40 $GtCO_2e$ in 2011 to 75 $GtCO_2e$ by 2060, of which a full two-thirds would be associated with the extraction and management of materials (Organisation for Economic Cooperation and Development, 2018, Figure 11, p. 17). Those emissions associated with energy supply, end users and transport would be largely removed by switching away from fossil fuels to zero-carbon energy sources such as renewables. But this still leaves several $GtCO_2e$ coming from agriculture (e.g. from livestock and fertiliser use) and industry (e.g. from steel-making with bituminous coal or from cement).

Between now and 2060, before full decarbonisation can be achieved, all emissions could be much reduced by far greater efficiency in the use of materials, far less wastage of them, and much more repair, reuse, remanufacturing and recycling of products and their constituent materials at the end of their lives, with the original design of products being oriented towards making these processes both feasible and economic.

The International Resource Panel (IRP) of the United Nations Environment Programme (UNEP) estimated that material efficiency strategies could reduce lifecycle GHG emissions from homes by 35% in G7 countries,[6] and 60% in China and India (International Resource Panel, 2020a, Figure 2, p. 3), and lifecycle GHG emissions from cars by 40% in G7 countries and 35% in China and India (International Resource Panel, 2020a, Figure 3, p. 4).

The result of such a transformation in material use has been called a 'circular economy', to distinguish it from the 'disposable economy' that is an all-too accurate description of current models of materials management, as has been noted.

Towards sustainability in materials use

The IRP has also projected material use, GHG emissions and various other environmental impacts through to 2060 on the basis of historical trends. In addition, it projected a Sustainability Scenario to 2060, that incorporated strong climate action and the implementation of stringent policies that greatly increased efficiency in the use of resources (resource efficiency) with the purpose of moving towards a more 'circular' economy.

Table 8.6 shows the key results of the two scenarios. With Historical Trends, global material extraction more than doubles by 2060 (as in the OECD projections above), and GHG emissions increase by 43% over the level in 2015. The increase of global pasture land for livestock (25%) and the associated area of agricultural land to grow their fodder reduces the forest area by 10% and other natural habitat by 20%. This scenario paints a world of rampant climate change, substantial biodiversity loss, and much further depletion of nature. It is not at all clear that 9–10 billion people could thrive in such a world.

In the IRP's Towards Sustainability scenario, which models strong policies for climate mitigation and increased resource efficiency, all the environmental indicators improve dramatically. GHG emissions fall by 90% from today's level, keeping global warming within the 2°C target. The implementation of resource efficiency and circular economy policies reduces global materials extraction 25% below the Historical Trends level (although that still entails a considerable increase from today's materials use). Changed diets reduce pasture land for livestock by 30% from the Historical Trends level, and below today's level, freeing up agricultural land so that forests and other natural habitats are able to increase by 11% above today's level.

A final remarkable result of the Towards Sustainability scenario is that global GDP is increased, rather than reduced, by the focus on decarbonisation and resource efficiency – by 2060 it is 8.8% above the level in Historical Trends. The Towards Sustainability scenario may not therefore describe a fully sustainable world, but it is much more likely to be a habitable one than the continuation of historical trends, as well as one in which the economic and social

---

6 Canada, France, Germany, Italy, Japan, UK, USA.

Table 8.6 Projections to 2060 of two scenarios, Historical Trends and Towards Sustainability

	'Historical Trends': Projected 2060 levels compared to 2015 in absence of urgent and concerted action	*'Towards Sustainability': Projected 2060 levels in comparison with 'Historical Trends'*
Global GDP		8% above Historical Trends
Global material extraction	More than doubles	25% lower than Historical Trends
Greenhouse gas emissions	Increases by 43%	Decrease by 90%
Areas of agricultural land	Increases by more than 20%	9% less than Historical Trends
Global pasture land	Increases by 25%	30% less than Historical Trends
Forest	Reduces by over 10%	Increases by 11% (forests and
Other natural habitat	Reduces by around 20%	other natural habitat)

Source: Results collated from International Resource Panel, 2019, Chapters 2 and 4.

aspirations of currently low-income countries are much more likely to be met. The economics of this 'sustainability transformation' are further explored in Chapter 11, while policies that can help bring it about are set out in Chapter 12.

Critical minerals for the clean energy transition

Many of the technologies required for decarbonisation require a wide range of metals and minerals, the use of which to date has been relatively small, and the production of which will need to be scaled up dramatically if these technologies are to replace fossil fuels over the next three decades.

Key minerals for the transition

In 2021 the IEA turned its attention to the issue of minerals for the clean energy transition, and produced a report on the materials that would be needed for different clean energy technologies. Figure 8.3 from that report shows that EVs and renewable power technologies (especially offshore wind) require a far greater mass (kg/vehicle or kg/MW) and variety of metals than their conventional or fossil fuel counterparts. Copper in particular is required in large quantities for all the technologies of low-carbon power generation and transport displayed in the figure. Lithium is important for the batteries of electric cars, as is nickel, which is also used in a number of power generation technologies. Manganese is needed for both kinds of vehicles and for wind turbines. Cobalt and graphite are currently critical for batteries for electric cars, while chromium and zinc are needed for wind turbines.

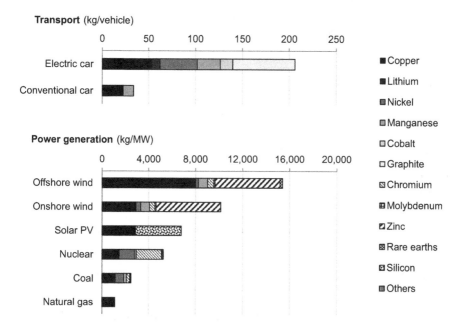

Figure 8.3 Important minerals (excluding steel and aluminium) for some clean energy technologies.

Source: International Energy Agency, 2021c, p. 6.

Table 8.7, from the same source, extends the list of both the technologies and minerals considered and indicates which minerals are important for which technologies. It can be seen that nickel and platinum group metals (PGMs) are important for hydrogen production, and rare earth elements (REEs) have a role in hydrogen too, as well as wind turbines and batteries. Copper, with its high conductivity, is crucial for the electricity networks, which will need to be much expanded as electricity takes over from fossil fuels in transport and heat provision, as well as power generation. Zinc has high importance for wind turbines, and is moderately important for hydropower, concentrated solar power (CSP) and bioenergy. Aluminium is important across a wide range of the technologies shown, as well as having many other uses.

Table 8.8 contains further mineral requirements related to fuel cells (absent from the list above), namely scandium and platinum, while gallium, indium, cadmium, tellurium and selenium are currently important elements for PV cells. Table 8.8 also compares the ratios of demand in 2013 and projected for 2035 for these minerals for these technologies, with the 2013 production level of these minerals. Taking lithium as an example, batteries and airframes took only 2% of lithium production in 2013, but these demands are expected to rise to 385% of the 2013 production in 2035. This means that lithium demand for

Table 8.7 Minerals for different clean energy technologies

	Copper	Cobalt	Nickel	Lithium	REEs	Chromium	Zinc	PGMs	Aluminium
Solar PV	Δ	⊗	⊗	⊗	⊗	⊗	⊗	⊗	Δ
Wind	Δ	⊗	*	⊗	Δ	*	Δ	⊗	*
Hydro	*	⊗	⊗	⊗	⊗	*	*	⊗	*
CSP	*	⊗	*	⊗	⊗	Δ	*	⊗	Δ
Bioenergy	Δ	⊗	⊗	⊗	⊗	⊗	*	⊗	*
Geothermal	⊗	⊗	Δ	⊗	⊗	Δ	⊗	⊗	⊗
Nuclear	*	⊗	*	⊗	⊗	*	⊗	⊗	⊗
Electricity networks	Δ	⊗	⊗	⊗	⊗	⊗	⊗	⊗	Δ
EVs and battery storage	Δ	Δ	Δ	Δ	Δ	⊗	⊗	⊗	Δ
Hydrogen	⊗	⊗	Δ	⊗	*	⊗	⊗	Δ	*

Source: International Energy Agency, 2021c, p. 45.

Notes: Δ: high; *: moderate; ⊗: low
REEs = rare earth elements; PGM = platinum group metals; CSP = concentrated solar power; aluminium use for electricity networks only.

these two technologies is expected to grow by a factor of nearly 20 between 2013 and 2035. Many other minerals in the table show spectacular growth. The only mineral for which demand for the associated technologies is projected to fall is tin (from 50% to 42% of 2013 production).

Companion metals

Many of the minerals listed in Tables 8.7 and 8.8 are not produced as the primary product of extraction and processing, but as by-products, co-products or 'companion metals'. Table 8.9 shows the proportion of particular minerals that are produced as 'companion metals'. Of the minerals most closely associated with clean energy technologies, mentioned in Tables 8.7 and 8.8, Al, Cr, Li, Mo, Ni, Cu, Zn, Sn, Ta, Ti and Pt are largely produced in their own right, but Sc, Co, Ga, Ge, Se, Cd, In, Pa, Re, Te[7] and the rare earth elements (Nd, Dy, Pr, Td) have a high percentage of their production obtained as companion metals.

The mixtures of minerals in their ores are complex, and principal metals in some ores may be companion metals in others. Thus Cu is a principal metal, but is also a companion to Au, Pt, Pb, Zn and Ni. About 40% of Co is a companion to Ni production, and slightly less as a companion to Cu, with small amounts also as a companion to Pt. Ga is mainly produced as a companion to Al, but is also a companion to Zn. The main REEs for renewable technologies

7 The full names of most of these elements can be found in Table 8.9, and all of them with their characteristics at https://ptable.com/?lang=en#Properties. Accessed October 10, 2023.

Table 8.8 Ratios of demand for particular minerals for particular technologies in 2013 and 2035 to production of these minerals in 2013

Metal	Demand 2013 / Production 2013 (%)	Demand 2035 / Production 2013 (%)	Related innovative technologies
Lithium	2	385	Lithium-ion batteries, lightweight airframes
HREE (Dy/ Tb)	85	313	Magnets, EVs, wind power
Rhenium	98	250	Superalloy
LREE (Nd/ Pr)	79	174	Permanent magnets (especially for EVs and wind power)
Tantalum	38	159	Microcapacitors, medical technology
Scandium	17	138	Solid oxide fuel cells
Cobalt	4	94	Lithium-ion batteries, synthetic liquid fuels
Germanium	39	81	Fibre optic, infrared technology
Platinum	0	60	Fuel cells, catalysts
Tin	50	42	Lead-free solders, wind turbines
Palladium	8	47	Catalysts, seawater desalination
Indium	29	45	Displays, thin layer photovoltaics
Gallium	25	37	Thin layer photovoltaics, integrated circuits, white LEDs
Silver	22	32	Lead-free solder, nanosilver, radio frequency identification devices (RFID), microcapacitors, high-temperature superconductors, concentrated solar panels
Copper	1	29	Electric motors, RFID
Titanium	4	18	Seawater desalination, medical implants

Source: International Resource Panel, 2020b, p. 111.

Notes: HREE and LREE are heavy and light rare earth elements respectively; Dy is dysprosium and Tb is terbium among the HREEs; Nd is neodymium and Pr is praseodymium among the LREEs; LEDs are light-emitting diodes.

(Pr, Nd, Dy and Tb) are mainly companions to Fe production (Nassar et al., 2015, Figure 2, p. 3). And so on. It should be noted that the proportions in this case relate to 2008, and may change over time as different ores are exploited, or mining or metallurgical processes change (Nassar et al., 2015).

This dependency of the production of some minerals on the production of a principal host mineral can be problematic, because the companion metal will only be co-produced if the economics of its production are profitable. Where prices for the companion metal are low, it may remain in the waste of the principal metal's production, which may be a cause for concern if the companion metal is toxic (e.g. cadmium). Considerable potential production

Table 8.9 Proportion of metals obtained as companions, by percentage

0–20%

Be (Beryllium)	B (Boron)	Mg (Magnesium)	Al (Aluminium)	Ti (Titanium)
Cr (Chromium)	Mn (Manganese)	Fe (Iron)	Ni (Nickel)	Sr (Strontium)
Nb (Niobium)	Sn (Tin)	Ba (Barium)	W (Tungsten)	Au (Gold)
Hg (Mercury)	Pb (Lead)	U (Uranium)		

20–40%

Cu (Copper)	Zn (Zinc)	Y (Yttrium)	Ta (Tantalum)	Pt (Platinum)

40–60%

Li (Lithium)	Mo (Molybdenum)

60–80%

Ag (Silver)	Ce (Cerium)

80–100%

Sc (Scandium)	V (Vanadium)	Ge (Germanium)	Ga (Gallium)	Co (Cobalt)
As (Arsenic)	Se (Selenium)	Zr (Zirconium)	Ru (Ruthenium)	Rh (Rhodium)
Pd (Palladium)	Cd (Cadmium)	Nd (Neodymium)	Sb (Antimony)	Te (Tellurium)
Hf (Hafnium)	Re (Rhenium)	Dy (Dysprosium)	Sm (Samarium)	Tl (Thallium)
Bi (Bismuth)	La (Lanthanum)	Pr (Praseodymium)	Ho (Holmium)	Ir (Iridium)
Eu (Europium)	Gd (Gadolinium)	Td (Terbium)	Os (Osmium)	In (Indium)
Er (Erbium)	Tm (Thulium)	Yb (Ytterbium)	Lu (Lutetium)	Th (Thorium)

Source: Adapted from Nassar et al., 2015, Figure 1, p. 2.

of critical minerals is forgone for this reason; Nassar et al. (2015, p. 7) find that "recovery efficiencies for some companion metals can be as low as 10%". Moreover, the financial dependence of companion metals on that of their principal metals means that the response of their production to increased demand can be very slow.

Concentration of production and processing

Another important consideration for the availability of materials for the clean energy transition is the concentration of their extraction and processing in different countries. Figure 8.4 from the 2020 EU study on Critical Raw Materials shows the countries which produce the highest proportions of selected materials. The major role played by China in the supply of many of these materials, with many of them at above 50% of global supply, and the REEs at 86%, is striking.

China's prominence is even more pronounced when it comes to the processing of extracted ores, as seen in Figure 8.5, which shows the shares of extraction and processing in major producing countries of some of the minerals of greatest interest for the clean energy transition (as well as for fossil fuels). China is among the top three extractors of copper, lithium and rare earths, but

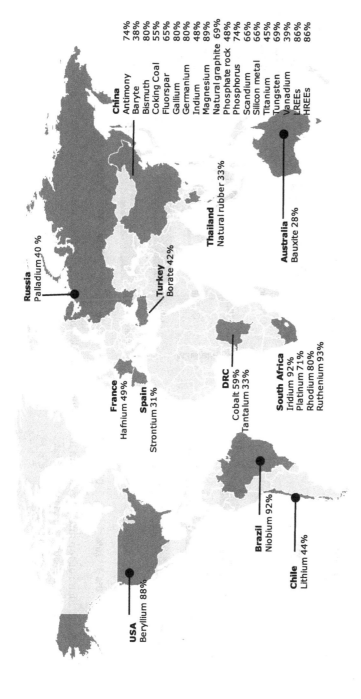

Figure 8.4 Countries which produce the highest proportion of selected materials.

Source: European Commission, 2020, p. 6.

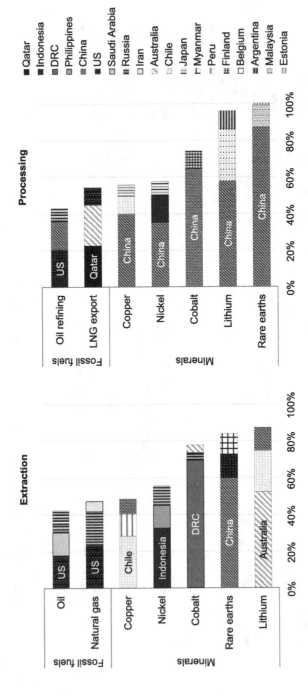

Figure 8.5 Share of top three countries in the production of selected minerals and fossil fuels.

Note: LNG is liquid natural (fossil methane) gas.

Source: International Energy Agency, 2021c, p. 13.

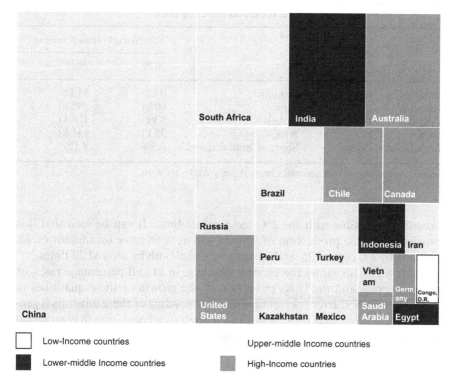

Figure 8.6 Share of value of production of non-energy minerals and metals of top 20 producing countries, by country and income group.

Source: International Resource Panel, 2020b, p. 65.

it is the top processor for all of them, with more than 50% of global processing of cobalt, lithium and rare earths.

Figure 8.6 shows that the value from non-energy minerals and metals production is similarly concentrated in relatively few countries, with the 20 countries shown representing 73% of the global production value of these commodities. Ten of these countries are upper middle-income countries, and six are high-income countries. The only low-income country in this list is the Democratic Republic of the Congo (DRC).

Growth in demand

Table 8.8 showed the projected growth of demand by 2035 of many clean energy minerals relative to their 2013 production. Table 8.10 shows this growth for five of these minerals by 2040, relative to their production in 2020, as projected in the IEA's Sustainable Development Scenario (SDS), which is

Table 8.10 Growth to 2040 of selected minerals in the IEA Sustainable Development Scenario, and of their reuse and recycling rates

	Growth index (2020 = 1)		*Recycled and reused volume*	
			2030	*2040*
Lithium	42	Lithium	0 kt	81 kt
Graphite	25	Copper	60 kt	395 kt
Cobalt	21	Cobalt	8 kt	108 kt
Nickel	19	Nickel	35 kt	611 kt
Rare earths	7	Share of total demand	0.3%	8.1%

Source: Adapted from International Energy Agency, 2021c, pp. 9, 16.

broadly compatible with the 2°C temperature limit. It can be seen that it is projected that the production of lithium will need to grow by a factor of 42, graphite by a factor of 25, and cobalt and nickel both by around 20 times.

Table 8.10 also shows the current recycling, in kt and percentage rates, of these minerals, and the IEA's projection of the growth of these quantities to 2040 in the SDS. Current percentage rates of recycling of these minerals is generally very low (around 1% or less), and though significant growth is projected from 2030 to 2040, the overall projected rate of recycling in 2040 remains low (8.1% of the total demand for lithium, nickel, copper and cobalt).

Nassar et al. (2015, Figure 4, p. 5) show that the rare earth elements which are more important for clean energy technologies (Pr, Nd, Dy and Tb) present a special challenge. They are exclusively companion metals, are very concentrated in production, and have very low recycling rates (<1%).

The growth in demand for these minerals that are critical for the clean energy transition has led to a substantial increase in the price of many of them. Between January 2021 and March 2022 the price of lithium increased more than seven-fold, while prices for cobalt, nickel and aluminium increased respectively by factors of 2.5, 2 and 1.8 (International Energy Agency, 2022, p. 113). For each of these metals the price in March 2022 was higher than at any time in the 2010s.

The price increases have had an inevitable impact on the costs of the technologies that incorporate these metals. For example, the cost of electric vehicle battery packs, the cathodes of which contain lithium, cobalt and nickel, had risen by 20% in January–February 2022, compared to 2020 (International Energy Agency, 2022, p. 114). If maintained, these price increases seriously threaten the falling costs of clean energy technologies that are described in Chapter 11.

The prices of critical minerals will only come down, and the increasing demand for them met, if there is a substantial increase in investment in mining for and processing them. 2021 saw an increase in investment by major mining companies over 2020 (International Energy Agency, 2022, p. 120), but the level is still substantially below that in 2012–2014, and the time lag between the investment and the opening of a new mine, which can easily be ten years, means that there is a real danger of the lack of supply of critical minerals being a bottleneck to the necessary and early large-scale deployment of clean energy technologies.

Conclusion

This chapter has explored three essential elements of the clean energy transition: digitalisation, shifts to a circular economy, and the need for many metals that have been hitherto hardly used. Any failure to fully exploit the opportunities in these areas, or to overcome the many barriers to delivering them, could act as a major roadblock to the necessary scale and timeline of GHG emissions reduction.

Digitalisation is already in full swing across the economy and is already being introduced extensively across both energy supply and energy demand sectors. Because digital technologies are powered by electricity, they are well suited to low-carbon economies, and there are enormous opportunities for big data, machine learning and AI, blockchain and the Internet of Things to facilitate both the supply of electricity and its more efficient use by buildings, industry and in transport. The substantial use of electricity by digital technologies themselves, especially data centres, will be a source of carbon emissions until electricity supplies themselves are decarbonised, which, as seen in an earlier chapter, is the first priority for decarbonisation more generally.

The use of materials – biomass, fossil fuels, metal ores and non-metallic minerals – has trebled since 1970 and without policy interventions is projected to double again by 2060. The extraction and processing of these materials is currently responsible for enormous environmental impacts, including about half of all GHG emissions. Without policy interventions, this could rise to two-thirds of a larger total by 2060. However, the implementation of strong policies of climate action, decarbonisation, resource efficiency and circular economy can both keep GHG emissions to a trajectory within 2°C, and greatly reduce the other environmental impacts of material use. Perhaps surprisingly, it could also increase the rate of economic growth, so that the world in 2060 in this scenario is both richer and has more nature than on current trends. Chapter 11 explains how it is that such win-win results for the economy and the environment can come about, while Chapter 12 explores the policy approaches that will be necessary for this to be achieved.

The production of many clean energy technologies is highly dependent on a number of minerals which, until fairly recently, have been used in relatively small quantities. The required scale up of these technologies, and therefore of the minerals required for them, for zero-carbon emissions in 2050 to be achieved is a formidable challenge. This is not so much because of the minerals' geological availability – it is likely that Earth's crust contains large quantities that have not yet been discovered, partly at least because until now there has not been a systematic search for them. Rather the challenge is that extracting these minerals is very demanding, in terms of the time taken to open a new mine, and the investment required for both the extraction and processing of the ores. And while undoubtedly the recycling rates of many of these minerals can be increased from current low levels, and this will help reduce the demand for new primary minerals, there are limits to the recycling rates that can be achieved, and the scale of the demand increase is such that it can only be met for the foreseeable future by the opening of a large number of new mines. There is a real risk that the supply of these minerals will increase the cost and slow down deployment of clean energy technologies, so that the clean energy transition is not fast enough to achieve the 1.5°C Paris temperature target.

Another potential source of delay of the clean energy transition relates to the governance of mining processes. These need to ensure that the social conditions of production, the environmental impacts, the share of revenues going to the host country, and mine closure and land restoration all contribute to the sustainable development of the country of production. This is what was referred to in the IRP's report on mineral resource governance (International Resource Panel, 2020b) as the 'sustainable development licence to operate' (SDLO). While mining practices in respect of some of these issues have improved in the last 30 years, much further improvement is likely to be required before many host country governments will welcome the mining and processing of their resources, without which the clean energy transition will be slowed, so that the target of zero emissions will not be achieved by 2050, if at all.

References

Bradley, P., Leach, M., & Torriti, J. (2011). *A review of current and future costs and benefits of demand response for electricity*. Centre for Environmental Strategy Working Paper 10/11, University of Surrey. www.surrey.ac.uk/sites/default/files/2018-03/10-11-bradley-deman-response.pdf

European Commission. (2020). *Study on the EU's List of Critical Raw Materials: Final Report*. European Commission, Brussels.

International Energy Agency. (2017). *Digitalization and Energy*. IEA, Paris.

International Energy Agency. (2021a). Energy Efficiency, 2021. IEA, Paris. https://iea.blob.core.windows.net/assets/9c30109f-38a7-4a0b-b159-47f00d65e5be/EnergyEfficiency2021.pdf

International Energy Agency. (2021b). *Net Zero by 2050: A Roadmap for the Global Energy Sector*. IEA, Paris.

International Energy Agency. (2021c). *The Role of Critical Minerals in Clean Energy Transitions*. IEA, Paris.

International Energy Agency. (2022). *World Energy Investment 2022*. IEA, Paris.

International Resource Panel. (2019). *Global Resources Outlook 2019: Natural resources for the future we want*. UN Environment Programme, Nairobi.

International Resource Panel. (2020a). *Resource Efficiency and Climate Change: Material Efficiency Strategies for a Low-Carbon Future*. Hertwich, E., Lifset, R., Pauliuk, S.,

Heeren, N. (Eds.) A report of the International Resource Panel. United Nations Environment Programme, Nairobi, Kenya.

International Resource Panel. (2020b). *Mineral Resource Governance in the 21st Century: gearing extractive industries towards sustainable development.* United Nations Environment Programme, Nairobi.

Nassar, N. T., Graedel, T. E., & Harper, E. M. (2015). By-product metals are technologically essential but have problematic supply. *Science Advances, 1*(3). https://doi.org/10.1126/sciadv.1400180

Navigant Consulting. (2015). *Peak Demand Reduction Strategy.* Advanced Energy Economy, Washington, DC. https://blog.advancedenergyunited.org/articles/new-report-reducing-peak-demand-saves-money-for-electricity-customers

Organisation for Economic Cooperation and Development. (2018). *Highlights: Global Material Resources Outlook to 2060: Economic drivers and environmental consequences.* OECD, Paris.

Rhodes, A. (2020). *Digitalisation and Energy.* An Energy Futures Lab Briefing Paper. Imperial College, London. https://spiral.imperial.ac.uk/bitstream/10044/1/78885/2/4709_EFL_Digitalisation_briefing_paper_WEB2.pdf. Accessed May 6, 2023.

U.S. Department of Energy. (2006). *Benefits of demand response in electricity markets and recommendations for achieving them: a Report to U.S. Congress.* US DOE, Washington DC. www.energy.gov/sites/prod/files/oeprod/DocumentsandMedia/DOE_Benefits_of_Demand_Response_in_Electricity_Markets_and_Recommendations_for_Achieving_Them_Report_to_Congress.pdf

Zakeri, B., Staffell, I., Dodds, P., Grubb, M., Ekins, P., Jääskeläinen, J., Cross, S., Helin, K., & Castagneto Gissey, G. (2022). *Energy Transitions in Europe – Role of Natural Gas in Electricity Prices.* SSRN. https://papers.ssrn.com/sol3/papers.cfm?abstract_id=4170906

9 Decarbonising buildings, transport, industry and business

Summary

Buildings, transport, and industry and business are major sources of greenhouse gas (GHG) emissions in their delivery of the products and services desired by people. While there is some scope for businesses and consumers to reduce their emissions through their own efforts, the scale of reduction involved will require the implementation of stringent policies.

Buildings globally account for around 40% of GHG emissions, the majority from their operation and use. Reaching 'real zero' will require a transformation in the energy efficiency of the building stock, and the systems through which they are heated and cooled. This chapter explores the extent of the energy demand reduction that will be necessary and the technologies through which zero-carbon heating and cooling can be delivered. Because of the diversity of the building stock, detailed analysis of the issue is only possible at a country level, which in this chapter is undertaken for the UK. Issues explored include technologies, finance, energy advice and labelling, and policy, and many of these are relevant to other European countries and elsewhere.

In respect of transport the great majority of transport emissions come from burning fuels made from oil, and the majority of those emissions come from road transport. Because there is no technology to capture tailpipe CO_2 emissions, the only way of eliminating them is to substitute away from oil products, using instead batteries, biofuels, synfuels, fuel cells, or hydrogen. The chapter looks in detail at the implications of these alternatives in the different modes of transport – road passengers and freight, rail, aviation and shipping. Supplying adequate quantities of these alternative fuels would be made easier by reducing car ownership and use in urban areas, by enabling increased walking, cycling, and public transport, and promoting car-sharing in compact, relatively densely populated cities.

With industry the second-largest direct emitter of GHGs after the energy sector, industrial decarbonisation is clearly critical to real zero.

DOI: 10.4324/9781003438007-10

The International Energy Agency (IEA) shows how emissions in this sector can be reduced by 95% by 2050, with major roles being played by material efficiency, hydrogen, electrification and carbon capture, utilisation and storage (CCUS). After looking at the decarbonisation of industry in general, the chapter focuses in on the three most energy-intensive sectors – iron and steel, chemicals and cement – to look at the options for their decarbonisation, finishing with options for the reduction of fluorinated gas (F-gas) emissions.

Introduction

Earlier chapters have made clear that the principal cause of greenhouse gas (GHG) emissions to the atmosphere is the human use of carbon-containing energy sources, principally fossil fuels. This chapter explores the contribution to GHG emissions of the three main 'sectors' that use fossil energy – buildings, transport and industry and business – and describes in detail the technologies that could reduce or eliminate these emissions, and the policies that are needed to implement them at scale. The next chapter does the same for agriculture, the main cause of non-CO_2 emissions.

Of course, in respect of buildings, it is true that 'buildings don't use energy, people do' (Janda, 2011), and this is true for transport and industry too. All the energy use in these sectors is the result of human decisions and behaviours to deliver certain kinds of products and services that people want.

To reduce their emissions from these activities consumers have a number of options: to reduce or change the consumption of the products and services that cause the emissions (e.g. turning down the thermostat in their homes, or taking a bus or train instead of using a car); to cause fewer emissions through their consumption of products and services (e.g. keeping warm through wearing more clothes or improving the energy efficiency of their home); to choose products and services that have been made in ways that do not cause emissions (e.g. renewable electricity or products that have been made with renewable electricity), or cause fewer emissions in their manufacture or use (e.g. energy-efficient appliances); to use products that last longer, or can be repaired or shared, or recycled (the circular economy approach described in the previous chapter).

Producers have many of the same options in manufacturing or delivering their products and services, using technologies that cause low or no emissions, and making products with less materials, that last longer and that can be repaired, remanufactured or recycled. All these options of consumption and production are included under the rubric of 'Responsible Consumption and Production', which is the 12th of the UN's Sustainable Development Goals (SDGs).[1]

---

1 https://sdgs.un.org/goals

However, both people in businesses and people as consumers face formidable problems and barriers in making these choices. Businesses face competitive pressures, and the low- or zero-carbon option may be more expensive and offer no extra consumer benefits. The information about the options, for both businesses and consumers, may be difficult to find, not reliable, or not available. Consumers may not be able to afford the low-carbon option, or it may be inconvenient to use (e.g. some public transport) or dangerous (e.g. lack of safe cycling infrastructure). Many businesses and consumers who perceive the importance of decarbonisation make heroic efforts to try to overcome these challenges, and do manage to reduce their emissions, but the emission reduction tends to be incremental and, at a societal level, nothing like the level that is required.

Removing the constraints to decarbonisation, making it profitable for businesses to reduce their GHG emissions, and expensive for them to keep on emitting, and making it attractive economically and socially to consumers to move to low- and zero-emission lifestyles, and increasing the cost of high-emission lifestyles, is the role of climate policy. Many examples of this have already been encountered in previous chapters in respect of energy supply. This chapter will discuss relevant policies for energy demand in buildings, transport, industry and business.

Buildings

Buildings are a major user of energy in their construction and throughout their lives. Globally,

> The buildings and construction sector accounted for 36% of final energy use and 39% of energy and process-related carbon dioxide (CO_2) emissions … [and] in 2018, buildings-related CO_2 emissions rose for the second year in a row, to an all-time high of 9.7 Gt CO_2.
>
> (GlobalABC/IEA/UNEP (Global Alliance for
> Buildings and Construction), 2020, p. 7)

The same source shows the whole-life energy use of buildings, from the extraction and manufacture of construction materials (e.g. stone, cement, bricks, steel, glass) through to demolition (GlobalABC/IEA/UNEP (Global Alliance for Buildings and Construction), 2020, Figure 3, p. 14). It shows that about two thirds of building energy use goes into the operation and use of the building (heating, cooling, etc.), with the final third roughly evenly split between materials provision (e.g. cement, steel, bricks), construction, maintenance, repair and refurbishment, and end-of-life. Carbon emissions arising from this last third, those mainly associated with the provision of these materials, are commonly called 'embodied emissions', despite the fact that the problem arises precisely because these emissions are not in fact 'embodied' in the materials, but emitted to the atmosphere.

Emissions from buildings in use arise from space heating and cooling, water heating, lighting, cooking and appliances and other equipment. Space heating is the largest source, followed by water heating and cooking. Emissions from residences account for a little more than half the total, and the majority of both residential and non-residential emissions come indirectly from their use of electricity. It was this source that was mainly responsible for the around 10% growth in CO_2 emissions from buildings globally between 2010 and 2018 (GlobalABC/IEA/UNEP (Global Alliance for Buildings and Construction), 2020, Figure 4, p. 16).

Table 9.1 shows end use energy in buildings, but for 2020, and then for 2030 and 2050 in the IEA's Net Zero Emissions (NZE) scenario. Demand for energy for space heating drops by two thirds because of increased efficiency of heat generation and buildings, and lower demanded indoor temperatures, and energy use for space cooling increases somewhat, as does that for appliances, despite efficiency gains, so that appliances become the single largest user of energy. 'Other' end uses drop dramatically with the disappearance of traditional biomass (2.7 billion people get access to clean cooking by 2030), leaving a residual demand for energy for desalination of water. Fossil fuels disappear from building energy use almost entirely, with the share of electricity in the energy mix doubling to two thirds by 2050. 'Renewables' is bioenergy, over 50% of which is used for cooking in emerging market and developing economies.

Going beyond 2050, another study suggests that by 2100, the dominant use of energy in buildings in developing regions becomes lighting and appliances and space cooling. Deep electrification of buildings' energy use is characteristic of all regions (Levesque et al., 2018).

Buildings in the UK

Because buildings, how they are owned and managed, and the weathers they have to cope with, differ so much around the world, there is no universal prescription for how their CO_2 emissions can be reduced, beyond such general statements as that they should be highly energy-efficient. Indeed, even within a single country like the UK, buildings are so diverse, and the equipment they contain and the people who occupy them, differ so much, it is not far from the truth to say that efforts to reduce emissions must be conducted more or less on a building-by-building basis, though within a supportive policy framework from both central and local government. The rest of this section therefore focuses on a single country, the UK, though many of its insights will be relevant to other northern hemisphere, temperate countries.

Direct GHG emissions from UK buildings, mainly from the use of fossil fuels for heating, were 85 $MtCO_2e$ in 2017 (83 $MtCO_2$, 2 $MtCO_2e$ non-CO_2 [methane and nitrous oxide]), 17% of the UK total. These direct emissions were split 77% from homes, 14% from commercial buildings, and 10% from public buildings. Direct emissions from buildings have fallen by 11% from

Table 9.1 Use of energy in buildings by end use and by fuel at different dates in the IEA Net Zero Emissions (NZE) scenario

	Use of energy in buildings (EJ)		
	2020	*2030*	*2050*
By end use			
Space heating	41.7	29.1	14.2
Water heating	16.9	16.0	14.8
Space cooling	6.35	7.7	9.12
Lighting	6.02	5.99	6.25
Cooking	10.8	17.4	14.0
Appliances	18.8	20.3	23.5
Other	26.4	2.54	4.13
By fuel			
Coal	3.96	1.09	0.004
Oil	13.0	9.18	1.18
Natural (fossil methane) gas	28.4	19.3	0.657
Hydrogen	0.002	1.9	2.55
Electricity	42.1	45.3	56.9
District energy	6.8	5.97	4.210
Renewables	7.2	16.1	20.5
Traditional use of biomass	25.1	0.000	0.000

Source: Adapted from International Energy Agency, 2021a, Figure 3.28, p. 143.

Note: 'Other' end use includes desalination and traditional use of solid biomass which is not allocated to a specific end use.

1990, largely due to energy efficiency improvements in buildings and growth in bioenergy use.

Buildings also consumed in 2017 66% of UK electricity, equivalent at then grid carbon intensities to a further 48 $MtCO_2e$ of emissions, called 'indirect emissions'. When including these emissions, buildings account for 26% of the UK total. Indirect emissions fell at an average rate of 8% per year from 2009 to 2017,[2] due to both reductions in demand and the decarbonisation of electricity generation.

Fossil methane gas meets around 75% of the UK's heating demand in buildings, oil 8%, and electricity most of the remainder. But most electricity consumption in homes stems from the use of appliances and lighting, and in non-residential buildings from cooling, catering and ICT equipment. (Data for the last three paragraphs from Climate Change Committee (2019b, pp. 69–70.)

Figure 9.1 shows energy use (horizontal axis) and CO_2 emissions (vertical axis), in both absolute terms and % share for different uses, from UK homes and workplaces from a slightly earlier year, which shows broadly the

2 This reduction then continued through to 2019.

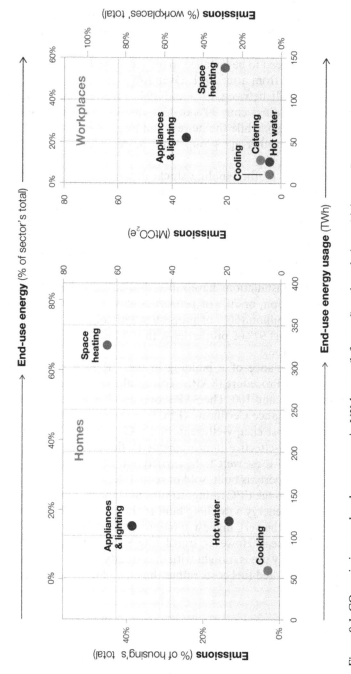

Figure 9.1 CO₂ emissions and end use energy in UK homes (left panel) and workplaces (right panel).

Source: Energy Research Partnership, 2016, Figure 2, p. 6.

same pattern. This includes both electricity and other energy use, and all the associated emissions.

Building techniques have changed dramatically over the years, as have the energy efficiency regulations that apply to new buildings, so that the age of a building has a significant influence over the level of its CO_2 emissions, and on the technologies that need to be installed to reduce these. In the UK only about 20% of dwellings date from after 1981. Over 20% date from before 1919, and another 16% were built between 1919 and 1944 (Traynor, 2020, Figure 6.4, p. 81). The oldest buildings emit 30% of CO_2 dwelling emissions (more than their share of dwellings), while the newest buildings emit less than 10% (less than their share). The other age groups emit more or less the same as their share of dwellings.

Given the diversity of the building stock, generalisations about heat loss from buildings will mask considerable variation, but the Energy Saving Trust estimates that, for a small detached building, 33% of energy loss is through the walls, 26% through the loft or roof, 18% through the windows, 12% as a result of ventilation and draughts, 8% through the floors and 3% through the doors.[3] It is therefore hardly surprising that government policy has tended to focus on loft and wall insulation (Climate Change Committee, 2019a, Figure 1.1, p. 29), although even these installations have fallen dramatically since 2012. By the end of 2020, 14.3 million, or 70% of properties with a cavity wall, had cavity wall insulation, 16.6 million (66% of properties with a loft) had got loft insulation, and 772,000 (just 9% of properties with solid walls – e.g. single bricks) had solid wall insulation.[4]

The energy performance of a building in the UK is assessed through a Standard Assessment Procedure (SAP).[5] This results in a building being given a SAP score between 1 and 100. The SAP scores are then related to the ratings on an Energy Performance Certificate (EPC), as shown in Table 9.2, with an A-rating being the most energy-efficient, and a G-rating the least. The EPC also recommends cost-effective improvements to the dwelling, and gives the potential EPC rating if these were to be carried out. EPCs are now mandatory in the UK when a property is built, sold or rented out.[6]

Table 9.2 also relates the EPC ratings to various standards for new buildings, and shows how much energy a dwelling built to these standards should use for heating. One built to the very highest standard (Passivhaus) should consume less than 15 kWh/m²/year. However, translating SAP ratings and EPC bands into estimates of energy use is fraught with uncertainty, not least because SAP/EPC is based on estimated fuel costs rather than a building's energy efficiency.

3 www.energysavingtrust.org.uk/sites/default/files/EnergySavingWeek2013infographic%20(4).pdf. Accessed May 5, 2023.
4 https://assets.publishing.service.gov.uk/government/uploads/system/uploads/attachment_data/file/970064/Detailed_Release_-_HEE_stats_18_Mar_2021_FINAL.pdf. Accessed May 5, 2023.
5 www.gov.uk/guidance/standard-assessment-procedure. Accessed May 5, 2023.
6 www.gov.uk/buy-sell-your-home/energy-performance-certificates. Accessed May 5, 2023.

Table 9.2 SAP scores, EPC ratings and building regulations

Building energy use (kWh/m²/annum)	Energy efficiency rating on the Energy Performance Certificate (range of SAP scores)	Notes
<15	A (>91)	Passive House Standard
<25 (A1)		EnerPHit Standard
>25 (A2)		
>50 (A3)		Low Energy Standard
>75 (B1)	B (81–91)	
>100 (B2)		Current UK building regulations
>125 (B3)		
>150 (C1)	C (69–81)	
>175 (C2)		
>200 (C3)		
>225 (D1)	D (55–68)	
>250 (D2)		
>300 (E1)	E (39–54)	
>340 (E2)		
>380 (F)	F (21–38)	
>400 (G)	G (1–20)	

Source: Traynor, 2020, Figure 2.1, p. 21.

Note: The energy usage unit is kWh/m²/annum, where the m² refers to the floor area of the building.

SAP only includes space heating, hot water and lighting, not cooking and other appliances, and also includes generated energy, such as that from solar PV. It is therefore difficult but not impossible to get F- and G-ratings with gas heating (which is cheaper than electricity), while getting an A-rating and most B-ratings requires the generation of energy (mostly from PV panels). So the two EPC bands at each end of the scale are more to do with fuel type and generation than with energy efficiency. So while these conversion factors from SAP/EPC to energy use may give a rough approximation for energy use in standard (i.e. average size, semi-detached, gas-heated) homes, emerging evidence suggests that A- and B-ratings underestimate, and F- and G-ratings overestimate, energy use.[7]

SAP ratings increased considerably between 1996 and 2015, as building regulations for the energy efficiency of buildings became more stringent and energy efficiency measures were implemented. SAP ratings in social housing went from about 49 to 63 over this period, while the owner-occupied/private rented sectors went from around 43/40 to around 58 (Traynor, 2020, Figure 1.3, p. 9).

---

7 Insights from personal communication with Professor Tadj Oreszczyn, UCL, August 1, 2022.

New, even stricter, building regulations were introduced in June 2022. However, in practice the energy performance of new buildings falls considerably short of their design specifications. The Climate Change Committee wrote in 2019:

> The way new homes are built and existing homes retrofitted often falls short of design standards. This is unacceptable. In the long run, consumers pay a heavy price for poor-quality build and retrofit. Greater levels of inspection and stricter enforcement of building standards are required, alongside stiffer penalties for non-compliance. … Closing the energy use performance gap in new homes (the difference between how they are designed and how they actually perform) could save between £70 and £260 in energy bills per household per year.
>
> (Climate Change Committee, 2019a, p. 9)

This was up to around 25% of average bills at that time. At 2022 gas prices, the savings could easily have been double those figures.

Climate Change Committee (2019a, Figure B1.1, p. 30) shows the distribution of SAP ratings across households in England and Scotland. Easily the most common are D-ratings, which apply to more than 12 million dwellings, followed by nearly 8 million C-rated dwellings. Very few dwellings are now F- and G-rated, and the best ratings, A and B, are also very few, perhaps because of the way the SAP score is calculated, as discussed above. These ratings show that there is clearly considerable scope for energy efficiency improvement in dwellings in England and Scotland and, by extension, Wales and Northern Ireland as well. The balance to be struck is between the desirability of reducing energy demand, to reduce the quantity of energy that needs to be supplied, and the upgrading of the electricity and gas infrastructure that might be required to supply it, and the cost of increasing the energy efficiency of the UK building stock. At the time of writing there were proposals going through the UK Parliament to require that: all landlords to ensure in respect of their rented dwellings that all new tenancies had at least a C-rated EPC from 2026, and all tenancies had this rating from 2028; all mortgage lenders averaged a C-rating across their housing portfolio; all owner-occupied homes would have at least a C-rating from 2035; and the same would apply to 'a significant proportion' of social housing. These requirements were always provided that the required measures for landlords and owner-occupiers were 'practical, cost-effective and affordable', with the meanings of these terms to be set by subsequent regulations.[8] A Future Homes Standard would require that new homes from 2025 were 'zero-carbon ready'.

Economidou (2011, Figure 3B1, p. 109) projected from 2010 how the costs of building renovation for increased energy efficiency might change through

---

8 https://publications.parliament.uk/pa/bills/cbill/58-02/0150/210150.pdf. Accessed May 5, 2023.

Table 9.3 Costs of different energy efficiency measures for different homes

Measure	Lowest cost/home, £	Average cost/home, £	Highest cost/home, £
1. Solid wall insulation: internal	6,800	7,900	8,900
2. Solid wall insulation: external	7,100	11,800	15,000
3. Cavity wall insulation	480	750	1,400
4. Party cavity wall insulation	300	325	350
5. Loft insulation	185	450	670
6. Underfloor insulation	3,500	5,800	8,300
7. Solid floor insulation	1,300	5,700	9,800
8. Replacement double-glazing (panes and frames)	3,900	6,400	10,700
9. Gas boiler replacement	1,600	2,000	2,400

Source: Department for Business, Energy and Industrial Strategy, 2017, Table 3, p. 15.

to 2050, given innovation and economies of scale, for different ambitions of increased energy efficiency. This suggested that a deep retrofit in 2020 costing €250/m² of a relatively modest-sized dwelling (80 m²) would have cost €20,000, with a moderate renovation costing about half as much. The projections suggest that by 2050 the cost of a deep retrofit for such a dwelling would have fallen to around €8,000.

It is interesting to compare these projections with the 2017 cost estimates of different measures from the UK's Department of Business, Energy and Industrial Strategy (BEIS). Table 9.3 reproduces the figures from the literature review in the source publication.

Table 9.3 shows that the average cost of a renovation requiring measures 2, 3, 5, 6, 8 and 9, which would count as a 'deep renovation', would cost around £27,200. Given that some of these estimations in 2017 probably dated from before 2017, and assuming some further cost reductions, this suggests that the €20,000/home projection for 2020 in Economidou (2011) was not far off the mark. For the UK in 2020, measures 8 and 9 can be largely excluded as by then 87% of UK homes had double-glazing,[9] and around 75% had condensing boilers.[10] Multiplying the total of measures 2, 3, 5, 6 by the 8.6 million homes

9 www.statista.com/statistics/292265/insulation-in-dwellings-in-england-uk-y-on-y/.   Accessed May 5, 2023.
10 https://assets.publishing.service.gov.uk/government/uploads/system/uploads/attachment_d ata/file/1055629/Energy_Report_2019-20.pdf. Accessed May 5, 2023.

without solid wall insulation suggests a national deep renovation bill of around £162 billion. Of course, some solid wall homes will not need all these measures, but there will also be many non-solid wall dwellings that do not have all the measures in Table 9.3, so this estimate probably serves as a first approximation. The estimate is considerably more than the government's estimated cost of £35–65 billion of increasing the SAP rating of all homes to a C (Her Majesty's Government, 2019), implying either that it considers that the costs in Table 9.3 will decline with implementation at scale, or that a C-rating falls short of the deep renovation envisaged by the £162 billion cost, or both.

The importance of reducing energy demand in homes through greater energy efficiency is well expressed by Figure 9.2. This shows that average household heat demand in winter is three times the summer level, and that there is more than a factor of four difference between the winter energy demands of the largest and oldest houses and those of the smallest and best insulated. Reducing the heat demand of all houses to that of the average would make a huge difference to the amount of heat that needed to be supplied in winter.

Increasing the energy efficiency of buildings is skilled work which, if done poorly, can be ineffective in reducing heat loss and, worse, can both damage the fabric of the building (e.g. through causing condensation and rot) and cause health problems if the building is inadequately ventilated. Given the diversity of the building stock, the range of possible energy measures to increase

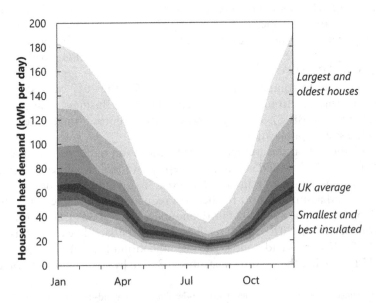

Figure 9.2 Household heat demand over the year for different types of houses.

Source: Staffell et al., 2019, Figure 6, p. 469.

building energy efficiency, the associated costs, and the need for expert installation, householders usually need professional advice on the best approach for any particular building. Such support, on an individual building basis, used to be provided in the UK by the Energy Saving Trust (EST), but is now only available in Scotland, funded by the Scottish Government.[11] The EST website still gives useful generic information about energy efficiency.[12]

While, because of the specific nature of the building stock in different countries, the energy efficiency issues in this chapter have largely reflected the situation in the UK, many of the points made, and the technologies discussed, will have much wider relevance. Indeed, the IEA, with reference to energy saving to facilitate decarbonisation, wrote: "Efficiency in buildings can deliver the largest share of avoided energy consumption to 2030" (International Energy Agency, 2021b, p. 61).

The relevance to net zero of reducing heat demand is shown by the dominance of fossil methane gas in UK household heating, which in 2010 heated more than 80% of UK households, with the balance supplied by electricity and other fuels, including oil and bioenergy (Staffell et al., 2019, Figure 8, p. 470). Real zero by 2050 demands that this fossil methane gas is entirely replaced by a zero-carbon alternative.

In its Net Zero Technical Report, the Climate Change Committee suggested how this transformation might be achieved, as shown in Figure 9.3. New houses, as already seen, are required by building regulations to be highly energy-efficient and, after 2025, must have low-carbon heating (i.e. not be heated by fossil methane gas). Existing buildings off the gas grid will be predominantly heated by heat pumps (see below), with a smaller role for bioenergy. In densely housed urban areas, the best low-carbon heating is likely to be a heating network (pipes that carry hot water to the different homes), with the heat supplied by large heat pumps or bioenergy. Some biomethane (also called biogas, from anaerobic digesters of food or farm waste) may be fed into the gas grid, but this will be nothing like sufficient to replace fossil methane gas. Challenging buildings will be those that are too far apart to be on heat networks, but without the space necessary for heat pumps.

In its Net Zero Balanced Pathway, the Climate Change Committee (CCC) modelled the solutions for low-carbon heat in buildings (Climate Change Committee, 2020, Figure 3.2.a, p. 110). There is surprisingly little CO_2 reduction from behaviour change (people like warm homes), or from improved efficiency in either residential or non-residential buildings, perhaps reflecting the fact that there are currently no government ambitions to go beyond the mandatory C-ratings for existing buildings, or the expense of doing so, as described above. In any case, as also seen above, fewer than 5 million homes have E-, F- and G-ratings, where the carbon savings of upgrading the dwelling

---

11 www.homeenergyscotland.org/contact-advice-support-funding/. Accessed May 5, 2023.
12 https://energysavingtrust.org.uk/energy-at-home/. Accessed May 5, 2023.

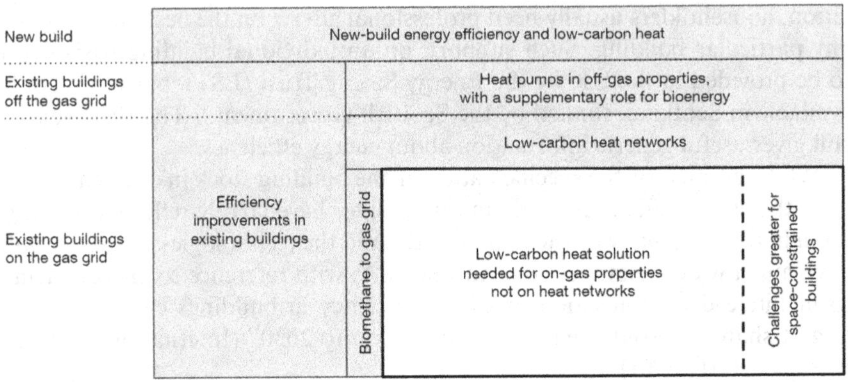

Figure 9.3 Low-carbon heating options for the UK housing stock.

Source: Climate Change Committee, 2019b, Figure 3.3, p. 74.

to a C-rating might be significant. In the CCC projections, low-carbon heat networks play a role in both residential and non-residential buildings, but the really big reductions come from low-carbon heat in homes, delivered largely by heat pumps.

It is still uncertain what proportion of low-carbon heat in homes currently on the gas grid, but for which heat networks are inappropriate, will be supplied by electricity, through heat pumps, or hydrogen, or some combination of the two. Heat pumps are likely to play by far the larger role. For example, in the CCC's Balanced Net Zero Pathway (BNZP) by 2050 there are heat pump installations in 27 million homes (see Figure 9.4). To reach this level, heat pumps need to start being installed at scale soon: the BNZP envisages that there are a total of 5.5 million heat pumps installed in homes by 2030, with annual sales in that year of 1 million per year. In public and commercial buildings, 52% of low-carbon heat in 2050 is supplied by heat pumps, with 42% from district heating and only 5% from hydrogen boilers. Globally, the International Energy Agency estimates that 90% of space and water heating could be supplied by heat pumps (International Energy Agency, 2021c).

Heat pumps work like fridges or freezers, which pump heat from a colder environment (the inside of the fridge or freezer) into a warmer environment (the room in which the fridge or freezer is located). For heating a building or providing hot water, heat pumps pump heat from outside (from the air, ground, or water) by heating water which is then piped into the building's radiators or hot water tank. Heat pumps are powered by electricity. Their great advantage over electric resistance heaters (e.g. electric fires or immersion heaters) is that the latter convert one unit of electricity into one unit of heat, whereas a heat pump can provide three units of heat for each unit of electricity input. If the electricity is derived from renewables, the heat generated is zero-carbon.

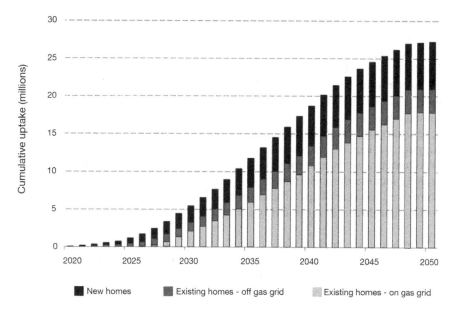

Figure 9.4 Cumulative uptake of heat pumps in residential buildings in the CCC's Balanced Net Zero Pathway.

Source: Climate Change Committee, 2020, Figure 3.2.c, p. 115.

The UK is currently ill-prepared for the transformation of its heating systems through heat pumps. In 2020 around 240,000 heat pumps were operational (Department for Business, 2020, p. 10). In 2020 the installation rate was around 35,000 heat pumps per year, which the government aims to ramp up to 600,000 per year by 2028, on the way to the installation of 1.7 million per year by 2035 (Her Majesty's Government, 2021, p. 10). There currently seems little prospect that these targets will be met. Even with 2022 gas and electricity prices, and the current government grant of £5,000 per installation, there is inadequate financial incentive for householders uncommitted to energy efficiency or climate change mitigation to bear the cost of £12,000–17,000 for an air source heat pump[13] (a ground source heat pump can cost twice that[14]). (The Domestic Renewable Heat Incentive in Great Britain, which made a per kWh payment for heat from some heat pumps, closed at the end of March 2022[15]).

13 www.edfenergy.com/heating/electric/air-source-heat-pump?gclid=Cj0KCQjwuaiXBhCCARIs AKZLt3lRJqK9SDic2dG-gpB_FpOI4yAynKI711YvsR5Rs3UqDj7SFuoygkAaAqUsE ALw_wcB. Accessed May 5, 2023.
14 www.greenmatch.co.uk/blog/2014/08/the-running-costs-of-heat-pumps. Accessed May 5, 2023.
15 www.gov.uk/domestic-renewable-heat-incentive. Accessed May 5, 2023.

The slow take up of heat pumps in the UK is puzzling at first glance, given the fact that heat pumps are a well-established heating technology in other Western European countries. However, a comparison between Finland and the UK (Martiskainen et al., 2021) reveals several reasons why Finland, with 3 million homes, has installed over a million heat pumps since 2000, while the UK has installed a quarter of that. One major reason has been the availability in the UK of relatively cheap (until recently) fossil methane gas (this is a reason sometimes also given for the UK having much less district heating than some other European countries[16]). Other obstacles were the lack of consistent policy support, the lack of consumer understanding of the technology and problems experienced with its operation. Other problems identified in a trade magazine in 2021 include the unsuitability of heat pumps for use in energy-inefficient homes and the lack of qualified installers.[17]

The first three recommendations from the CCC were:

1 Overhaul the compliance and enforcement framework so that it is outcomes-based (focussing on performance of homes once built), places risk with those able to control it, and provides transparent information and a clear audit trail, with effective oversight and sanctions. Fund local authorities to enforce standards properly across the country.
2 Reform monitoring metrics and certification to reflect real world performance, rather than modelled data (e.g. SAP). Accurate performance testing and reporting must be made widespread, committing developers to the standards they advertise.
3 Review professional standards and skills across the building, heat and ventilation supply trades with a nationwide training programme to upskill the existing workforce, along with an increased focus on incentivising high 'as-built' performance. Ensure appropriate accreditation schemes are in place (Climate Change Committee, 2019a, p. 17).

Unfortunately these recommendations are as valid in 2023 as they were in 2019. Once they have been implemented, and householders and businesses can be reasonably assured that building energy efficiency upgrades and low-carbon heating equipment will deliver energy savings and low-carbon heat as specified, the next problem is how both the energy efficiency measures and heat pumps, which together in some cases could cost £30,000–40,000 per home, are to be funded.

The Green Finance Institute draws a clear distinction in this respect between owner-occupied homes, private rented homes, and social rented housing.

---

16 www.bioregional.com/news-and-opinion/why-cant-we-get-district-heating-right-in-the-uk. Accessed May 5, 2023.
17 www.building.co.uk/news/major-challenges-in-persuading-homeowners-to-install-heat-pumps-government-admits/5112066.article. Accessed May 5, 2023.

Table 9.4 summarises the financial and non-financial barriers facing the relevant actors in these sectors.

While these barriers can be removed, there is no magic bullet that will do so within one of these sectors, let alone across them all. Each solution needs to be designed for the particular context in which it will be implemented, although obviously learning can take place from similar projects. The Green Finance Institute publication describes a number of 'Demonstrator Solutions', which include:

- improved valuation of homes to take account of energy efficiency investments;
- metered evidence of savings from such investments;
- building renovation 'passports' that clearly show what energy improvements have been carried out and the potential for further measures;
- a TrustMark platform to give information about reliable technologies and supply chains, and routes to finance;
- Residential Retrofit Principles, and a certification scheme, for green buildings and retrofits to increase the confidence of lenders and borrowers;
- 'Pay As You Save' financing that provides up-front loans that are paid back over a long period as energy bills are reduced;
- Green Equity Release for older homeowners, to allow them to use some of the value of their homes to improve their energy efficiency;
- Green Equity Loans, or additions to mortgages, for homeowners against the value of their property, to be paid back as energy bills are reduced;
- Salary Sacrifice Scheme for employees, with repayments on energy efficiency loans repaid by employers out of gross income;
- 'Green Leases' to allow landlords to increase rents by some fraction of the predicted savings from an energy efficiency investment; and
- a range of other green financial instruments based on performance guarantees, such as ISAs, bonds and crowd funding.

None of these financial 'solutions' are risk- or problem-free, and all would need government policy support, regulation or legislation to ensure the protection and gain the confidence of owner-occupiers, tenants and landlords. Until such policy support is provided with long-term commitment, ambitious home energy efficiency building renovations are likely to remain the niche activity of enthusiasts.

In 2015 the then UK Government commissioned a review of Consumer Advice, Protection, Standards and Enforcement for home energy efficiency and renewable energy measures in the UK. At the heart of the Bonfield Review, published a year later (Bonfield, 2016), was a recommendation to establish a quality mark based on a Consumer Charter, and a Code of Conduct and Codes of Practice, to improve standards of performance in the sector and give householders and social housing providers the confidence to improve their homes by making them more energy-efficient and installing renewable energy

Table 9.4 Barriers to investment in energy efficiency in buildings for different tenure types

Type of household	Financial barrier	Non-financial barrier
Owner occupier	High up-front costs and lack of access to capital	Low awareness about the link between energy efficiency and climate change
	Lack of confidence in extent of bill savings	Lack of professional advice from small builders, plumbers etc.
	May move before the investment pays back	Lack of good quality information and support about products, suppliers etc.
	Investment may not be (fully) reflected in the property value	Duration, hassle and complexity of projects
	Lack of financing package, e.g. through a mortgage	Institutional disconnect, e.g. between leaseholder and freeholder
Private rented	Split incentive – landlords make the investment but tenants get the savings on their bills	Low awareness of Minimum Energy Efficiency Standard (MEES) legislation
	High up-front costs and lack of appropriate financial products to access capital	Duration, hassle and complexity of projects, plus tenancy issues and concern about empty periods
	Improvements may not be reflected in increased rents or property values	Lack of good quality information and support about products, suppliers etc. to meet MEES
	Freeholders cannot increase service charges to leaseholders for property improvements, and have no incentive to undertake them	Uncertainty about regulation, especially in respect of future MEES requirements
Social rented	Limited funds and access to capital	Supply chain constraints on delivering renovation at scale
	Bureaucratic approval processes and short-term planning horizons	Lack of capacity in project development and delivery expertise
	Higher interest rates for Housing Associations	Reluctant 'Right to Buy' private leaseholders on social housing estates may impede central management of projects and achievement of economies of scale

Source: Adapted from Tables 1, 3 and 5 in Green Finance Institute, 2020.

measures. The recommendation was not followed up but remains as valid as when it was issued. Without such a framework for consumer confidence, it seems unlikely that these crucial markets for the decarbonisation of the buildings sector will develop to the scale required by the UK's net-zero target.

Transport

In 2018 global CO_2 emissions from the transport sectors were around 8 $GtCO_2$, of which 74.5% came from road vehicles (45.1% passenger, 29.4% freight), 11.6% came from aviation and 10.6% came from shipping.[18] This rose by about 0.5 $GtCO_2$ in 2019, before falling to around 7 $GtCO_2$ in 2020 because of the Covid pandemic. But the IEA estimates that transport emissions rebounded in 2021, and that passenger travel will nearly double by 2050, with the global car fleet increasing from 1.2 billion in 2020 to 2 billion in 2050, and that freight in 2050 will be 2.5 times that in 2020 (International Energy Agency, 2021a, p. 131). Clearly all transport modes will have to use very different technologies if transport CO_2 emissions are to reduce to zero by 2050.

Figure 9.5a shows the trajectory of CO_2 emissions for different transport modes in the IEA Net Zero Emissions (NZE) scenario. The technologies for decarbonising light-duty vehicles (LDVs), namely batteries and to a lesser extent fuel cells, are available in markets today, so emissions from this mode start to fall steeply more or less immediately. Clearly a requirement for this to happen is a combination of a rapid roll-out of charging points, and a massive increase in the extraction of minerals for battery production on a hitherto unprecedented scale. The latter issue was discussed in the previous chapter.

Figure 9.5b shows the maturity of the technologies for the 'hard-to-abate' transport sectors of heavy trucks, shipping and aviation. Low-carbon options in shipping and aviation are well behind heavy trucks in terms of technological maturity. Market uptake for heavy trucks and shipping has now begun, even though the relevant technologies cannot yet be described as 'mature', but in aviation decarbonisation technology is still mainly at the prototype stage. Given the length of time it takes technologies to get established, and the long lives of both ships and aircraft, it will clearly be some years before new low-carbon technologies in these sectors will be widely established.

Heavy trucks, for which batteries struggle with the power and range requirements, rely more heavily on fuel cells, a less mature technology, so that decarbonisation of this mode starts later than for LDVs. Given the difficulties of decarbonising aviation, the NZE scenario assumes that government policy acts to halve the annual growth in demand for flights from 6% per year (the pre-pandemic rate, likely to resume when Covid concerns are past) to 3% per year through a shift to high-speed rail for short-haul flights. The NZE scenario projects that CO_2 emissions from aviation rise back to their pre-pandemic level

---

18 https://ourworldindata.org/co2-emissions-from-transport. Accessed May 5, 2023.

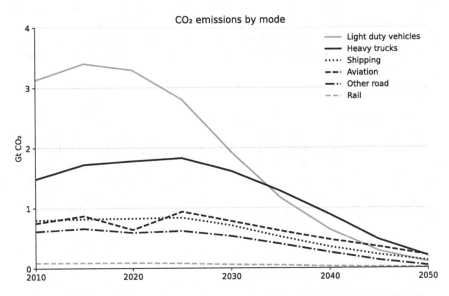

Figure 9.5a CO_2 emissions trajectory in transport sectors in the IEA's NZE scenario.

of around 1 $GtCO_2$ by 2025, and then start to fall slowly to 210 Mt in 2050, by when jet kerosene from oil only supplies a little over 20% of aviation fuel to an aircraft fleet that is significantly more efficient, with the other 80% coming from biofuel (biojet kerosene, 45%) and synthetic hydrogen-based fuels (30%). With the IEA assumptions about technology improvements and cost reduction, this is estimated to increase the price of a mid-haul (1,200 km) flight by only around US$10 per passenger (International Energy Agency, 2021a, pp. 135–136).

In shipping, CO_2 emissions declined from 880 $MtCO_2$ in 2019 to 830 $MtCO_2$ in 2020, largely due to the pandemic, and in the NZE scenario decline further from the mid-2020s through efficiency measures (slow steaming and wind-assisted ships) and in the longer term to 120 $MtCO_2$ by 2050, the residual emissions indicating the difficulty of producing alternatives to fossil fuels. In 2050 60% of fuel for shipping is ammonia and hydrogen, with biofuels providing nearly 20%. The 20 largest ports, which handle more than half the world's cargo, could become hubs for the production of hydrogen (perhaps from fossil methane gas with carbon capture and storage (CCS) and ammonia. The use of batteries in this sector is limited by considerations of power density, even with considerable projected improvements in battery technology, to journeys of less than 200 km (International Energy Agency, 2021a, pp. 136–137).

Figure 9.6a shows the fuels used by different transport modes in that scenario. Striking is that currently fossil fuels account for about 90% of transport fuels. By 2050 this will need to fall to 10% or less for net zero to be achieved,

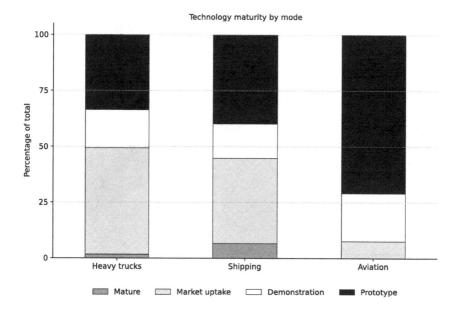

Figure 9.5b Technological maturity in hard-to-abate transport sectors.

Source: International Energy Agency, 2021a, Figure 3.21, p. 132.

with their place being taken by electricity, biofuels and hydrogen. For real zero, the use of fossil fuels in transport will have to be removed completely, as there is no way that their emissions can be captured and stored from vehicle tailpipes, aircraft or boats.

Another striking fact to emerge from Figure 9.6a is that, despite the projected growth in both passenger and freight transport, the energy use in a decarbonised transport system in 2050 is more than a quarter less than it was in 2020. This is because, as discussed in Chapter 3, battery electric vehicles (BEVs), which dominate sales of LDVs (cars and vans) and other road vehicles after 2030, are far more energy-efficient than the internal combustion engine (ICE) vehicles they are replacing, so that LDV emissions in 2050 are less than half those in 2020, despite an increase in LDV transport (Figure 9.6b).

As discussed in Chapter 6, it should be remembered in the context of biofuels that they will only deliver significant reductions in carbon emissions if they are produced in ways that emit considerably lower CO_2 per MJ of energy than the diesel or gasoline for which they are substituting. As was seen in Chapter 6, this is very strongly driven by the land use changes, direct and indirect, to which the production of the biofuels gives rise.

The European Union (EU) sought to take this issue into account by defining 'sustainability criteria' in its Renewable Energy Directive of 2009, which both

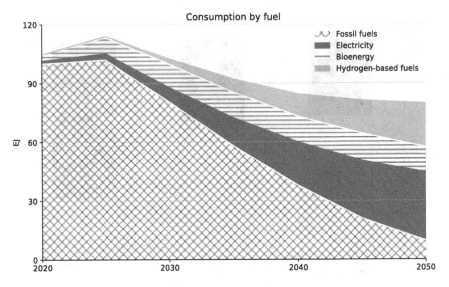

Figure 9.6a Fuel use in transport to 2050 in the IEA's NZE scenario.

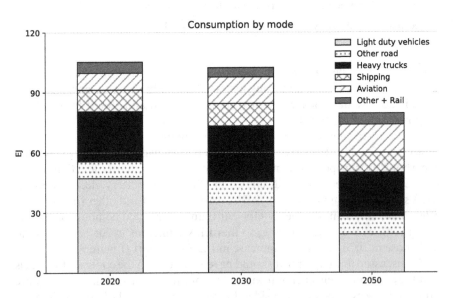

Figure 9.6b Consumption of fuel by different transport modes.
Note: LDVs = light-duty vehicles (cars and vans); Other road = two/three wheelers and buses.
Source: International Energy Agency, 2021a, Figure 3.22, p. 133.

set a target for a 10% share of biofuels in the EU consumption of transport fuels, to be reached by 2020, but also mandated that, to be counted in this target, the fuels by 2012 had to save at least 35% of the CO_2 emissions of the fossil transport fuels they replaced, to rise to 50% in 2017.[19] The Directive also stipulated that growing crops for biofuels should not result in deforestation or impinge on land of high biodiversity. While it did not seek to take account of indirect land use change in these targets, it expressed the intention to devise a methodology to do so in the future.

Njakou Djomo & Ceulemans (2012, p. 400), on the basis of the numbers cited in Chapter 6, felt that "none of the biofuel crops [cited in the paper] was likely to achieve the 35% (increasing to 50% in 2017) reduction required by the EU directive by 2012". However, in assessing the performance of EU Member States against this Directive, the European Environment Agency (EEA) found that the target of the 10% share in biofuels had in fact been met,[20] with Sweden and Finland having the lowest biofuel carbon intensity, of 15.4 gCO_2e/MJ and 12.1 gCO_2e/MJ respectively.[21] This is about a fifth of the CO_2 intensity of diesel and petrol (gasoline). The EEA also found that another target, that Member States between 2010 and 2020 should reduce the CO_2 intensity of their biofuels by 6%, had been missed by all but 11 Member States, with the average being 5.5%, excluding indirect land use change (iLUC) (including iLUC the reduction was only 3.3%).

The European Union revised the Renewable Energy Directive in 2018,[22] and increased to 14% in 2030 the share of transport fuel in the EU that should come from renewable sources. It mandated that to count against this target the biofuels should save 50–70% of the GHG emissions of the fossil fuel they were replacing, depending on when the biofuel plant was commissioned (50% if from plants operating before 2015, 70% if from plants starting operation after 2026) (Article 29.10). In addition, it listed a number of categories of land likely to be of high biodiversity which could not be used for biofuel production to comply with the Directive (Article 29.3–29.7), and it placed a declining limit on the volume of this renewable transport fuel that could be counted against this target if it came from sources deemed at high risk of indirect land use change. By 2030 this limit is zero (Article 26.2).

Another possibility for road freight transport is the electrification of roads, through overhead power lines, induction loops under the road surface or electric rails embedded in the road, all of which could both drive the vehicle directly or charge its battery. There are currently experiments with these systems in Sweden and Germany, and advocates argue that this is the most energy-efficient and cost-effective means of road freight decarbonisation, but at

---

19 Article 17.2 of the Directive, https://eur-lex.europa.eu/LexUriServ/LexUriServ.do?uri=OJ:L:2009:140:0016:0062:en:PDF. Accessed May 5, 2023.

20 www.eea.europa.eu/ims/use-of-renewable-energy-for. Accessed May 5, 2023.

21 www.eea.europa.eu/ims/greenhouse-gas-emission-intensity-of. Accessed May 5, 2023.

22 https://energy.ec.europa.eu/topics/renewable-energy/bioenergy/biofuels_en. Accessed May 5, 2023.

€2 million per km the investment needed to install them widely (e.g. on the motorway network) would be very large indeed.[23]

The emissions from rail transport, the most energy-efficient passenger transport mode, are currently low. In the NZE scenario the share of passenger rail transport in total transport in 2050 doubles to 20%, but its emissions fall almost to zero as the 55% of rail transport that currently uses diesel from oil is substituted by electricity (90% of energy demand in 2050) and, to a much lesser extent, by hydrogen (5%) for those lines which are not used enough for electrification to be economic (International Energy Agency, 2021a, p. 137).

For LDVs McKinsey has identified and compared four zero-emission vehicle (ZEV) technologies (biofuel and synthetic fuels, internal combustion engines, fuel cells using hydrogen, and battery electric vehicles) across a number of dimensions. It can be seen from Table 9.5 that no vehicle type is best across all the criteria listed. The H_2 fuel cell (H_2FC) and BEVs are zero emissions at the tailpipe, and overall if the H_2 is produced, or the battery is charged, using renewable electricity. But both are currently more expensive than bio/synfuel and H_2-ICEs, and both (and H_2-ICEs) also need extensive refuelling or charging infrastructure. In those countries with fast-growing BEV markets, the charging infrastructure issue is being addressed – as of July 2022 the UK had 42,000 charging points in over 15,500 locations (more than petrol stations),[24] but there are still issues of long charging times and non-standard charging plugs to be addressed.

Other disadvantages are, for BEVs, the weight and limited range of batteries, and the charging time, and for H_2FCs (and H_2-ICEs) the required size of the hydrogen fuel tank. But BEVs are easily the most energy-efficient technology 'from well to wheel' (i.e. taking all the energy losses from energy generation, conversion and transmission/distribution into account). Bio/synfuels are very inefficient, and therefore relatively costly in terms of fuel, but can use existing internal combustion engines and refuelling infrastructure. How these relative advantages and disadvantages of the different options play out in the future will depend on how their costs come down with innovation. The IEA's NZE scenario suggests that BEVs will dominate the car and light-van markets, H_2FCs will be widely used for heavier road transport (buses and trucks), and the bio/synfuels will be used for aviation and shipping. But technology pathways are not predetermined, and breakthroughs or bottlenecks could mean that the future actually turns out very differently.

The brief summary of projected transport developments to 2050 from the IEA's NZE scenario perhaps does not convey the profound change in transport patterns and behaviours that these technological developments entail. For example, global car ownership in the NZE scenario increases from 0.15

---

23 www.cleanenergywire.org/factsheets/electric-highways-offer-most-efficient-path-decarbonise-trucks. Accessed May 5, 2023.
24 www.edfenergy.com/electric-cars/charging-points. Accessed May 5, 2023.

Table 9.5 Comparison between different zero-emission vehicle technologies

	Bio/synfuel	Hydrogen internal combustion engines (H_2-ICE)	Hydrogen (H_2) fuel cell	Battery electric
Emissions				
CO_2 intensity	CO_2 intensity depends on source of biomass/ carbon	Zero/minimal CO_2 if using green/blue H_2	Zero/minimal CO_2 if using green/blue H_2	CO_2 intensity depends on grid mix; zero CO_2 if using renewable power
Air quality	NO_x[1] and particulate-matter emissions similar to diesel	No significant NO_x emissions with SCR[2] aftertreatment	Zero emissions	Zero emissions
Cost and performance issues				
Efficiency (well-to-wheel)	−20%	−30% for renewable H_2 production	−35% for renewable H_2 production	75–85%+ depending on transmission and charging losses
Powertrain capital expenditure	Same as today's combustion engines	H_2 engines with similar capex as diesel ICE, but H_2 tank required	High capex for fuel cells and batteries, but more scalable than BEV[3]	High capex if large batteries required (medium for smaller/lighter segments)
Constraints (space/ payload)	Same size and weight as today's combustion engines	Engine with same size as today, but H_2 tank needed	More space needed than combustion engine for fuel cell and H_2 tank	Higher weight than combustion engine; payload constraints subject to use case
Uptime/ refuelling	<15 minutes, depending on tank size	<15–30 minutes, depending on tank size	<15–30 minutes, depending on tank size	3+ hours, depending on ability for fast charging
Infrastructure costs	Can use existing infrastructure	H_2 distribution and refuelling infrastructure as required	H_2 distribution and refuelling infrastructure as required	Charging infrastructure and grid upgrades required

Source: McKinsey, www.mckinsey.com/industries/automotive-and-assembly/our-insights/how-hydrogen-combustion-engines-can-contribute-to-zero-emissions. Accessed May 5, 2023.

Notes:
[1] Nitrogen oxides.
[2] Selective catalytic reduction.
[3] Battery electric vehicle.
'Capex' is the capital expenditure needed to build or buy a vehicle, as opposed to 'opex' which is the expenditure needed to operate (refuel, charge and maintain) it.

cars per person in 2020 to only 0.2 cars per person in 2050, despite the global economy having doubled in size.

The relationship between income and car ownership is complex. The take up of technologies over time is often modelled as an S-curve, with slow take up by early adopters, then an acceleration as the technology becomes mainstream, and then a slowdown and flattening out as the technology approaches saturation. Dargay et al. (2007, Figure 10, p. 6) shows the relationship between income and vehicles (cars, vans, buses, trucks, i.e. all 4-wheelers) per 1,000 people over 1960–2002 in a number of countries (India, China, South Korea, Japan), and then projects this to 2030. The relationship seems to follow a shallow S-shaped curve ('Gompertz function') that starts at low levels with low incomes, but then rises quicky for an income of between US$3,000 and US$20,000 (1995 dollars, purchasing power parity [PPP]), but then saturates before 1,000 vehicles per 1,000 people.

Of the countries mentioned by Dargay et al. (2007, Figure 10, p. 6) their vehicles per 1,000 people in 2021/2022 were as follows: USA 868, Japan 624, South Korea 485, China 219, India 55.[25] If saturation of 1,000 vehicles per 1,000 people comes about in 2050, then a population of 9 billion people, on current patterns of car ownership, would own 9 billion vehicles by that date (compared to about 1.4 billion today). If most of those vehicles were electric, the new mineral requirements would be far higher than those projected in Chapter 7, and almost certainly unachievable.

The IEA only gives its estimate of the number of cars per person in 2050 (0.2, or 200 per 1,000 people), rather than the number of vehicles (in the UK around 80% of vehicles are cars[26]), but still the assumptions about cars per person seem low in light of the trends in Dargay et al. (2007, Figure 10, p. 6). Achieving them may well need widespread changes in behaviour and urban design. In respect of the former the IEA notes a possible "shift to cycling, walking, ridesharing or taking buses for trips in cities that would otherwise be made by car", which would

> represent a break in familiar or habitual ways of life and as such would require a degree of public acceptance and even enthusiasm, ... [and] would also require new infrastructure, such as cycle lanes ..., clear policy support and high quality urban planning.
>
> (International Energy Agency, 2021a, p. 67)

It seems that such a shift will be absolutely required if the demand for metals and minerals is not to become a serious bottleneck to the clean energy transition.

A review of city-level interventions that are effective in reducing both car ownership and car use in European cities found that such policies need a

---

25 https://en.wikipedia.org/wiki/List_of_countries_by_vehicles_per_capita#cite_note-:6-5. Accessed May 5, 2023.

26 www.racfoundation.org/motoring-faqs/mobility#a1. Accessed May 5, 2023.

combination of both sticks (measures that make car use in cities more difficult and/or more expensive) and carrots (measures that actively encourage a shift in travel mode). Among the most effective sticks were congestion and parking charges, and traffic and parking restrictions in city centres and residential streets, while carrots included infrastructure to make walking and cycling safer, easier and more pleasant, improved public transport, car-sharing schemes and organisational travel planning (Kuss & Nicholas, 2022). An often-cited poster child for 'the sustainable city' is the German city of Freiburg, which has been describing itself as a 'green city' since 2008, and in which the main means of mobility for one third of its citizens is cycling, and the proportion of car drivers is now only 21%.[27] Many other European cities are now moving in this direction, with pedestrianisation of city centres, an expansion of cycling infrastructure, and restrictions on car use such as, in the UK, the creation of Low Traffic Neighbourhoods (LTNs). These have been introduced in many London Boroughs and make walking and cycling safer by restricting car use in residential or busy shopping streets.[28] Like all such changes they can be controversial and receive political push-back from drivers. However, such trends of establishing 'Living Streets'[29] that are less dominated by motor vehicles now seem well established, in Europe at least.

In lower- and low-income countries the overwhelming priority is to channel the current phase of rapid urbanisation into the creation of compact cities with a good mix of local amenities, where travel needs can relatively easily be served by walking, cycling and public transport, rather than sprawling conurbations in which car use is a practical necessity. The celebrated comparison between Atlanta (USA) and Barcelona (Spain) with very similar populations in 1990 (2.5 and 2.8 million respectively) showed that the built-up area of compact Barcelona occupied 262 km^2, while that of the sprawling Atlanta spread over 4,280 km^2 (Bertaud & Richardson, 2004).

Industry

The industrial sector in 2019 produced GHG emissions from fuel combustion (7.1 $GtCO_2e$ direct and about 5.9 $GtCO_2$ indirect from electricity and heat generation), industrial processes (4.5 $GtCO_2e$) and the use of products (0.2 $GtCO_2e$), as well as from waste (2.3 $GtCO_2e$). In total, therefore, industrial direct GHG emissions in 2019 amounted to 14.1 $GtCO_2e$, or 20 $GtCO_2e$ including indirect emissions. This makes industry (24% direct emissions) second to the energy sector in direct GHG emissions, and the largest emitter (34% in 2019) after indirect emissions are allocated (IPCC, 2022).

27 www.dw.com/en/the-green-city-of-freiburg-is-this-germanys-future/a-60438622. Accessed May 5, 2023.
28 https://madeby.tfl.gov.uk/2020/12/15/low-traffic-neighbourhoods/. Accessed May 5, 2023.
29 www.livingstreets.org.uk/about-us/our-work-in-action/walkable-london-2022. Accessed May 5, 2023.

Table 9.6 Global combustion and process CO_2 emissions from industry in 2016, in absolute terms and as a percentage of total global emissions

	Million tonnes CO_2	*Percentage (%)*
Other industry	3473	9.7
Iron and steel	2977	8.3
Cement and lime	2297	6.4
Non-ammonia chemicals	1419	4.0
Ammonia for fertilisers	420	1.2
Non-ferrous metals	296	0.8
Pulp and paper	282	0.8

Source: Bataille, 2020, Figure 2, p. 2.

In terms of CO_2 only, industry in 2016 emitted 11.2 $GtCO_2$, or about 31% of global CO_2 emissions, through direct combustion of fossil fuels or process emissions (i.e. not including emissions from the generation of the electricity it uses) (Bataille, 2020). Table 9.6 shows the main CO_2-emitting sectors. The process emissions come mainly from the cement industry (from the decomposition of limestone to lime), but also from iron and steel and chemicals.

Of these industrial emissions around 70% came from just three 'heavy industry' sectors – steel, chemicals and cement – and the great majority of production (70–90%) of these goods comes from emerging market and developing economies (with China alone producing 60% of the world's steel and cement in 2020). The IEA's Net Zero Emissions scenario sees a dramatic reduction in CO_2 emissions in these sectors, from 8.4 $GtCO_2$ in 2020 to less than 500 $MtCO_2$ in 2050, although cement production remains at about its current level, while steel demand increases by 12% and primary chemicals by 30%, the latter mainly in emerging market and developing economies (International Energy Agency, 2021a, Figure 3.15, p. 121). This emission reduction comes about through a wholesale change in heat and process technologies.

There are various pathways through which this almost total decarbonisation of these three sectors, and other energy-intensive industries (e.g. non-ferrous metals, pulp and paper, ceramics) might be achieved (Bataille et al., 2018, Figure 1, p. 963). The first priority is to make better use of materials, recycling more and wasting less of them, sometimes called moving towards a 'circular economy', as discussed in Chapter 8. Then there is the decision as to whether to keep or change existing processes. Decarbonising existing processes will entail either using carbon capture, utilisation and storage (CCUS) technologies (see Chapter 7) or moving to an alternative, zero-carbon, heat source. Changing the existing processes will also entail using different fuels. An example of a changed process would be to move from producing steel in blast furnaces with metallurgical coal to the direct hydrogen reduction of iron, as discussed further below.

In its NZE scenario, the IEA modelled the uptake of such measures to achieve 95% decarbonisation in heavy industry at least cost. With no interventions CO_2 emissions from heavy industry are projected to increase by 39%, because of increased output (activity), but with a range of measures can be reduced by 95% in 2050. CCUS makes the greatest contribution (around 40%), followed by material efficiency (around 20%). Other contributions, in order of their scale, are electrification, hydrogen, bioenergy and other fuel shifts, and energy efficiency (International Energy Agency, 2021a, Figure 3.16, p. 123). However, a relatively small proportion of these emissions reductions are delivered by mature technologies that are already being deployed at scale. The great majority are still at the stages of initial market uptake, demonstration or prototype. This indicates the scale of technology development and deployment that is still required for the NZE projections to be realised.

In contrast to the IEA NZE projection, Madeddu et al. (2020) assessed to what extent the industrial sector in the EU plus UK could be decarbonised exclusively through the direct use of electricity. Their modelling included sectors which are responsible for 92% of the CO_2 emissions from EU+UK industry, which itself accounts for 30% of European CO_2 emissions. Fuel combustion comprises 70% of this industry's final energy use (excluding chemical feedstocks), to provide heat at different temperatures, with the other 30% coming from electricity, mainly used for cooling and mechanical power (e.g. motors, compressors). The study electrified these sectors in three stages: Stage 1 largely installed widely available electric boilers and heat pumps for low- and medium-temperature applications across a wide range of sectors. Stage 2 involved the deployment of more technologically advanced, but still available, technologies specific to particular products and processes. Stage 3, in contrast, involved the use of technologies of lower maturity and therefore of higher uncertainty, in order to achieve maximum possible electrification (Madeddu et al., 2020, Figure 2B, p. 6).

The modelling suggested that Stage 1 enabled 23% of the sectors' useful industrial energy to be added to the 19% of useful energy already supplied by electricity. Stage 2 added a further 8%, such that the sectors ceramics and glass, paper, wood, food, textiles, machinery, and transport equipment were completely electrified, while non-ferrous metals and secondary steel were 97% and 98% electrified respectively. After Stage 3, the useful energy of cement and steel also becomes almost completely decarbonised (cement still has substantial emissions from the cement-making process that are not affected by electrification) and 60% of total useful energy is supplied by electricity, with the great majority of the remaining 40% consisting of energy feedstocks for the chemical industry. The full Stage 3 electrification process results in a 20% fall in final energy use (final energy is useful energy plus losses in the appliance or process using it), indicating the generally greater energy efficiency of electrical generation of heat over combustion of fossil fuels.

The extent of decarbonisation achieved by this electrification process depends crucially, of course, on the carbon intensity of the electricity that is being used. At the 2015 level of EU electricity carbon intensity, around 300 gCO_2/kWh, the decarbonisation even at Stage 3 would be relatively modest, and Stages 1 and 2 would result in an emissions *increase* of around 10%. But the EU and UK both have ambitious plans for the almost total decarbonisation of their power systems well before 2050. Madeddu et al. (2020, Figure 5, p. 10) show that emissions from the modelled sectors would be reduced by 43% from Stage 1, 54% from Stages 1 and 2, and 78% from all three stages, if the CO_2 intensity of European electricity decreased to 12 gCO_2/kWh, which is more or less what is planned.

The differences between the industrial decarbonisation outcomes of the IEA's NZE scenario and that of Madeddu et al. (2020) are instructive. A large part of the hydrogen and bioenergy in the IEA scenario, which achieved much bigger cuts in CO_2 emissions, is needed to decarbonise the feedstocks that remain after the electrification scenario. In addition, the CCUS in the IEA scenario (used in relation to the production of hydrogen and bioenergy) is projected to probably be cheaper than the Stage 3 electrification options (although the electrification study did not estimate costs).

Further insights on the cost-related balance between electrification and CCUS in these heavy industry sectors are shown in Figure 9.7, which shows the electricity price at which different decarbonisation options become competitive, in three sectors: cement, steel and ethylene (where ethylene is one of the major products of the very diverse chemicals industry [see below]). The relevant electricity price differs according to whether the industry is on a greenfield or brownfield site. The latter are cheaper for CCS because of the relative ease of putting in the necessary infrastructure. Thus, on brownfield, CCS is cheaper than electricity at electricity prices above around US$25/MWh, and at slightly less than twice that on greenfield sites. For steel, CCS is cheaper than electricity-based hydrogen at electricity prices above around US$20/MWh (brownfield) and US$40/MWh (greenfield). For ethylene, electricity prices have to be lower still before electricity can compete with CCS.

All the above routes to decarbonisation (Stage 3 electrification, CCS and electricity with hydrogen) rely heavily on relatively immature technologies. The above modelling exercise therefore emphasises the importance of technological development to achieve deep decarbonisation. This will only come about through a relentless focus on innovation and investment in deployment of new technologies, as discussed in Chapter 11. To minimise the costs of this, and the 'stranded' energy-using assets (plant and infrastructure that cannot be used because of the CO_2 emissions that they generate), the investment in new low-carbon technologies will have to be carefully scheduled to substitute for upgrades and retrofits to high-carbon technologies.

The lifetimes of energy-using plant and infrastructure (assets) are typically around 40 years. This means that investments in high-carbon assets made before and up to 2020 will typically last until 2060 and emit CO_2 until 2060, along the path of the outer curve in Figure 9.8. This is clearly incompatible with the need for almost total decarbonisation by 2050. However, these

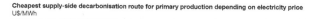

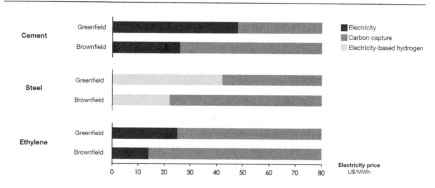

Figure 9.7 Use of electricity, hydrogen or CCUS in the decarbonisation of three heavy industry products.

Source: Energy Transitions Commission, 2018, Exhibit 2, p. 17.

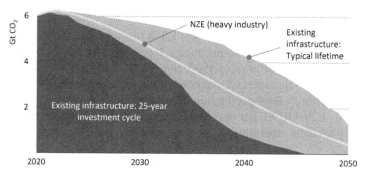

Figure 9.8 The need for timely investment in low-carbon infrastructure.

Source: International Energy Agency, 2021a, Figure 3.17, p. 124.

assets typically require substantial upgrading or retrofitting after 25 years: an example is blast furnaces for steel-making, which need relining after this time. If at this point the 25-year-old high-carbon plant is decommissioned, and the investment put instead into low-carbon plant (e.g. an electric arc furnace, or even direct reduction by hydrogen, for steel-making, as discussed in Vogl et al., 2021), then emissions could be brought down along the NZE (heavy industry) line in Figure 9.8, which is the pathway in the IEA's NZE scenario. In this way the use of existing high-carbon infrastructure consistent with the NZE can be maximised, and the wasted ('stranded') investment in high-carbon plant and infrastructure minimised.

Table 9.7 Main energy inputs to steel production and their applications

Energy input	Application as energy	Application as energy and reducing agent
Coal	Blast furnace (BF), sinter and coking plant	Coke production, BF pulverised coal injection
Electricity	EAF, rolling mills and motors	-
Natural gas	Furnaces, power generators	BF injection, DRI production
Oil	Steam production	BF injection

Source: World Steel Association fact sheet, 2021, https://worldsteel.org/wp-content/uploads/Fact-sheet-Energy-use-in-the-steel-industry.pdf. Accessed May 5, 2023.

Note: EAF stands for electric arc furnace, DRI stands for direct reduced iron.

This section ends with a brief overview of the three main CO_2-emitting industrial sectors, and those that are most difficult to decarbonise (steel, chemicals and cement), with the options for decarbonisation beyond electrification.

Steel

Global crude steel production in 2020 was 1,860 Mt, up from about 1,400 Mt in 2010.[30] The energy required to produce a tonne of steel (its energy intensity) has fallen by about 60% since 1960, but it has remained fairly constant since around 2005, so that steel CO_2 emissions since then have increased along with steel production.[31] Steel production currently emits around 1.85 tCO_2 per tonne of steel, or 2.6 GtCO_2 per year (excluding the emissions from generating the electricity used by steel production), amounting to 7–9% of global CO_2 emissions.[32]

Table 9.7 shows the energy inputs to steel production and their application. The main production routes today are from iron ore using a blast furnace-basic oxygen furnace (BF-BOF, about 75% of production), or from steel scrap using an electric arc furnace (EAF, about 25% of production); 89% of a BF-BOF's energy input comes from coal, 7% from electricity, 3% from fossil methane gas and 1% from other gases and sources. For EAFs, the energy input from coal accounts for 11%, from electricity 50%, from fossil methane gas 38% and 1% from other sources.[33] BF-BOF steel uses 2.8 times as much energy, and 11 times

30 www.iea.org/reports/iron-and-steel. Accessed May 5, 2023.
31 https://worldsteel.org/wp-content/uploads/Fact-sheet-Energy-use-in-the-steel-industry.pdf. Accessed May 5, 2023.
32 https://worldsteel.org/publications/policy-papers/climate-change-policy-paper/. Accessed May 5, 2023.
33 https://worldsteel.org/publications/policy-papers/climate-change-policy-paper/. Accessed May 5, 2023.

as much material, per tonne of steel produced than EAF (Li et al., 2018). This means that EAF steel tends to be far less carbon-intensive that BF-BOF steel. The amount depends on the carbon intensity of EAF electricity, but one estimate puts the weighted average of the carbon intensity of BF-BOF steel at 2.88 times that of EAF steel (Madias, 2016).

In addition to providing energy, the coal in the BF-BOF route plays a crucial non-energy role as a reducing agent, turning the iron oxide in the iron ore into iron. In direct iron reduction (DRI) this reducing role can be played by fossil methane gas (which emits less CO_2) or hydrogen (which emits no CO_2).

In its roadmap for steel decarbonisation (International Energy Agency, 2020), the IEA identifies four main routes:

- Increasing material efficiency (using less steel to get the same results) and energy efficiency (so that all steel companies get closer to the performance of the most energy-efficient).
- Producing more steel from scrap steel in EAFs (the electrification route for steel-making, though this is limited by the availability of scrap steel) or by fuel-switching from coal to DRI based on fossil methane gas (which emits less CO_2 than coal, but is far from carbon-neutral).
- Using CCUS to prevent the CO_2 reaching the atmosphere (this and material efficiency are major elements in the IEA's NZE scenario, as noted above).
- Using low-carbon (green or blue) hydrogen to provide the energy and reducing agent for primary steel production (H-DRI).

There are a number of innovative low-carbon steel-making projects identified by the IEA,[34] of which one of the most promising is the H-DRI Hybrit project in Sweden,[35] which is aiming to reach full-scale industrial production in 2026. Both the BF-BOF and H-DRI processes start with iron ore concentrate, converted to iron ore pellets (by non-fossil fuels in the Hybrit case). Hybrit then uses hydrogen, generated from the electrolysis of water using renewable energy, to provide the energy and reducing agent to produce a form of iron called 'sponge iron' which, with the addition of some scrap steel, can be used to produce steel. This process has zero CO_2 emissions. The car and truck maker Volvo announced that it would start using the first steel from this process in its trucks from 2022.[36]

34 www.iea.org/reports/iron-and-steel. Accessed May 5, 2023.
35 www.hybritdevelopment.se/en/. Accessed May 5, 2023.
36 www.volvotrucks.com/en-en/news-stories/press-releases/2022/may/volvo-trucks-first-in-the-world-to-use-fossil-free-steel-in-its-trucks.html. Accessed May 5, 2023.

Chemicals

Today's chemical industry produces a huge range of products that have become essential for modern life. In their world organic chemical industry mass balance for 2017, Saygin & Gielen (2021, Figure 1, p. 2) found that the total products of this kind in use had a mass of 2.933 Gt, with an addition to the stock of 176 Mt per year. The quantity of hydrocarbons used for chemical feedstock (mainly fossil methane gas, but a considerable use of coal in China) is considerably larger than the hydrocarbon inputs for energy into the chemicals sector. In mass terms plastics dominate the output mix, with 2017 production of 390 Mt, but the industry also produced in 2020 175 Mt of ammonia (NH_3) and in 2019 more than 98 Mt of methanol, as an intermediate for the production of other organic chemicals (Saygin & Gielen, 2021). Most ammonia is used to manufacture fertiliser.

The CO_2 emissions from the chemical sector shown in Table 9.6 are only from the energy and process inputs to chemical production, but the sector is also associated with GHG (nitrous oxide, N_2O) emissions from the use of ammonia-based fertilisers and from the incineration of waste plastics. The former are attributed to agriculture and both this and waste are discussed in Chapter 10.

In value terms the global chemical sector has grown by about 140% in the last 20 years, with easily the fastest growth in China, which in 2020 had nearly 45% of the world market.[37] While market growth has slowed, the sector is still projected to grow by 3.3% per year through to 2024.[38] Options for emission reduction from the sector are to move away from coal in China, and elsewhere substitute fossil methane gas with electricity, biomass or low-carbon hydrogen. Where carbon storage is available, CCUS is another option.

One high-level roadmap for the electrification of the chemical industry through to 2050[39] shows various ways electricity can be used to decarbonise the chemicals sector. Power-to-heat uses electricity instead of fossil fuels to provide heat, and Power-to-gas provides hydrogen in order to synthesise some basic chemicals (e.g. ammonia, methanol). There is also a direct Power-to-chemicals route involving hydrogen, but this is too technical for discussion here – see Jiang et al. (2019) for further information.

Cement

Of the sectors discussed here, this is perhaps the most challenging to decarbonise, because of the large quality of process CO_2 emissions involved in

37 https://cefic.org/a-pillar-of-the-european-economy/facts-and-figures-of-the-european-chemical-industry/profile/. Accessed May 5, 2023.
38 https://report.basf.com/2021/en/_assets/downloads/frc-forecast-basf-ar21.pdf. Accessed May 5, 2023.
39 www.tno.nl/media/5813/electrification_of_chemical_industry.pdf. Accessed May 5, 2023.

decomposing limestone to lime. These could be captured and used or stored underground, if the use opportunities (e.g. in the chemicals sector, with hydrogen) and storage were available. 'Breakthrough innovations' identified by the Energy Transitions Commission include the development of high-temperature electric kilns and furnaces, and different, low-carbon cement and concrete chemistries (Energy Transitions Commission, 2018, p. 33). Bio-based building materials (e.g. cross-laminated timber[40]) could also be developed or more widely used, depending on the availability of the resource, but like all biomass this is likely to be limited.

But one of the most effective means of decarbonising the cement, and other energy-intensive industrial sectors, is to make more efficient use of the materials that are produced. This is the materials efficiency option that is shown to make a considerable contribution to decarbonisation in the IEA's NZE scenario (as discussed above). The Energy Transitions Commission has estimated that achieving greater materials efficiency through a systematic move to a more circular economy would involve:

> making better use of existing stocks of materials through greater and better recycling and reuse and (ii) reducing the materials requirements in key value chains (e.g. transport, buildings, consumer goods, etc.) through improved product design, longer product lifetime, and new service-based and sharing business models (e.g. car sharing).
>
> (Energy Transitions Commission, 2018, p. 16)

The Commission's modelling of these measures in a 'circular economy scenario' suggested that they: could reduce CO_2 emissions from four key industry sectors (steel, cement, plastic and aluminium) by 40%: emissions from aluminium would fall from 1.3 to 0.8 $GtCO_2$, from plastics from 2.2 to 0.9 $GtCO_2$, from steel from 2.8 to 1.9 $GtCO_2$ and from cement from 2.9 to 2.0 $GtCO_2$ (Energy Transitions Commission, 2018, Exhibit 1, p. 16). Improved building design could reduce emissions from the cement sector by 34%.

Because the heavy industries that have been the main focus of this section are heavy users of energy, and therefore have a strong financial incentive to use energy as efficiently as possible, the opportunities for further energy efficiency in these sectors are limited, though they are still significant in many lower- and low-income countries. However, this is not true with regard to other industrial sectors, in which energy efficiency still offers considerable potential to reduce energy use, and therefore lower the cost of the low-carbon transition by requiring less low-carbon energy to be provided.

---

40 www.vox.com/energy-and-environment/2020/1/15/21058051/climate-change-building-materi als-mass-timber-cross-laminated-clt. Accessed May 5, 2023.

Figure 9.9 shows a range of policies that can be implemented by businesses or governments to increase energy efficiency across the board. The policies can be applied to specific pieces of equipment, to whole industrial processes, to particular sites or factories, to whole companies, to specific industries or sectors or across the economy as a whole. The possible types of policies include regulations on technologies, targets or management systems; voluntary actions by businesses on targets or management systems; economic instruments, usually implemented by government; information programmes; and government action on energy efficiency in its own buildings, on providing training and in considering energy efficiency in its procurement of goods and services.

A particularly common approach in business is *benchmarking*, a process whereby businesses or regulation identify best performance or best practices in relation to technologies or companies, and then regulate or adopt voluntary commitments to reach these standards over a period of time.

An obvious issue that arises with voluntary actions by companies is that most businesses operate in a competitive market context, which places clear constraints on their abilities to adopt measures that involve costs but do not contribute to profits. A similar issue arises with a government's abilities to impose policies like carbon taxes on highly traded sectors, like chemicals or steel, that raise the costs of businesses in their countries, but not those in other countries. This runs the clear risk of damaging the competitiveness of businesses at home and increasing imports from businesses abroad, thereby also increasing carbon emissions from those businesses abroad. Given that carbon emissions are equally damaging wherever they are emitted in the world, this 'carbon leakage' to other countries undermines the whole purpose of the carbon tax. Where the industry from which imports have been increased is less efficient than the one at home that has been adversely affected by the carbon tax, the policy may be counter-productive in that it might actually increase emissions globally. It is to prevent such perverse outcomes, while still being able to use instruments like carbon taxes to put a price on carbon emissions, that the EU has agreed, and the UK is proposing to consult on, a Carbon Border Adjustment Mechanism (CBAM), to ensure that imports into these areas face similar carbon prices to those within it.[41]

F-gases

F-gases (fluorinated gases) are a group of gases that contain the element fluorine (F). Some of them are powerful GHGs, with a global warming potential thousands of times that of CO_2. The F-gases controlled under the UN Framework Convention on Climate Change are hydrofluorocarbons (HFCs)

---

41 www.consilium.europa.eu/en/press/press-releases/2022/03/15/carbon-border-adjustment-mechanism-cbam-council-agrees-its-negotiating-mandate/. Accessed May 5, 2023. www.icaew.com/insights/tax-news/2022/jul-2022/uk-government-commits-to-consultation-on-carbon-border-adjustment-mechanism. Accessed May 5, 2023.

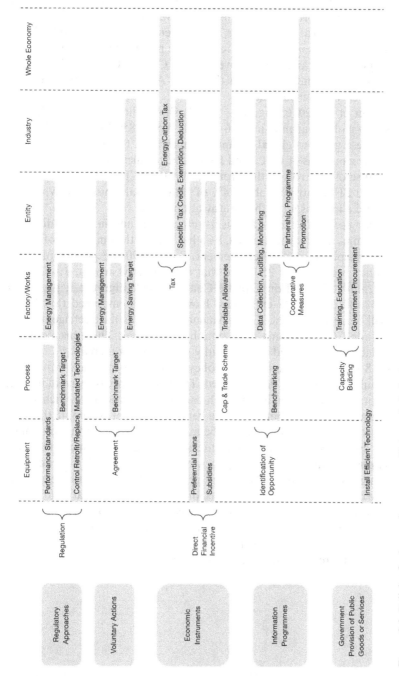

Figure 9.9 Policies for increasing energy efficiency in industry.

Source: IPCC, 2014, Figure 10.15, p. 781.

(around 90% of emissions), perfluorocarbons (PFCs), sulphur hexafluoride (SF_6), and nitrogen trifluoride (NF_3). They also include chlorofluorocarbons (CFCs) and hydrochlorofluorocarbons (HCFCs), but these two gases deplete the ozone layer as well as being GHGs, and their emissions are controlled by the Montreal Protocol; their emissions are therefore not included in the emissions reported to the UN Framework Convention on Climate Change (UNFCCC). HFCs are now also controlled under the Kigali Amendment to the Montreal Protocol, which was signed in 2016.

De Richter et al. (2016, Figure 1, p. 451) shows the relative contribution of F-gases to global warming in 2016 as 12%. About 85% of the total contribution of these gases to global warming comes from the Montreal Protocol gases, but the emissions of CFCs and HCFCs have fallen by 98% since the late 1980s, because of controls in the Montreal Protocol. The other halocarbons plus NF_3 and SF_6 contribute about 4% of total global emissions. While this doesn't sound a lot, it is more than the aviation sector.

The major use of all these gases is as refrigerants and in air conditioners and heat pumps, and as the use of CFCs has fallen, the emissions of HFCs have risen sharply in partial replacement. Since 1990 the growth of F-gas emissions has been around 250%, so that they are now around 1.4 $GtCO_2e$ (Intergovernmental Panel on Climate Change, 2022, Figure 2.3, pp. 2–14), about 2.4% of total GHG emissions. However, it is envisaged that future growth will be curbed by the Kigali Amendment. Substitute refrigerants for HFCs include CO_2, ammonia and some hydrocarbons. These are commercially available and reported to be 'rapidly gaining market share'.[42]

Table 9.8 lists the minimum set of policy priorities if industry, transportation and buildings are to be decarbonised. The first priority is to avoid making matters worse through 'locking-in' high-carbon infrastructure, which would become 'stranded' (i.e. become unusable before the end of its life) if the Paris Agreement temperature targets are to be met. This means not building high-carbon or carbon-intensive infrastructure across electricity supply (e.g. new fossil fuel power plants), industry (e.g. new or reconditioned blast furnaces for steel-making) and buildings. The second priority is to decarbonise electricity supply, one of the most important means for which is the removal of fossil fuel subsidies, which will of course incentivise decarbonisation of other sectors as well. Thirdly there are the broad 'system' actions that need to be taken across each of the sectors, that will then put national policymakers in a good position to design more detailed policies according to their national contexts and priorities.

---

42 https://eia-international.org/wp-content/uploads/EIA-Kigali-Amendment-to-the-Montreal-Protocol-FINAL.pdf. Accessed May 5, 2023.

Table 9.8 Policy priorities for electricity supply, industry, transportation and buildings

Avoid lock-in	Electricity supply	Industry	Transportation	Buildings
Prepare to enable transformation to zero emissions	Avoid fossil fuel subsidies	Avoid CO_2-intensive industrial infrastructure	Avoid false dichotomies Avoid delay	Avoid inefficient buildings
	Avoid new fossil fuel infrastructure		Avoid unsustainable solutions	Avoid new gas connections for buildings
Sustain deep reductions	Plan a just transition	Develop zero-emission steel	Expand electric vehicles	Minimise embodied emissions
	Prepare electricity system for high shares of renewables	Develop zero-emissions cement	Shift modes in transport	Build only zero-emissions buildings
	Expand renewables	Electrify industry	Reduce transport demand	Increase retrofitting rate

Source: United Nations Environment Programme, 2022, Figure 5.1, p. 39.

Conclusion

The decarbonisation of the energy used in buildings, transport and industry must be at the heart of any strategy for 'real zero'.

In order for buildings to emit no GHGs, they will need to be heated and cooled, and other building services like lighting and hot water provided, predominantly by zero-carbon electricity. In order to limit the quantity of electricity that will need to be produced to achieve this, buildings in many countries, including the UK, will need to be made much more energy-efficient. The dominant zero-carbon heating technology in a country like the UK is likely to be heat pumps, which can produce two to three times as much heat as the electricity they use, but for them to heat effectively they need to operate in relatively energy-efficient houses and other buildings. The conclusion is inescapable that building decarbonisation in the UK and similar countries will require policies that generate the very large investments required to refurbish the building stock to the necessary level of energy efficiency, and install very large numbers of heat pumps, as well as train current plumbers and electricians to be expert heat pump installers.

In zero-carbon road transport, the dominant technology is likely to be battery electric vehicles (BEVs) for passenger cars and urban delivery and commercial vans. The extent to which batteries will be able to power heavier road freight vehicles over longer distances depends on battery development, but it seems likely that fuel cells using hydrogen or some synfuel (e.g. methanol) will have at least some role to play. Electric highways are also a possibility for road freight. Rail will use electric power, and in countries with adequate rail capacity rail may increase its share of freight transport. Both shipping and aviation are harder to decarbonise, with batteries likely to be unable to power them over long distances. Shipping may come to use hydrogen, either in fuel cells or as ammonia, while aviation seems more likely to use biofuel or synfuel.

Whatever the solution, and barring completely unforeseen technological developments (which are, of course, always possible), decarbonised shipping and aviation are likely to be considerably more expensive than the fossil-fuelled versions we have become used to. This will doubtless have implications for international travel and trade. Finally, the dominance of BEVs for passenger cars does not mean that a simple one-for-one switch from internal combustion engines to BEVs will be unproblematic. The first task is obviously the roll-out of adequate numbers of charging points, but a more important issue is the sheer quantity of new minerals that will be required for the billions of motor cars that will be demanded on current trends, and the investment in mining that this will need, even with a considerable development of recycling. The pressure to supply these minerals would be considerably eased if towns and cities, where most people will live by 2050, could move away from individual private motorised transport towards car-sharing, public transport, and active travel (walking and cycling) for short journeys. Such trends are already under way in some cities, and if pronounced could have considerable benefits in terms of reduced congestion, as well as lowering the demand for minerals and the energy required for car manufacture.

Much of industry can be electrified, though the amount of energy it will need will largely depend on the extent to which countries can move towards a circular economy, with products lasting longer, and being easier to repair, re-manufacture and recycle (see Chapter 8), processes that often require less energy than making new products, at their end-of-life. Those industrial processes that are most difficult to decarbonise through electricity, including steel-making from iron ore, and the production of chemicals and cement, will require a mixture of other technologies, including hydrogen, carbon capture and storage (CCS) and synfuels. Many of the technologies under consideration for these sectors are still at a relatively early stage of development, and are inevitably more expensive at the moment than their fossil fuel counterparts. Bringing these technologies into the commercial mainstream will require the industries to be protected by the countries into which they are first introduced, and then rolled out globally as factories and equipment using fossil fuel-based technologies come to the end of their lives. This is likely to require an unprecedented level of global cooperation.

References

Bataille, C. (2020). Physical and policy pathways to net-zero emissions industry. *WIREs Climate Change*, *11*(2). https://doi.org/10.1002/wcc.633

Bataille, C., Åhman, M., Neuhoff, K., Nilsson, L. J., Fischedick, M., Lechtenböhmer, S., Solano-Rodriquez, B., Denis-Ryan, A., Stiebert, S., Waisman, H., Sartor, O., & Rahbar, S. (2018). A review of technology and policy deep decarbonization pathway options for making energy-intensive industry production consistent with the Paris Agreement. *Journal of Cleaner Production*, *187*, 960–973. https://doi.org/10.1016/j.jclepro.2018.03.107

Bertaud, A. & Richardson, H. (2004). Transit and density: Atlanta, the United States and Western Europe. Chapter 17. In: Bae, C. & Richardson, H. (Eds.) *Urban sprawl in Western Europe and the USA*. Routledge, London/New York.

Bonfield, P. (2016). *Each Home Counts: An Independent Review of Consumer Advice, Protection, Standards and Enforcement for Energy Efficiency and Renewable Energy*. Her Majesty's Government. www.gov.uk/government/publications/each-home-counts-review-of-consumer-advice-protection-standards-and-enforcement-for-energy-efficiency-and-renewable-energy. Accessed May 6, 2023.

Climate Change Committee. (2019a). *UK housing: Fit for the future?* CCC, London.

Climate Change Committee. (2019b). *Net Zero Technical Report*. CCC, London.

Climate Change Committee. (2020). *The Sixth Carbon Budget: the UK'S Path to Net Zero*. CCC, London.

Dargay, J., Gately, D., & Sommer, M., 2007. Vehicle ownership and income growth, worldwide: 1960–2030. *Energy Journal.* 28, 143–170. doi:10.2307/41323125

Department for Business, Energy and Industrial Strategy. (2020). *Heat Pump Manufacturing Supply Chain Research Project*. BEIS, London.

Department for Business, & Energy and Industrial Strategy. (2017). *What Does It Cost to Retrofit Homes?* BEIS, London.

De Richter, R. K., Ming, T., Caillol, S., & Liu, W. (2016). Fighting global warming by GHG removal: Destroying CFCs and HCFCs in solar-wind power plant hybrids producing renewable energy with no-intermittency. *International Journal of Greenhouse Gas Control*, *49*, 449–472. https://doi.org/10.1016/j.ijggc.2016.02.027

Economidou, M. (2011). *Europe's Buildings under the Microscope*. Buildings Performance Institute Europe, Brussels.

Energy Research Partnership. (2016). *Heating Buildings*. ERP, London. https://erpuk.org/wp-content/uploads/2017/01/ERP-Heating-Buildings-report-Oct-2016.pdf. Accessed May 6, 2023.

Energy Transitions Commission. (2018). *Mission Possible*. ETC, London. www.energy-transitions.org/publications/mission-possible/. Accessed May 6, 2023.

GlobalABC/IEA/UNEP (Global Alliance for Buildings and Construction), IEA and the UNEP (2020). *GlobalABC Roadmap for Buildings and Construction: Towards a zero-emission, efficient and resilient buildings and construction sector*. IEA, Paris.

Green Finance Institute. (2020). *Financing Energy Efficient Buildings: The Path to Retrofit at Scale*. GFI, London. www.greenfinanceinstitute.co.uk/wp-content/uploads/2020/06/Financing-energy-efficient-buildings-the-path-to-retrofit-at-scale.pdf. Accessed M1y 6,2023.

Her Majesty's Government. (2019). *Green Finance Strategy*. HMG, London.

Her Majesty's Government. (2021). *Heat and Buildings Strategy*. HMG, London.

International Energy Agency. (2020). *Iron and Steel Technology Roadmap: Towards more sustainable steelmaking*. IEA, Paris.

International Energy Agency. (2021a). *Net Zero by 2050*. IEA, Paris.

International Energy Agency. (2021b). *Energy Efficiency, 2021*. IEA, Paris. https://iea. blob.core.windows.net/assets/9c30109f-38a7-4a0b-b159-47f00d65e5be/EnergyEfficie ncy2021.pdf

International Energy Agency. (2021c). *Heat Pumps*. IEA, Paris.

IPCC. (2014). *Industry*. Fischedick M., J. Roy, A. Abdel-Aziz, A. Acquaye, J. M. Allwood, J.-P. Ceron, Y. Geng, H. Kheshgi, A. Lanza, D. Perczyk, L. Price, E. Santalla, C. Sheinbaum, & K. Tanaka. In: Edenhofer, O., R. Pichs-Madruga, Y. Sokona, E. Farahani, S. Kadner, K. Seyboth, A. sadler, I. Baum, S. Brunner, P. Eickemeier, B. Kriemann, J. Savolainen, S. Schlömer, C. von Stechow, T. Zwickel and J. C. Minx (Eds.) *Climate Change 2014: Mitigation of Climate Change. Contribution of Working Group III to the Fifth Assessment Report of the Intergovernmental Panel on Climate Change*. Cambridge University Press, Cambridge, United Kingdom and New York, NY, USA.

IPCC. (2022). *Climate Change 2022: Mitigation of Climate Change. Contribution of Working Group III to the Sixth Assessment Report of the Intergovernmental Panel on Climate Change*. P. R. Shukla, J. Skea, R. Slade, A. Al Khourdajie, R. van Diemen, D. McCollum, M. Pathak, S. Some, P. Vyas, R. Fradera, M. Belkacemi, A. Hasija, G. Lisboa, S. Luz, & J. Malley (Eds.). Cambridge University Press, Cambridge, UK and New York, NY, USA. doi: 10.1017/9781009157926.001

Janda, K. (2011). Buildings don't use energy: People do. *Architectural Science Review, 54*(1), 15–22. https://doi.org/10.3763/asre.2009.0050

Jiang, H., Kulkarni, A. P., Rego De Vasconcelos, B., & Lavoie, J.-M. (2019). Recent advances in power-to-X technology for the production of fuels and chemicals. *Frontiers in Chemistry, 1*, 392. https://doi.org/10.3389/fchem.2019.00392

Kuss, P. & Nicholas, K. (2022). A dozen effective interventions to reduce car use in European cities: Lessons learned from a meta-analysis and transition management. *Case Studies in Transport 10*(3), 1494–1513. https://doi.org/10.1016/j.cstp.2022.02.001

Levesque, A., Pietzcker, R. C., Baumstark, L., de Stercke, S., Grübler, A., & Luderer, G. (2018). How much energy will buildings consume in 2100? A global perspective within a scenario framework. *Energy, 148*, 514–527. https://doi.org/10.1016/j.ene rgy.2018.01.139

Li, X., Sun, W., Zhao, L. & Cai, J. (2018). Material metabolism and environmental emissions of BF-BOF and EAF steel production routes. *Mineral Processing and Extractive Metallurgy Review, 39*(1), 50–58. https://doi.org/10.1080/08827 508.2017.1324440

Madeddu, S., Ueckerdt, F., Pehl, M., Peterseim, J., Lord, M., Kumar, K. A., Krüger, C., & Luderer, G. (2020). The CO_2 reduction potential for the European industry via direct electrification of heat supply (power-to-heat). *Environmental Research Letters, 15*(12), 124004. https://doi.org/10.1088/1748-9326/abbd02

Madias, J. (2016). Electric arc furnace. In: Cavaliere, P. (Ed.) *Ironmaking and Steelmaking Processes – Greenhouse Emissions, Control, and Reduction*, pp. 267–281, Springer.

Martiskainen, M., Schot, J., & Sovacool, B. K. (2021). User innovation, niche construction and regime destabilization in heat pump transitions. *Environmental Innovation and Societal Transitions, 39*, 119–140. https://doi.org/10.1016/j.eist.2021.03.001

Njakou Djomo, S., & Ceulemans, R. (2012). A comparative analysis of the carbon intensity of biofuels caused by land use changes. *GCB Bioenergy, 4*(4), 392–407. https://doi.org/10.1111/j.1757-1707.2012.01176.x

Saygin, D., & Gielen, D. (2021). Zero-emission pathway for the global chemical and petrochemical sector. *Energies, 14*(13), 3772. https://doi.org/10.3390/en14133772

Staffell, I., Scamman, D., Dodds, P. E., Ekins, P., Shah, N., & Ward, K. R. (2019). The role of hydrogen and fuel cells in the global energy system. *Energy and Environmental Science, 12*(2). https://doi.org/10.1039/c8ee01157e

Traynor, J. (2020). *EnerPHit*. RIBA, London. https://doi.org/10.4324/9780429347863

United Nations Environment Programme (2022) *Emissions Gap Report 2022*, UNEP, Nairobi.

Vogl, V., Olsson, O., & Nykvist, B. (2021). Phasing out the blast furnace to meet global climate targets. *Joule, 5*(10), 2646–2662. https://doi.org/10.1016/j.joule.2021.09.007

10 Climate, agriculture and waste

Summary

The world's current food system is a major contributor to climate change. It is directly responsible for about a quarter of the world's greenhouse gases (GHGs), and contributes to much more as the major driver of deforestation. However, in many countries it is also likely to be negatively affected by climate change, and many countries thus affected already suffer from food insecurity which could be made worse. An astonishing proportion of food that is grown is not eaten. It is either lost because of inadequate infrastructure in developing countries to get it to market, or, in richer countries, it is bought and then thrown away.

Stopping climate change requires that food production greatly reduces its contribution to climate change. The main emission reduction strategies in agriculture centre on reducing emissions from livestock, by reducing meat consumption to healthy levels where it is currently excessive, and by producing their feed without associated deforestation. The main way to reduce the threat of climate change to agriculture and food security is to reduce the extent of climate change by reducing the emissions that contribute to it. The world's food system needs to be transformed and made more resilient by introducing more sustainable agricultural methods, and reducing food loss and waste. None of the measures needed to achieve these outcomes are likely to be easy to introduce.

Introduction

Everyone needs to eat. In many ways it is remarkable what the world's farmers, and the wider food system, have achieved: the sustenance of 8 billion people. Not all of these people eat enough. The World Health Organization (WHO)'s 2021 Fact Sheet on Malnutrition classifies 462 million as 'underweight',

DOI: 10.4324/9781003438007-11

Table 10.1 Global greenhouse gas emissions from food production

	Global greenhouse gas emissions from food production (%)
Total contribution of the food system	26
Supply chain	**18**
Retail	3
Packaging	5
Transport	6
Food processing	4
Livestock and fisheries	**31**
Wild catch fisheries	1
Livestock & fish farms	30
Crop production	**27**
Crops for animal feed	6
Crops for human food	21
Land use	**24**
Land use for human food	8
Land use for livestock	16

Source: Our World in Data, https://ourworldindata.org/emissions-by-sector#agriculture-forestry-and-land-use-18-4. Accessed April 28, 2023.

149 million of whom were children. It finds that far more people – 1.6 billion – are overweight or obese.[1]

The world's current food system is a major contributor to climate change. Its food supply is also under significant threat from it. This chapter examines both of these aspects of food and climate change: how food can reduce its contribution to climate change, and become more resilient to the climate change that is in prospect.

Food's contribution to climate change

Table 10.1 shows the 26% contribution of the food system in 2017 to global GHG emissions. Of this total, 24% was from land use, 27% from crop production, 31% from livestock and fisheries and the rest from various aspects of the food supply chain. Pendrill et al. (2022) showed that agriculture also was associated with at least 90% of deforestation between 2011 and 2015, a further major driver of CO_2 emissions.

The main GHG emissions from agriculture are not CO_2, but methane (CH_4, from livestock and crop production) and nitrous oxide (N_2O, from denitrification of fertilisers). The Intergovernmental Panel on Climate Change (IPCC) (2019, p. 12) estimated that the 6.2 $GtCO_2e$ emitted annually from agriculture

1 www.who.int/news-room/fact-sheets/detail/malnutrition. Accessed April 29, 2023.

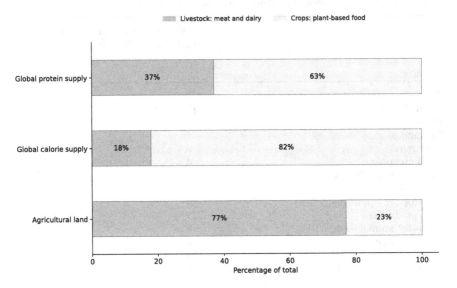

Figure 10.1 Global land use for food production.

Source: Our World in Data, https://ourworldindata.org/land-use. Accessed April 28, 2023.

(average for 2007–2016) comprised 4 GtCO$_2$e (65%) from CH$_4$, and 2.2 GtCO$_2$e (35%) from N$_2$O. No global data were available in this source for agricultural CO$_2$ emissions, but according to Figure 2.7 for agriculture and fisheries these were around 0.7 GtCO$_2$e (1.7% of total GHGs).

It is clear from Table 10.1 that over 40% of agriculture emissions come from livestock. Figure 10.1 shows that livestock also occupy 77% of agricultural land, but they supply only 18% of calories and 37% of protein for human consumption. It is clear that if livestock numbers could be significantly reduced, and more protein for humans provided by crops, a substantial amount of land could be released for nature.

This is important, because agriculture is not only a major contributor of GHG emissions. It is also the principal cause of biodiversity loss and water stress – the 2019 *Global Resource Outlook* from UNEP's International Resource Panel (IRP) estimated that land use, mainly for agriculture, is the driver of 80% of biodiversity loss and 85% of water stress (International Resource Panel, 2019). Reducing the global livestock herd, and eating more plant-based food, would therefore free land for nature and ecosystems, as well as reducing GHG emissions. Other environmental impacts of agriculture include eutrophication caused by the overuse of fertilisers in certain areas. While humans are therefore absolutely dependent on the products of agriculture, what we produce for food, and the way we produce it, is not environmentally sustainable.

There can be no global blueprint for a sustainable agriculture because agriculture is too dependent on place-based characteristics, such as soil quality, the slope of land, the weather and climate, and, of course, local and national cultures of farming and food consumption. There are, however, some broad principles for a sustainable agriculture. These are set out in the 2014 publication of the Food and Agriculture Organization of the United Nations (FAO, 2014). The publication first lists the "unprecedented confluence of pressures" facing global agriculture (poverty, inequalities, hunger and malnutrition; inadequate diets and unsustainable consumption patterns; land scarcity, degradation and soil depletion; water scarcity and pollution; loss of living resources and biodiversity; climate change; and stagnation in agricultural research). It then articulates its five principles for sustainable agriculture: improved efficiency in the use of natural resources; action to conserve, protect and enhance these natural resources; protection and improvement of rural livelihoods, equity and social well-being; enhanced resilience of people, communities and ecosystems (not least to climate change); and responsible and effective governance mechanisms.

These rather abstract principles are then illustrated in the enumeration of practices through which they may be, and, on a small scale, have been, implemented. The detail of these is outside the scope of this book. The tragedy is that, as with so much good practice that could address the twin crises of climate and nature, as well as providing for human needs, these practices remain everywhere the exception rather than the rule, so that the problems of agriculture – of sustainably feeding the world's large and growing human population – largely remain.

This is as true in the UK as anywhere else, where the problems, and their solutions, were brought into sharp focus by the publication in 2021 of the UK National Food Strategy (Dimbleby, 2021). Dimbleby (2021, Figure 1.7, p. 21) shows that between 2008 and 2018 GHG emissions from the UK food system fell by only 13%, less than half the rate in the wider UK economy, but GHG emissions from agriculture did not fall at all.

Defra (Department for Environment, 2020, p. 18) estimates UK farmland at 17.3 million hectares, around 71% of the UK land area. Figure 10.2 shows that UK diets require roughly the same amount of land outside the UK as farmland within it. Of UK farmland, 85% is used to rear animals (a larger proportion than globally, which is 77%, as shown in Figure 10.1). These animals are, as shown in Table 10.1, the single largest source of greenhouse gas (GHG) emissions.

In 2018 UK agriculture accounted for around 10% of total UK GHG emissions, but 70% of total nitrous oxide (N_2O) and 49% of total methane (CH_4) emissions (Department for Environment, 2020, p. 13). CO_2 emissions from agriculture were just 1.6% of UK GHGs. Table 10.2 shows how these GHGs from agriculture have changed since 1990. Unlike CO_2 emissions more generally, it can be seen that every category of GHGs from agriculture increased between 1990 and 2018.

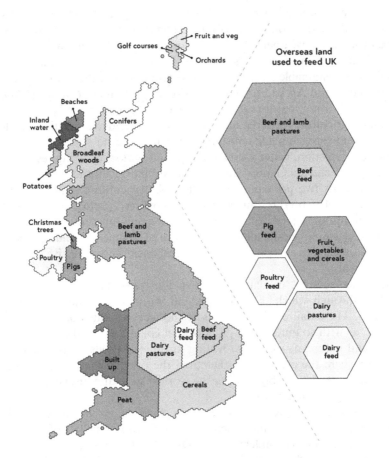

Figure 10.2 Land use in the UK, and land for food production for the UK outside it.
Source: Dimbleby, 2021, Figure 9.3, p. 90.

Table 10.2 Greenhouse gas emissions from UK agriculture, 1990–2018

	Million tonnes carbon dioxide equivalent	
	1990	*2018*
Carbon dioxide emissions	5.7	6.5
Methane emissions	25.4	30.3
Nitrous oxide emissions	14.3	17.2
UK agriculture: total GHG emissions	45.4	54.0

Source: Department for Environment, Food and Rural Affairs, 2020, Figure 1.1, p. 13.

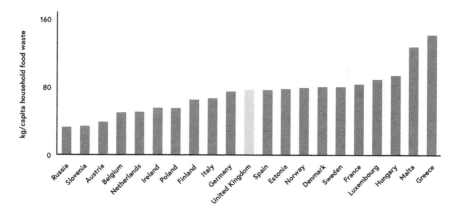

Figure 10.3 Food waste in selected European countries.

Source: Dimbleby, 2021, Figure 11.1, p. 108.

N_2O emissions could be reduced through more efficient management of fertilisers, including manure, and CH_4 emissions could be reduced through reducing the 'feed conversion ratio'[2] of farmed animals (Department for Environment, 2020, p. 18) (i.e. the weight in kg of feed required to increase the weight of an animal by 1 kg). Average feed conversion ratios are 8 for cattle, 3.9 for pork, 1.9 for chickens and 1.3 for salmon.[3]

It is truly extraordinary that, given that food needs to be bought, over a quarter of food sold in the UK (as globally, see later in this chapter) is not eaten, but is thrown away: "The biggest contributors to food waste are households (70%), followed by manufacturers (18%), the hospitality and food industry (10%) and then retailers (2%)" (Dimbleby, 2021, p. 107). Figure 10.3 shows that this is not something that is unique to the UK, but occurs throughout Europe, to a greater or lesser extent.

This food waste not only 'wastes' the GHG emissions that were incurred in its production. It also contributes further to GHG emissions by creating methane when it is put into landfill. The UK Waste and Resources Action Programme estimated that the GHGs associated with the 9.5 million tonnes of UK food waste in 2018 were around 25 $MtCO_2e$ (including the emissions from both the production of wasted food and the disposal of food waste), around

---

2 Feed conversion ratio (FCR) is the conventional measure of livestock production efficiency: the weight of feed intake divided by weight gained by the animal. Lower FCR values indicate higher efficiency. FCRs are typically 6.0–10.0 for beef, 2.7–5.0 for pigs, 1.7–2.0 for chicken and 1.0–2.4 for farmed fish and shrimp. See www.tabledebates.org/research-library/feed-conversion-efficiency-aquaculture-do-we-measure-it-correctly. Accessed April 29, 2023.

3 www.statista.com/statistics/254421/feed-conversion-ratios-worldwide-2010/. Accessed April 29, 2023.

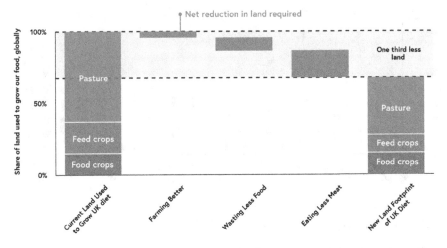

Figure 10.4 Possible land 'saving' from farming better, wasting less food and eating less meat.

Source: Dimbleby, 2021, Figure 11.2, p. 110.

5% of total UK GHG emissions. The food that could have been eaten (6.4 Mt) cost over £19 billion (over £500 per household), and would make the equivalent of over 15 billion meals.[4]

Looking to the future, the UK National Food Strategy (Dimbleby, 2021) estimated that better farming (adoption of best agricultural practices and crop genetics) could increase UK crop yields by 15%, which, combined with a 50% reduction in food waste, and a reduction in meat eating by 30%, could feed the UK on a third less land (see Figure 10.4). This would both reduce UK agricultural emissions and free up land, both to regenerate nature and increase biodiversity, and to store carbon taken from the air, one of the means of greenhouse gas removal (GGR), or carbon dioxide removal (CDR), discussed in Chapter 7.

Table 10.3 shows how the reduction in land use from UK farming, and better farming, could lead to reduced GHG emissions by 2035. The great majority of the emissions reductions come from changes in land use that take up and store carbon from the atmosphere, most notably through forests and peatlands.

This becomes even more obvious in the Climate Change Committee's projections of GHG emissions from agriculture and land use change through to 2050 (Climate Change Committee, 2020). In its Balanced Net Zero Emissions Pathway, the carbon reductions from agriculture itself are relatively modest,

---

4 https://wrap.org.uk/taking-action/food-drink/actions/action-on-food-waste). Accessed April 28, 2023.

Table 10.3 Possible GHG emissions reductions in 2035 from changes in food production, consumption and waste, and alternative land uses

	GHG savings from measures to reduce agriculture and land use emissions, 2035 ($MtCO_2e$)
Peatlands	6
Low-carbon farming practices & machinery	5
Diet change, food waste & other land release	8
Forestry – land	3
Forestry – other	12
Agro-forestry & hedges	1
Energy crops – land	2
Energy crops – other	3

Source: Climate Change Committee, 2020, Figure 3.6a, p. 164.

Note: 'Other' forestry and energy crops is the additional savings elsewhere in the economy by displacing fossil fuels with biomass material. These are annual savings compared to emissions in the baseline in 2035.

with most coming from diet change and reduced food waste, and the residual GHG emissions are quite sizeable (over 30 $MtCO_2e$), even by 2050 (Climate Change Committee, 2020, Figure 3.6b, p. 166). In contrast, Climate Change Committee (2020, Figure 3.6d, p. 171) shows how the changes in land use made possible by freeing up land from agriculture can lead to negative emissions of nearly 20 $MtCO_2e$ by 2050, with the major contribution coming from afforestation (new forests) and the restoration of peatlands.

Such changes will not happen by themselves. They will need to be stimulated and incentivised by strong and sustained public policy, as shown in Table 10.4.

However, reductions in emissions from agriculture do not need to wait for these policies. Sizeable reductions are possible now, as shown by a case study in Ortiz et al. (2021, p. p. 86). Newhouse Farm is an 800 ha, predominantly arable, family-owned farm in Hampshire. Between 2000 and 2020 the farm became carbon negative through a 17% reduction in nitrogen fertiliser use (with no loss of wheat yields), a 25% reduction in fuel consumption by reducing tillage and installing biomass boilers, and a 1% increase in soil organic matter. Other such experiences may be found in the 'farmer stories' of the UK's Nature-Friendly Farming Network.[5]

The impacts of climate change on food production

The percentage of the world's land area devoted to agriculture has remained roughly constant at 37–39% since the early 1990s.[6] However, food supply (measured in calories per person per day) has increased faster than the human population, and was 9% higher in 2018–2020 than in 2000–2002 (Food

5 www.nffn.org.uk/category/farmer-stories/. Accessed April 29, 2023.
6 https://databank.worldbank.org/reports.aspx?source=2&series=AG.LND.AGRI.ZS. Accessed October 7, 2023.

Table 10.4 Timeline of measures for reducing emissions from agriculture and land use change and forestry

	Time horizon
Ensure the post-CAP framework incentivises on-farm emissions reduction, including a strong regulatory baseline.	Short term (2020–2030)
Targeted investment in R&D and innovation to deliver productivity improvements in crops and livestock.	Medium-term (2030–2040)
Support the commercialisation of low-carbon agriculture vehicles and infrastructure.	Long term (2040–2050)
Introduce policies to support the shift toward healthier and less carbon-intensive diets.	Long term (2040–2050)
Mandate the availability of non-meat options in the public sector by 2022.	Short term (2020–2030)
Announce policies and measures to achieve a 20% reduction in food waste by 2025.	Short term (2020–2030)
Ensure existing schemes for woodland creation incentivises further afforestation.	Short term (2020–2030)
Ensure the post-CAP framework promotes transformational land use change and measures for deep emissions reduction including afforestation and peatland restoration.	Short term (2020–2030)
Introduce measures to address non-financial barriers to alternative land uses.	Short term (2020–2030)
Targeted investment in R&D and innovation to deliver productivity improvements in tree and energy crops.	Medium term (2030–2040)

Source: Climate Change Committee, 2019, Figure 7.4, p. 207, and Figure 7.9, p. 228.
Note: CAP stands for the Common Agricultural Policy of the European Union.

and Agriculture Organization, 2021b, p. 28). At a global average of 2,950 kilocalories per person per day, the 11 billion tonnes of food produced in 2020 was enough to feed the whole world's population. That a tenth of this population went hungry is to do with the food's distribution, and the inability of poor people to buy enough food, rather than due to shortages of production itself (Food and Agriculture Organization, 2021a).

Population growth, and the transition as people get richer to higher-meat diets will require substantial increases in food production in the future. It is estimated that 50–60% more food will need to be produced to feed the projected population of around 10 billion people in 2050. In order not to reduce further the space for nature, this extra food will need to be grown without increasing the land area for agriculture and, for climate change reasons, with greatly reduced GHG emissions.[7] This would be challenging at the best of times, especially as there is evidence of a slowdown in the rate of growth, or even stagnation,

7 www.wri.org/insights/how-sustainably-feed-10-billion-people-2050-21-charts. Accessed April 28, 2023.

of yields for some major crops in some areas (Ray et al., 2012). However, the Food and Agriculture Organization (FAO) (2018, p. 21) projects, in a baseline without climate change, that improvements in agricultural technology will increase crop yields in 2050 by an average of 38%.

Climate change has long been recognised as a threat to food production, and therefore to food security, with effects including soil erosion, land degradation, drought, salinisation, water stress, flooding, and biodiversity loss. Already in 2009 the FAO was writing:

> Climate change will *affect agriculture and forestry systems* through higher temperatures, elevated carbon dioxide (CO_2) concentration, precipitation changes, increased weeds, pests and disease pressure. … Although the countries in the Southern hemisphere are not the main originators of climate change, they may suffer the greatest share of the damage in the form of declining yields and greater frequency of extreme weather events (droughts and floods). … All current quantitative assessments show that *climate change will adversely affect food security*.
>
> (Food and Agriculture Organization, 2009, pp. 29–30, emphasis in original)

In fact, there was already evidence in 2009 that climate change was indeed reducing food productivity. Research in 2007 found that at least 30% of year-to-year fluctuations in agricultural yield were due to climate variability (Lobell & Field, 2007). In Europe, while there has been considerable growth in crop yield (tonnes of crop per hectare) over the last 50 years, there has been an apparent recent fall in the yield growth rate compared to previous decades. Recent research has found that weather factors and temperature in particular are the most likely cause. Up to 30% of the long-run increase over time in the yield of European crops has been cancelled out by the adverse impact of weather, although this impact varies markedly across countries and crops (Agnolucci & De Lipsis, 2020).

This effect has been observed more widely. Ortiz-Bobea et al. (2021, Figure 5b, p. 311) have found that climate change has already slowed agricultural productivity growth in some parts of the world. It is particularly worrying that, while the effects of recent warming trends vary by crop and region, and can be positive or negative, the most pronounced adverse impacts tend to be in countries, such as those in sub-Saharan Africa, that still have relatively low agricultural productivity. This is particularly clear in the case of barley, maize, millet, pulses, rice and wheat. Moreover, it seems that the countries with the worst level of food security (as measured by the daily per person intake of calories) are also worst affected by rising temperature (Agnolucci et al., 2020). This seems to bear out the finding from the IPCC (2014) that the higher average global temperatures and more extreme weather events associated with climate change will amplify variability in food production.

The IPCC's assessment (with high confidence) in 2022 was that

Climate change will increasingly put pressure on food production and access, especially in vulnerable regions, undermining food security and nutrition. Increases in frequency, intensity and severity of droughts, floods and heatwaves, and continued sea level rise will increase risks to food security. ... At 2°C or higher global warming level in the mid-term, food security risks due to climate change will be more severe, leading to malnutrition and micro-nutrient deficiencies, concentrated in sub-Saharan Africa, South Asia, Central and South America and Small Islands.

And with medium confidence: "Global warming will progressively weaken soil health and ecosystem services such as pollination, increase pressure from pests and diseases, and reduce marine animal biomass, undermining food productivity in many regions on land and in the ocean" (IPCC, 2022b, SPM B.4.3, p. 14).

Table 10.5 from the FAO summarises the likely negative impacts on food security, especially in low-income countries.

Another change which will be brought about by the increased heat and rainfall associated with climate change is increased land degradation – due to the loss of soil nutrients and organic matter, salinisation and saltwater intrusion – with negative effects on crop yields (IPCC, 2018). In addition, accelerating sea level rise will compound these negative impacts by increasing salinisation through saltwater intrusions and permanently inundating crop land (IPCC, 2021). Even without climate change, land degradation has been severe, with the IPCC writing (with medium confidence):

About a quarter of the Earth's ice-free land area is subject to human-induced degradation. Soil erosion from agricultural fields is estimated to be currently 10 to 20 times (no tillage) to more than 100 times (conventional tillage) higher than the soil formation rate. Climate change exacerbates land degradation, particularly in low-lying coastal areas, river deltas, drylands and in permafrost areas ... including through increases in rainfall intensity, flooding, drought frequency and severity, heat stress, dry spells, wind, sea-level rise and wave action, and permafrost thaw with outcomes being modulated by land management.

(IPCC, 2019 pp. 7, 10)

Recent modelling of soil loss in wheat and maize fields shows large variations between tropical climate regions and regions with a large proportion of flat and dry land, with losses ranging from less than 1 tonne/hectare in Central Asia to 100 tonnes/hectare in South East Asia. The strong impact of climate and topography on simulated water erosion is clearly shown in the five largest wheat- and maize-producing countries. In Brazil, China and India, where a large proportion of cropland is in tropical areas, water erosion is relatively

Table 10.5 Impacts of climate change on food security

Dimension of food security	Climate change effects on food security	Time horizon
Availability	Global mean crop yields of rice, maize and wheat projected to decrease 3–10% per degree of warming Impacts on livestock through reduced feed quantity/quality, pest and disease prevalence, physical stress; meat, egg and milk yield and quality decrease 5–10% decrease in potential fish catch in tropical marine ecosystems	Slow onset, long term
Access	Increasing food prices Relocation of production with impacts on prices, trade flows and food access	Slow onset, long term
Utilisation	Reduced food safety due to higher rates of microbial growth at increased temperatures Reduced nutritional quality of crops due to decreases in leaf and grain nitrogen, protein and macro and micro-nutrient concentrations associated with increased carbon dioxide concentrations and more variable and warmer climate	Slow onset, long term
Stability	Damage to crops and livelihoods from extreme events (heatwaves, droughts, floods, storms, etc.) Short-term disruptions of trade through effects on transport systems	Extreme events, short term

Source: Food and Agriculture Organization, 2018, Table 2.1, p. 16.

high with annual median values of 10 t/ha, 6 t/ha, and 37 t/ha, respectively. In Russia and the United States annual median values are much lower at 1 t/ha, and 2 t/ha, respectively (Carr et al., 2020, p. 5269).

Historically, poor management of lands in Europe and the US has been largely remedied through policy interventions, with chemical fertilisers and other management measures such as irrigation able to offset a massive amount of degradation. For example, one study has shown that, with no inputs of fertiliser, US yields of corn over the past 100 years would have fallen from around 7 to a little over 1 tonne per hectare, due to soil degradation. However, fertiliser inputs have enabled yields to be broadly maintained, although at an annual cost to farmers of over half a billion dollars (Jang et al., 2021). Such interventions have very real consequences in terms of greenhouse gas emissions, water pollution, and energy use. These results also have worrying implications for poorer parts of the world which are currently experiencing land degradation, but do not have the resources to compensate for this with fertilisers. And the results become more worrying still if the soil erosion and wider land degradation is exacerbated by climate change.

FAO's projection of changes due to climate change for 2050 in agricultural production (changes from its baseline which, as noted above, assumes considerable improvements in productivity due to technological progress) is more optimistic than Ortiz-Bobea et al. (2021)'s modelling of losses in agricultural productivity already brought about by climate change. Results in North and Central America, Europe, the Russian Federation, Japan, Korea, Australia and South America except Brazil, all show increases in production from climate change (Food and Agriculture Organization, 2018, Figure 2.1, p. 20). Part of this may be due to the CO_2 fertilisation effect, which is the increased rate of photosynthesis and reduced leaf transpiration in plants caused by higher atmospheric concentrations of CO_2. However, this has an effect on the nutritional quality of the food produced. Ebi & Loladze (2019, p. e283) note after a review of the relevant literature:

> Higher CO_2 concentrations increase photosynthesis in C3 plants (e.g. wheat, rice, potatoes, barley), which can increase crop yields. But those increases come at the cost of lower nutritional quality as plants accumulate more carbohydrates and less minerals (e.g. iron and zinc), which can negatively affect human nutrition.

This could exacerbate the micro-nutrient deficiencies which were already being suffered by 2 billion people in 2003.

The global food system is used to adapting to short-term changing climatic conditions. Many aspects of land management for food production have changed in recent decades including spatial shifts in crop distribution in response to increased temperatures. The overall result of these changes has been greatly increased food yields in many parts of the world, and land managers may be expected to adapt their strategies for changes in the climate. On the other hand, food prices in the last 15 years have shown extreme volatility, with extreme price spikes in 2007–2008 and 2011–2012, followed by a precipitate fall in 2012 (Global Panel on Agriculture and Food Systems and Nutrition, 2016).

Given the multiple stresses on food production, and the increasing demand from growing populations and changing diets, it is perhaps remarkable that, thus far, the global agricultural system has been fairly robust and major food shortages (as opposed to difficulties for poor people to buy the food available) have been rare. One major factor in potentially smoothing out localised food shortages is food trade, the value of which increased by a factor of three between 2000 and 2016 (Food and Agriculture Organization, 2018, p. 2). As a result of climate change, those countries that are already net food importers are expected to import more (Food and Agriculture Organization, 2018, p. 20).

If world markets continue to operate effectively, if technological progress increases agricultural production as the FAO projects, and if climate change leads to the increases in production, as also projected by the FAO, then food

Table 10.6 Top ten grain exporting countries, 5-year average 2012–2013 to 2016–2017

	United States	Ukraine	Brazil	Argentina	Russian Federation
Thousand metric tonnes (2012/ 2013–2016/2017)	71,063	28,188	27,114	25,343	24,260
	Canada	Australia	India	Thailand	Viet Nam
Thousand metric tonnes (2012/ 2013–2016/2017)	23,046	17,858	15,918	9,817	7,310

Source: Janetos et al., 2017, Figure 2, p. 3.

shortages and accompanying price spikes may be avoided. But that is quite a lot of 'ifs'.

Table 10.6 shows the production of the top ten grain exporting countries. Reassuringly, North America, Ukraine, the Russian Federation, Argentina and Australia are all projected by FAO to increase their agricultural production with climate change. However, there are many reasons why this may be optimistic. One, of course, is the unpredictability of human affairs generally (this text is being written while the war between Ukraine and Russia is raging, and this is seriously affecting agricultural production and exports in Ukraine, and Russian food exports may be curtailed by sanctions). Another is the uncertainty of country reactions when their harvests are affected by adverse weather, or when food prices rise for other reasons.

While there are many influences on food prices (including crop yield, weather variations, international trade, the financialisation of and speculation in food commodity markets, and land management practices), technology and trade have buffered quite a few impacts thus far over a period of time when the world has largely embraced more or less open trading systems. Now that the world seems to be moving towards more protectionist policies at a time when climate change is intensifying, these stabilising effects may start to fail. For example, in the last food price spike Argentina, Russia, and Thailand simultaneously closed their borders to food export (Janetos et al., 2017, pp. 5–6).

The risk of food shortages would be especially high if more than one of the 'breadbasket' regions shown in Table 10.6 were to experience simultaneous crop failure, due to climate change or any other reason. Gaupp et al. (2020, p. 54) analysed this risk in respect of climate change and showed "an increasing risk of simultaneous failure of wheat, maize and soybean crops" but decreased risk for rice. If climate change were indeed to result in simultaneous failure in two or more major breadbasket regions, then the food security prospects of the net-importing, poorer parts of the world could be particularly dire.

Waste and wastewater

Increases in waste generation are closely related to income and population growth, and urbanisation, as a result of which total global emissions from waste almost doubled from 1970 to 2010, and increased by 13% over 2000–2010 (IPCC, 2014, p. 385). In 2010 they totalled around 1.4 $GtCO_2e$, which was around 3.0% of total GHG emissions from all sources. IPCC (2014, Figure 5.19, p. 382) shows the relationship between GHG emissions from waste per capita and broad country income levels, indicating that high-income countries generate up to 1 tCO_2e per person from waste.

In 2010 90% of GHGs from waste were from methane (CH_4), mainly emitted from waste disposal on land or in landfill sites and from wastewater, with nitrous oxide (N_2O) accounting for 8% (IPCC, 2014). In 2005, 56% of CH_4 emissions were from landfill, and 44% from wastewater, while all the N_2O emissions were from wastewater (IPCC, 2007, Table 10.3, p. 596).

GHG emissions from wastewater can be reduced mainly through anaerobic digestion and the subsequent use for energy of the CH_4 generated. IPCC's AR6 report estimated that GHG emissions from wastewater could be reduced by 0.2 $GtCO_2e$ per year (IPCC, 2022a, Table 12.3, p. 12–23).

There are many ways to reduce GHG emissions from solid waste. These are reviewed in detail in the IPCC's fourth assessment report in 2007, which had a whole chapter on the contribution of waste management and wastewater to GHG emissions (IPCC, 2007, Chapter 10). Figure 10.5 illustrates some of the major technologies for reducing GHG emissions from solid waste, indicating the level and cost of the technology, and whether the technology requires energy inputs (negative energy balance) or generates it (positive energy balance).

The gathering of methane from landfill sites is now commonplace in developed countries and typically attracts support as a renewable energy source, to substitute for fossil-based heat and electricity. It is also increasingly practised in developing countries as they have got richer, and the measure attracted some support under the Clean Development Mechanism (CDM) of the Kyoto Protocol. The European Union now has strict targets for the reduction of bio-degradable waste going to landfill, which will require alternative management options, including recycling, composting, incineration (see below), anaerobic digestion and mechanical-biological treatment (MBT), to recover recyclables and reduce the organic carbon content of the waste. Landfill sites can also serve as a long-term sink for carbon, as typically at least 50% of the organic carbon in a landfill site remains in the landfill (IPCC, 2007, p. 601).

Incineration with energy recovery is increasingly practised in industrial and richer developing countries, with Japan incinerating 78% of its residual (i.e. non-recycled) waste.[8] In the early 2000s around 600 incinerators in 35 countries

---

8 www.tokyoreview.net/2019/07/burning-problem-japan-waste-recycling/. Accessed April 29, 2023.

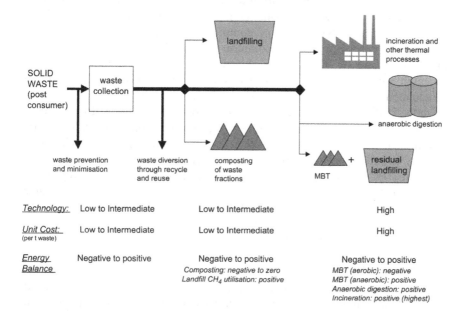

Figure 10.5 Technologies for the management of solid waste.

Source: IPCC, 2007, Figure 10.7, p. 600.

Note: MBT stands for mechanical-biological treatment.

burned around 130 Mt of waste (IPCC, 2007, p. 601), but by 2019 this had grown to 1,200 incineration facilities with a capacity of 310 Mt waste per year. Calculations indicate that burning 261 Mt waste could produce 283 TWh (1 EJ) of power and heat, that might otherwise be produced by fossil fuel plants (IPCC, 2022a, pp. 6–48).

The incineration of 1 tonne (t) solid waste produces 0.7–1.0 t of CO_2, and these emissions (perhaps 200 $MtCO_2$ per year) are therefore less than 2% of total emissions from waste. However, the proportion of emissions that is from biogenic waste (organic matter, and therefore the carbon in it has been drawn down from the atmosphere) is accounted as renewable and zero-carbon. This biogenic proportion is estimated as 33–50% of total incinerated waste. Only the fossil carbon element is therefore accounted as net emissions (Johnke, 2003). Most of the fossil carbon component of waste is plastic, 100 Mt of which were incinerated in 2015 (IPCC, 2022a, pp. 11–48). It has been calculated that fitting carbon capture and storage technology to just European incinerators could capture and store 60–70 Mt of CO_2 per year and, with some of this CO_2 being biogenic, this proportion could count as 'negative emissions' (IPCC, 2022a, pp. 6–48).

Both this accounting of emissions from incineration, and indeed incinerators themselves, are controversial. Critics claim that biogenic CO_2 emissions are not really net zero,[9] while opposition to incinerators also comes from community groups who worry about non-CO_2 emissions. Unregulated waste incineration can produce a wide range of common combustion air pollutants, such as nitrogen oxides, but also less common ones, such as heavy metals, dioxins and furans, some of which can be very damaging to both human health and the environment. This is a huge issue in some developing countries, where the unregulated dumping and burning of waste is still widespread (Ferronato & Torretta, 2019), but somewhat outside the scope of this book. In developed countries, although waste incineration has been problematic in terms of pollution in the past, which has left a residual suspicion of the technology in many communities, regulation of emissions from incineration is now very stringent (IPCC, 2007, p. 607). Nevertheless, although emissions from well-managed modern incinerators should be very low, studies have failed to give them a completely clean bill of health (Tait et al., 2020), so that community opposition to them in their localities may well continue.

Food loss and waste

A 2011 report to the FAO found that "roughly one-third of edible parts of food produced for human consumption globally was lost or wasted, corresponding to about 1.3 billion tonnes of food per year" (Food and Agriculture Organization, 2019, Box 3, p. 12). The FAO's most recent report defines *food loss and waste* "as the decrease in quantity or quality of food along the food supply chain", *food losses* "as occurring along the food supply chain from harvest/slaughter/catch up to, but not including, the retail level", while *food waste* "occurs at the retail and consumption level" (Food and Agriculture Organization, 2019, p. xii). The FAO's conceptual framework for food loss and waste is shown in Figure 10.6, where food loss and waste comprises only the boxes labelled as such. On the basis of these definitions, FAO estimates that *food losses* globally are around 14%, where the physical proportions of food loss are estimated for each type of food, and then weighted by their economic value before being aggregated. FAO's re-estimation of *food waste* is ongoing.

However, food loss may also be measured in purely physical or calorific terms. Food and Agriculture Organization (2019, Box 4, p. 15) shows the proportion of food loss by region according to three different metrics: physical quantity (tonnes), economic value and calorific value. While the aggregate difference between these three metrics at the world level is not great, the differences between them for particular regions is more marked, with the economic loss being greatest for North America and Europe (more than 15%), the physical loss being greatest for central and southern Asia (more than 20%), and the calorific loss being largest for other world regions (more than 15% in sub-Saharan Africa). FAO give the reasons for these losses as "inadequate harvesting time,

---

9 www.energyjustice.net/incineration/climate. Accessed April 29, 2023.

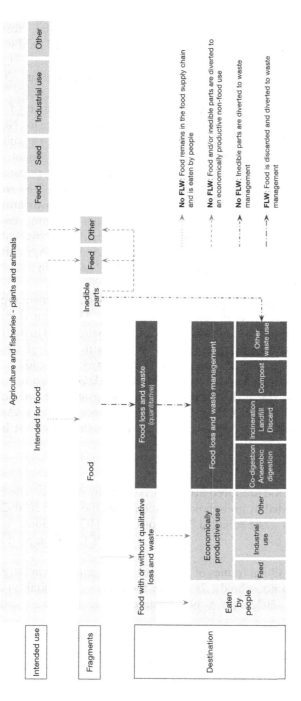

Figure 10.6 The FAO's conceptual framework for food loss and waste.

Note: "Industrial use" includes biofuels, fibres for packaging material, creating bioplastics, making traditional materials (e.g. leather, feathers [e.g. for pillows]) and rendering fat, oil or grease into a raw material to make soaps, biodiesel or cosmetics. It does not include anaerobic digestion, which is intended to manage waste. "Other" includes uses such as fertiliser and ground cover. The length of the bars is not representative of the total volume or value of the products concerned.

Source: Food and Agriculture Organization, 2019, Figure 2, p. 7.

climatic conditions, practices applied at harvest and handling, and challenges in marketing produce" as well as inadequate storage. Food loss can be reduced by the provision of cold storage and good transport infrastructure, efficient trade, and good processing and packaging facilities. The causes of food waste at the retail level are limited shelf life, aesthetic requirements (colour, shape, size), and demand variability (Food and Agriculture Organization, 2019, p. xiii).

Studies of consumer food waste over the last 10 years are relatively few – the FAO's 2019 meta-analysis reviewed 20 studies, 19 of which were for North America and one for Norway. The results, shown in Food and Agriculture Organization (2019, Figure 8, p. 39), of the eight studies of meat and animal products showed that over 2012–2017 around 14–33% were wasted (i.e. bought but not consumed). An earlier review of consumer food waste (Parfitt et al., 2010) found that total UK consumer food waste was around 25%. Consumer waste results from poor purchase and meal planning, excess buying (from over-large portions and package sizes), confusion over labels (best before and use by) and poor in-home storage (Food and Agriculture Organization, 2019, pp. xiii–xiv). 'Buy-one-get-one-free' offers can also contribute to food waste.[10]

Food loss and waste is a major topic in the IPCC's sixth assessment report (AR6), with 126 mentions spread across its 2900 pages. The major climate-related concern from food loss and waste is not the emissions that arise from waste management, although its anaerobic decomposition will of course contribute to methane (CH_4) emissions if it finds its way into landfill, but the fact that its production has involved considerable GHG emissions, water use and land use right down the supply chain, from agriculture (including from fertilisers and livestock) through transport, processing, packaging and retail, which could have been avoided had the wasted food not been produced.

The IPCC's AR6 estimate is that

> roughly 20–40% of food produced worldwide is lost to waste before it reaches the market, or is wasted by households, … and during the period 2010–2016 global food loss and waste equalled 8–10% of total GHG emissions.
>
> (IPCC, 2022a, pp. 5–26)

Globally, over 900 million tonnes of food waste were generated in 2019, of which 61% came from households, 26% from food service (hospitality) and 13% from retail (IPCC, 2022a, pp. 5–39). Reduced food loss and waste would

---

10 www.theguardian.com/business/2014/apr/06/buy-one-get-one-free-food-waste-supermarkets. Accessed April 28, 2023.

reduce GHG emissions just from the avoided agricultural production by 0.5–0.7 GtCO$_2$e per year (IPCC, 2022a, Table 12.3, pp. 12–19), but the full GHG emission reductions, taking account of changes in land use and emissions reductions along the full food value chain (i.e. from production through to consumption) could increase this emission reduction to 2.1–3.7 GtCO$_2$e per year. Other benefits would include "reducing environmental stress (e.g. water and land competition, land degradation, desertification), safeguarding food security, and reducing poverty" and would also "free up several million km^2 of land" (IPCC, 2022a, pp. 7–84).

Conclusions

Food production faces both ways on climate change. On the one hand, current methods of food production are responsible for up to a quarter of GHG emissions, and also account for 70% of global water use and have huge impacts on biodiversity. On the other, climate change is likely to have major negative impacts on food production in many parts of the world, while more food will be needed to feed growing populations, often in the same parts of the world where climate change is making its production more challenging.

At the same time very large quantities of food are either lost or wasted.

For all these reasons, systems of food production will need to be transformed. More sustainable forms of agriculture will need to be introduced almost everywhere. Diets need to change to reduce meat consumption in those countries that eat too much of it for their health. The land freed up for crops for human consumption will allow for a less chemically intensive agriculture that has lower impacts on ecosystems. Food loss can be reduced through development processes that allow farmers readier access to market. Food waste by consumers can be reduced by a range of incentive and regulatory policies.

None of these means of transformation are easy to implement. Table 10.7 shows that all parts of society have a role to play. But for these changes to be introduced there will need to be far greater social awareness of their need, engendering far greater political will by richer countries both to put their own agricultural houses and practices in order, and to help developing countries to adopt farming systems that reduce soil erosion, build soil fertility, have lower impacts on nature and allow farmers to get their produce to market in good condition. It is a huge agenda that goes well beyond climate change. But stopping climate change will not be achieved without these issues being effectively addressed.

Table 10.7 Potential solutions and barriers to food systems transformation by actor group

	Major transformation gaps	Potential solutions	Barriers
National governments	Absence of national strategy and clear measurable targets Lack of data and capacity Lack of key performance indicators to monitor progress Weak evaluation of externalities and incorporation into national accounting	Science-based national food systems transformation strategy and corresponding national coordination and accountability mechanism Open government data Integrate low carbon into national food and dietary guidelines Strengthen national land monitoring system for carbon reduction	Unbalanced power across ministries (and objectives) Lack of multisectoral coordination Acceptability of measures versus success at next elections
Cities and local governments	Carbon reduction is not part of the local and city government mandate Lack of awareness of carbon footprint of food systems	Strengthen coordination between national and city governments/local plans and policies Strengthen coordination between urban and rural areas Align public procurement with healthy and sustainable diets	Local economic development versus carbon reduction National versus local/ city interests
Private sector	Lack of commitments Lack of capacity Lobby against taxes and environmental regulations	Monitor and disclose progress towards environmental commitments Remove 'best before' label from fresh fruits and vegetables	Economic profitability versus social and environmental objectives
Civil society	Lack of knowledge and incentives Small number of platforms which enable involvement in decision-making Lack of resources (NGOs)	Social campaigns and social movements Mainstream low carbon into teaching curriculums NGOs develop score cards for companies	Budget constraints Well-being, cultural norms and preferences versus social and environmental goals
Academia	Science not fully aligned with societal needs Interdisciplinary approaches required but difficult to implement	Build strong science-policy interface between governments and academia Independent monitoring of progress towards targets related to food policy	Disciplinary funding structures and research traditions Independence/ separation of academia from policy processes

Source: United Nations Environment Programme, 2022, Table 6.5, p. 63.

References

Agnolucci, P., & De Lipsis, V. (2020). Long-run trend in agricultural yield and climatic factors in Europe. *Climatic Change, 159*(3), 385–405. https://doi.org/10.1007/s10 584-019-02622-3

Agnolucci, P., Rapti, C., Alexander, P., De Lipsis, V., Holland, R. A., Eigenbrod, F., & Ekins, P. (2020). Impacts of rising temperatures and farm management practices on global yields of 18 crops. *Nature Food, 1*(9), 562–571. https://doi.org/10.1038/s43 016-020-00148-x

Carr, T. W., Balkovič, J., Dodds, P. E., Folberth, C., Fulajtar, E., & Skalsky, R. (2020). Uncertainties, sensitivities and robustness of simulated water erosion in an EPIC-based global gridded crop model. *Biogeosciences, 17*(21), 5263–5283. https://doi.org/10.5194/bg-17-5263-2020

Climate Change Committee. (2019). *Net Zero Technical Report*. CCC, London.

Climate Change Committee. (2020). *The Sixth Carbon Budget: The UK'S Path to Net Zero*. CCC, London.

Department for Environment, Food and Rural Affairs (2020). *Agricultural Statistics and Climate Change*. https://agri-epicentre.com/wp-content/uploads/2021/01/Agric ultural-Statistics-and-Climate-Change-DEFRA-2020.pdf. Accessed May 6, 2023.

Dimbleby, H. (2021). *National Food Strategy: The Plan*. www.nationalfoodstrategy.org/. Accessed May 6, 2023.

Ebi, K. L., & Loladze, I. (2019). Elevated atmospheric CO_2 concentrations and climate change will affect our food's quality and quantity. *The Lancet Planetary Health, 3*(7), e283–e284. https://doi.org/10.1016/S2542-5196(19)30108-1

Ferronato, N., & Torretta, V. (2019). Waste mismanagement in developing countries: A review of global issues. *International Journal of Environmental Research and Public Health, 16*(6), 1060. https://doi.org/10.3390/ijerph16061060

Food and Agriculture Organization. (2009). *How to Feed the World in 2050*. FAO, Rome. www.fao.org/fileadmin/templates/wsfs/docs/expert_paper/How_to_Feed_th e_World_in_2050.pdf

Food and Agriculture Organization. (2014). *Building a Common Vision for Sustainable Food and Agriculture: Principles and Approaches*. FAO, Rome.

Food and Agriculture Organization. (2018). *The State of Agricultural Commodity Markets 2018. Agricultural trade, climate change and food security*. FAO, Rome.

Food and Agriculture Organization. (2019). *State of Food and Agriculture 2019: Moving forward on food loss and waste reduction*. FAO, Rome.

Food and Agriculture Organization. (2021a). *The State of Food and Agriculture 2021. Making agrifood systems more resilient to shocks and stresses*. FAO, Rome. https://doi.org/10.4060/cb4476en

Food and Agriculture Organization. (2021b). *World Food and Agriculture – Statistical Yearbook 2021*. FAO, Rome. https://doi.org/10.4060/cb4477en

Gaupp, F., Hall, J., Hochrainer-Stigler, S., & Dadson, S. (2020). Changing risks of simultaneous global breadbasket failure. *Nature Climate Change, 10*(1), 54–57. https://doi.org/10.1038/s41558-019-0600-z

Global Panel on Agriculture and Food Systems and Nutrition. (2016). *Managing Food Price Volatility: Policy Options to Support Healthy Diets and Nutrition in the Context of Uncertainty*. www.glopan.org/food-price-volatility/

International Resource Panel. (2019). *Global Resources Outlook 2019: Natural resources for the future we want*. UN Environment Programme, Nairobi.

IPCC. (2007). *Waste Management*. Bogner, J., M. Abdelrafie Ahmed, C. Diaz, A. Faaij, Q. Gao, S. Hashimoto, K. Mareckova, R. Pipatti, T. Zhang. In: B. Metz, O. R. Davidson, P. R. Bosch, R. Dave, L. A. Meyer (Eds.) *Climate Change 2007: Mitigation. Contribution of Working Group III to the Fourth Assessment Report of the Intergovernmental Panel on Climate Change*, Cambridge University Press, Cambridge, United Kingdom and New York, NY, USA.

IPCC. (2014). *Climate Change 2014: Mitigation of Climate Change. Contribution of Working Group III to the Fifth Assessment Report of the Intergovernmental Panel on Climate Change*. Edenhofer, O., R. Pichs-Madruga, Y. Sokona, E. Farahani, S. Kadner, K. Seyboth, A. Adler, I. Baum, S. Brunner, P. Eickemeier, B. Kriemann, J. Savolainen, S. Schlömer, C. von Stechow, T. Zwickel and J. C. Minx (Eds.). Cambridge University Press, Cambridge, United Kingdom and New York, NY, USA.

IPCC. (2018). *Global Warming of 1.5°C. An IPCC Special Report on the impacts of global warming of 1.5°C above pre-industrial levels and related global greenhouse gas emission pathways, in the context of strengthening the global response to the threat of climate change, sustainable development, and efforts to eradicate poverty*. Masson-Delmotte, V., P. Zhai, H.-O. Pörtner, D. Roberts, J. Skea, P.R. Shukla, A. Pirani, W. Moufouma-Okia, C. Péan, R. Pidcock, S. Connors, J. B. R. Matthews, Y. Chen, X. Zhou, M. I. Gomis, E. Lonnoy, T. Maycock, M. Tignor, and T. Waterfield (Eds.). Cambridge University Press, Cambridge, UK and New York, NY, USA, pp. 93–174. https://doi.org/10.1017/9781009157940.004

IPCC. (2019). Summary for Policymakers. In: P. R. Shukla, J. Skea, E. Calvo Buendia, V. Masson-Delmotte, H.- O. Pörtner, D. C. Roberts, P. Zhai, R. Slade, S. Connors, R. van Diemen, M. Ferrat, E. Haughey, S. Luz, S. Neogi, M. Pathak, J. Petzold, J. Portugal Pereira, P. Vyas, E. Huntley, K. Kissick, M. Belkacemi, J. Malley (Eds.) *Climate Change and Land: An IPCC Special Report on Climate Change, Desertification, Land Degradation, Sustainable Land Management, Food Security, and Greenhouse Gas Fluxes in Terrestrial Ecosystems*. https://doi.org/10.1017/978100 9157988.001

IPCC. (2021). *Climate Change 2021: The Physical Science Basis. Contribution of Working Group I to the Sixth Assessment Report of the Intergovernmental Panel on Climate Change*. Masson-Delmotte, V., P. Zhai, A. Pirani, S.L. Connors, C. Péan, S. Berger, N. Caud, Y. Chen, L. Goldfarb, M. I. Gomis, M. Huang, K. Leitzell, E. Lonnoy, J. B. R. Matthews, T. K. Maycock, T. Waterfield, O. Yelekçi, R. Yu, and B. Zhou (Eds.). Cambridge University Press, Cambridge, United Kingdom and New York, NY, USA, pp. 3–32, doi:10.1017/9781009157896.001

IPCC. (2022a). *Climate Change 2022: Mitigation of Climate Change. Contribution of Working Group III to the Sixth Assessment Report of the Intergovernmental Panel on Climate Change*. P. R. Shukla, J. Skea, R. Slade, A. Al Khourdajie, R. van Diemen, D. McCollum, M. Pathak, S. Some, P. Vyas, R. Fradera, M. Belkacemi, A. Hasija, G. Lisboa, S. Luz, J. Malley (Eds.). Cambridge University Press, Cambridge, UK and New York, NY, USA. doi: 10.1017/9781009157926.001

IPCC. (2022b). Summary for Policymakers In: *Climate Change 2022: Impacts, Adaptation, and Vulnerability. Contribution of Working Group II to the Sixth Assessment Report of the Intergovernmental Panel on Climate Change*. H.-O. Pörtner, D.C. Roberts, M. Tignor, E.S. Poloczanska, K. Mintenbeck, A. Alegría, M. Craig, S. Langsdorf, S. Löschke, V. Möller, A. Okem, B. Rama (Eds.). Cambridge University Press, Cambridge, UK and New York, NY, USA, pp. 2411–2538, doi:10.1017/ 9781009325844.025

Janetos, A., Justice, C., Jahn, M., Obersteiner, M., Glauber, J., & Mulhern, W. (2017). *The Risks of Multiple Breadbasket Failures in the 21st Century: A Science Research Agenda.* Boston University, Boston. www.bu.edu/pardee/files/2017/03/Multiple-Brea dbasket-Failures-Pardee-Report.pdf

Jang, W. S., Neff, J. C., Im, Y., Doro, L., & Herrick, J. E. (2021). The hidden costs of land degradation in US maize agriculture. *Earth's Future, 9*(2). https://doi.org/ 10.1029/2020EF001641

Johnke, B. (2003). *Emissions from Waste Incineration.* IPCC, Geneva.

Lobell, D. B., & Field, C. B. (2007). Global scale climate–crop yield relationships and the impacts of recent warming. *Environmental Research Letters, 2*(1), 014002. https:// doi.org/10.1088/1748-9326/2/1/014002

Ortiz, M., Baldock, D., Willan, C., & Dalin, C. (2021). *Towards Net Zero in UK Agriculture: Key information, perspectives and practical guidance.* www.ucl.ac.uk/ bartlett/news/2021/apr/uk-agriculture-pathways-net-zero. UCL/HSBC, London.

Ortiz-Bobea, A., Ault, T. R., Carrillo, C. M., Chambers, R. G., & Lobell, D. B. (2021). Anthropogenic climate change has slowed global agricultural productivity growth. *Nature Climate Change, 11*(4), 306–312. https://doi.org/10.1038/s41558-021-01000-1

Parfitt, J., Barthel, M., & Macnaughton, S. (2010). Food waste within food supply chains: quantification and potential for change to 2050. *Philosophical Transactions of the Royal Society B: Biological Sciences, 365*(1554), 3065–3081. https://doi.org/ 10.1098/rstb.2010.0126

Pendrill, F., Gardner, T. et al. (2022). Disentangling the numbers behind agriculture-driven deforestation, *Science, 377*(6611), September 9. https://doi.org/10.1126/scie nce.abm9267

Ray, D. K., Ramankutty, N., Mueller, N. D., West, P. C., & Foley, J. A. (2012). Recent patterns of crop yield growth and stagnation. *Nature Communications, 3*(1), 1293. https://doi.org/10.1038/ncomms2296

Tait, P. W., Brew, J., Che, A., Costanzo, A., Danyluk, A., Davis, M., Khalaf, A., McMahon, K., Watson, A., Rowcliff, K., & Bowles, D. (2020). The health impacts of waste incineration: A systematic review. *Australian and New Zealand Journal of Public Health, 44*(1), 40–48. https://doi.org/10.1111/1753-6405.12939

United Nations Environment Programme. (2022). *Emissions Gap Report 2022: The Closing Window – Climate crisis call for transformation of societies.* UNEP, Nairobi.

11 Economics of climate change mitigation

Summary

Earlier chapters have shown that carbon emission reduction can be brought about by a wide range of technologies. This chapter explores how the costs of these technologies have changed and may change in the future, and what the effects of their large-scale introduction will be on economic growth.

The most important economic characteristic of the clean energy transition is innovation, the process that creates new technologies and that leads to their cost reduction so that they can achieve large-scale deployment. This chapter first explores the nature of this process throughout the journey from the laboratory to the mainstream, and presents the extraordinary cost reductions in low-carbon technologies which have been brought about.

However, the large-scale global deployment of these technologies, to reach zero emissions by mid-century, will require huge levels of investment in low-carbon technologies. Such investment has increased greatly in recent years, but is still well short of where it needs to be. The issue is not shortage of capital *per se*, but a shortage of projects with the right risk-return ratio for it to invest in. There are now many initiatives in the financial sector to try to direct more capital into zero-carbon projects and technologies, but what is required to bring this investment about is greater zero-carbon ambition from businesses, more transparency in how they intend to achieve it, and government that is prepared to intervene strategically in markets to ensure that businesses can deliver large emissions reductions and remain profitable.

Investment is required not just in technologies, but also in the skills that are needed to manufacture and operate them effectively. The evidence suggests that the transition to clean energy will lead to a net increase in jobs, but there are challenges to ensure that those who lose their jobs in declining sectors are retrained to be able to take jobs in the low-carbon economy. With innovation and technical progress, and investment in

DOI: 10.4324/9781003438007-12

physical and human capital, it is likely that decarbonising economies will grow, and evidence to this effect is presented in this chapter. It is therefore surprising that a current of academic thought has arisen that posits that decarbonisation to zero emissions will require a reduction in economic output, in rich countries at least. This chapter argues that not only is this not necessary, but also that the suggestion itself is potentially very damaging politically in a world in which increased material living standards are still a major aspiration for most people.

Reducing and then eliminating fossil fuel use, and improving the diets of affluent people, will result in a number of benefits apart from greenhouse gas (GHG) emissions reduction, called 'co-benefits'. The largest of these is the health benefit from the improvement in outdoor air quality that would come from reducing fossil fuel use, and in indoor air quality from moving towards cooking and heating with cleaner fuels. Active travel in cities also has health benefits, as does reducing excess meat and dairy consumption. Afforestation and better land management that stores carbon should also improve ecosystems and lead to increased biodiversity. This chapter finishes by exploring the size and extent of these benefits.

Introduction

An obvious, and important, question that arises from the discussion in previous chapters of the many technologies that can reduce GHG emissions is: what will introducing them at the necessary scale to get to zero emissions actually cost? To answer that question it is necessary first to explore how technologies are created and the processes of innovation that take them to success and deployment at scale in markets.

Technology and innovation

The apparently simple question about the 'cost' of a technology can in fact mean a number of different things. The most obvious is: how much money needs to be spent to bring about mass deployment of these technologies? This is often referred to as the 'investment' needs of decarbonisation. However, because the costs of deploying technologies over time can change dramatically, as will be seen below, and because the deployment will need to take place over many years, it is not just the current costs of the technologies that need to be calculated. Their future costs over the relevant time period also need to be estimated, and, again, as will be seen, this is no simple task.

However, it is misleading simply to refer to these expenditures on technologies as 'costs'. As investments they will produce a range of economic benefits. They will employ people, they will generate incomes and, perhaps, exports. They will thus contribute to the gross domestic product (GDP) of the country in question, and, potentially, to GDP growth. At this macroeconomic level, the cost of decarbonisation is therefore computed as the difference in GDP at a certain date between a scenario with high carbon emissions and a scenario with low, or zero, carbon emissions. These are the issues which will be explored in this chapter, but it all starts with the costs of decarbonisation technologies now, and estimates of how they will develop in the future.

The Intergovernmental Panel on Climate Change (IPCC, 2022a, Figure SPM 7, p. SPM-50) estimated both the potential carbon saving and the estimated 2030 costs for the whole range of low- or zero-carbon technologies in energy, transport, buildings, industry, agriculture, forestry and other land use (AFOLU), waste, wastewater and fluorinated gases (F-gases). Many of these technologies have been discussed in earlier chapters. Taken from around 175 underlying sources, the costs are expressed in terms of US$/tCO$_2$e/year not emitted, compared to the cost of a reference high-carbon technology which would otherwise have been likely to be used. The costs and emission reduction potential for the main technologies are discussed here.

Energy
Wind and solar are the big hitters, able to save more than 2 GtCO$_2$e/year each by 2030, at a *negative* cost (i.e. the cost is *less* than that of the fossil fuel technology). Another 1–2 GtCO$_2$e/year can be saved by these technologies at a cost of less than US$50/tCO$_2$e/year. Bioelectricity, geothermal, nuclear and reducing upstream oil and gas emissions each save around 1 GtCO$_2$e/year, with costs that go up to around US$100/tCO$_2$e/year for each technology except bioelectricity, which is more expensive. Hydropower, reducing coal mine methane, Carbon Capture and Storage (CCS) and bioenergy with CCS (BECCS) save well under 1 GtCO$_2$e/year, with hydropower and coal mine methane reduction being fairly cheap (< US$50/tCO$_2$e) but CCS and BECCS being expensive (US$50–200/tCO$_2$e).

Buildings
Demand reduction (e.g. turning down thermostats, wearing more clothes in winter) and more efficient lighting, appliances and equipment save money (i.e. have *negative* cost) and together can save more than 1 GtCO$_2$e/year. So can very energy-efficient new buildings, but the costs there can go up to US$200/tCO$_2$e saved. On-site renewables save around 0.5 tCO$_2$e/year at a similar cost of high-performing buildings. Home retrofits (most relevant in colder climates) save around 0.25 tCO$_2$e/year, again at a cost of up to around US$200/tCO$_2$e/year.

Transport
All the main technologies listed for this sector are estimated to actually *save* money by 2030. They include shifting to electric vehicles (EVs), bikes or e-bikes, and adopting fuel-efficient heavy-duty vehicles, ships and aircraft. In total these negative-cost measures can save in excess of 1 GtCO$_2$e/year. Biofuels come in at a cost of up to US$100/tCO$_2$e/year, and save up to 0.5 GtCO$_2$e/year. The zero-carbon fuels for aviation and shipping (e.g. synfuels, bio-kerosene, hydrogen, ammonia) are not mentioned, but would be significantly more expensive.

Industry

Industry is shown as having no negative-cost mitigation opportunities, apart from small no-cost reductions in F-gases and methane from solid waste. Its cheapest opportunities, at less than US$20/tCO$_2$e/year, lie in energy efficiency, which in 2030 can save over 1 GtCO$_2$e/year at that cost. At under US$50/tCO$_2$e/year, material efficiency and enhanced recycling can save around 1.5 GtCO$_2$e/year. Fuel-switching (e.g. to hydrogen) has about the same opportunity at this cost, but can save 2 GtCO$_2$e/year at under US$100/tCO$_2$e. Changing feedstocks and industrial processes (e.g. in cement and chemicals) can save another 0.5 GtCO$_2$e/year at US$50–100/tCO$_2$e.

AFOLU

CO$_2$ sequestration, and reducing emissions of methane and N$_2$O in agriculture can save around 2 GtCO$_2$e/year, at a cost below US$50/tCO$_2$e, making this the most cost-effective option after wind and solar. Ecosystem restoration, afforestation, reforestation and improved sustainable forest management can save nearly 1.5 GtCO$_2$e/year, but at higher cost, up to US$100/tCO$_2$e. And as was discussed in Chapter 7, there is always a chance that the CO$_2$ from such measures can be released if the forests are later degraded or destroyed (e.g. by wildfires during heat waves).

While the IPCC does not make this calculation, a quick summation down the bars of Figure SPM 7 in IPCC (2022a, p. SPM-50), suggests that over half of current GHG emissions (59 GtCO$_2$e) could be reduced by technologies that, by 2030, will cost less than US$100 per tonne CO$_2$e.

The IPCC is at pains to stress the uncertainties involved in such cost estimates, which could turn out to be higher or lower. In addition, the estimated costs obviously depend on the assumptions made about the technologies which the low-carbon technologies will replace, and how the costs of both sets of technologies will develop between now and 2030. The costs will also vary between countries according to their technological capacities and economic and political contexts. The renewable electricity costs, for example, may not include the costs of incorporating them into electricity systems, which may become significant after 2030 as their proportion in electricity systems increases beyond about 20%.

However, the estimates above also do not contain the avoided costs from climate change, which is of course the main purpose of introducing them. Nor do they include the other benefits, called 'co-benefits' which substituting these technologies for fossil fuel technologies would bring about. The main co-benefit is the reduction of local air pollution associated with burning fossil fuels, but the co-benefits also include reduced traffic congestion, and more potential green space in cities (from the shift away from cars), less pollution of rivers from reduced use of fertilisers, and reduced health problems from obesity (from the adoption of healthier diets). Estimates of these co-benefits are given later in this chapter.

One of the major uncertainties of the cost estimates of the various technologies above, as has already been noted, is how they will develop over time.

This is well illustrated by how the costs of several of these technologies have in fact evolved over the past couple of decades.

The IPCC has shown the extent to which the cost of key renewables technologies has fallen over 20 years: by more than a factor of six for solar photovoltaics (PV), more than half for onshore wind and concentrated solar power (CSP), and more than half for offshore wind since 2015 (IPCC, 2022a, p. SPM-13). All four of these technologies now have costs in the range of fossil fuel generation costs, with PV costs falling from around US\$600/MWh in 2000 to around US\$50/MWh in 2020. It is especially encouraging that batteries for EVs have experienced an even faster cost reduction than PV, with their cost coming down from around US\$1400/kWh in 2005 to less than US\$100/kWh in 2020. However, enthusiasm over these cost reductions, and the very fast rates of increase of adoption of these technologies that they have stimulated, must be tempered by the very small proportions of the global electricity supply that these technologies still account for (see Figure 4.1). There is still a very long way to go before these technologies comprise even a significant share, let alone the majority, of their markets globally.

The technology cost reductions cited in the previous paragraph have not come about by accident. They are the result of government policies that supported research into and development of these technologies, and then created early markets for them, following which private sector companies deployed them at scale. In this way, public policy, aided in the later stages by markets, has driven these technologies through what are called Technology Readiness Levels (TRLs), which are described in Table 11.1.

The idea of TRLs was introduced by NASA (the US National Aeronautical and Space Administration), originally in the context of space technology, to describe the various stages through which a technology must pass, from a basic conception in research, through to operational deployment. TRLs are now

Table 11.1 Technology readiness levels (TRLs)

Research	TRL 1 – Basic principles observed and reported.
	TRL 2 – Technology concept and/or application formulated.
	TRL 3 – Analytical and experimental proof of concept.
Development	TRL 4 – Component or technology validated in laboratory.
	TRL 5 – Component or technology validated in an industrially relevant environment.
	TRL 6 – System or sub-system model or technology prototype demonstrated in relevant environment.
Deployment	TRL 7 – System prototype demonstration in operational environment.
	TRL 8 – Actual system completed through test and demonstration.
	TRL 9 – Actual system proven through successful operation.

Source: Adapted from NASA, 2012, www.nasa.gov/directorates/heo/scan/engineering/technology/technology_readiness_level. Accessed April 24, 2023.

used to describe this journey for technologies in general. Universities and corporate research laboratories tend to concentrate on TRLs 1–4. At that point the technology tends to face what is sometimes called 'the Valley of Death', often requiring substantial sums of money for it to be developed and demonstrated in a real-life industrial environment. Those technologies that are successful at that stage, passing through TRLs 5 and 6, are then ready for deployment, first in niche markets and then, if successful there, at scale.

This very much describes the journeys of the technologies shown in IPCC (2022a, Figure SPM 7, p. SPM-50), and described at the beginning of this chapter. At each stage of this innovation journey, the technology may experience cost reduction through learning: 'learning by research' in TRLs 1–4, and 'learning by doing' thereafter. 'Learning by doing' becomes particularly important as the technology is deployed at scale. The relationship between the unit cost and the cumulative deployment of a technology is called a 'learning (or experience) curve', which gives the percentage reduction in the cost of a technology with a doubling of its cumulative deployment.

But it would be a mistake to imagine that this journey of technologies through the TRLs is brought about by technological innovation and development alone. On the contrary, large-scale technological deployment requires innovation in a wide range of social and economic systems as well, including business organisation, business models and supply chains, consumer preferences and behaviours, financing, regulation and standard-setting in markets, institutions of research and governance (e.g. patenting) and infrastructure (e.g. smart grids, charging points for EVs) (Grubb et al., 2017). Public policy has an absolutely critical role to play in stimulating and setting the direction of innovation in all these areas such that it results in low- and zero-carbon emission outcomes, rather than a continuing dependency on fossil fuels.

The Fraunhofer Institut 2023 Photovoltaics Report shows the cost development for a large PV system between 2010 and 2021 (Fraunhofer Institut, 2023, p. 44[1]). In 2010 the cost was EUR 0.315/kWh. By 2021 the cost had fallen to EUR 0.041/kWh – a fall of 87%, with a year-on-year decrease in the cost of electricity of 17% (Fraunhofer Institut, 2023, p. 44). Over the same period global cumulative installed capacity increased more than 20-fold, from about 40 GW to 940 GW.[2]

The International Renewable Energy Agency (IRENA) (2022, Figure 1.2, p. 32) gives the costs of a range of power generation technologies over 2010–2021, derived from its large database of actually installed electricity projects. These costs are shown in Table 11.2 and have a number of points of interest.

1 www.ise.fraunhofer.de/content/dam/ise/de/documents/publications/studies/Photovoltaics-Report.pdf, Accessed April 24, 2023.
2 www.statista.com/statistics/280220/global-cumulative-installed-solar-pv-capacity/#:~:text=Global%20cumulative%20installed%20solar%20PV%20capacity%202000%2D2021&text=Global%20cumulative%20solar%20photovoltaic%20capacity,installed%20in%20that%20same%20year. Accessed April 24, 2023.

Table 11.2 Unit cost developments in a range of power generation technologies, 2010–2021

	Mean cost (2021 US$/kWh)	
	2010	*2021*
Bioenergy	0.078	0.067
Geothermal	0.050	0.068
Hydropower	0.039	0.048
Solar photovoltaic	0.417	0.048
Onshore wind	0.102	0.033
Offshore wind	0.188	0.075
Concentrated solar power	0.358	0.114
Fossil fuel electricity	0.075	0.067
Fossil fuel cost range	Approx. US$0.03–0.18/kWh	

Source: Adapted from International Renewable Energy Agency, 2022, Figure 1.2, p. 32.

First, Table 11.2 shows that not all power generation technologies have experienced a reduction in costs since 2010. The unit costs of both geothermal and hydropower have increased since 2010. Second, it shows that fossil fuel generation has also come down in costs, though this may not be the case in 2022, with the great increase in oil and gas prices. Third, it shows that the cost reductions in PV have been exceptional. The costs of wind and concentrated solar power (CSP) have come down less quickly. Fourth, it shows that in 2021 all the renewable generation technologies are either below the mean cost of fossil fuel electricity (solar PV, onshore wind, hydropower), or well within that band (offshore wind, geothermal, CSP).

The cost reductions cited above have both driven, and been driven by, the very fast rates of deployment of these technologies. The innovation that has brought about these cost reductions has been critical in giving the world any chance at all of reducing carbon emissions to the extent necessary to avoid devastating climate change. The hope must be that similar rates of deployment of a wide range of the technologies shown in IPCC (2022a, Figure SPM 7, p. SPM-50), and described above, will bring about similar cost reductions in the future, through 2030 and beyond.

However, there is no certainty that cost reductions for technologies will occur along with their cumulative deployment, as Table 11.2 makes clear – the costs of both geothermal and hydropower shown there actually increased over 2010–2021. However, to estimate the costs of getting to net zero by 2050, some estimate of future cost development must be made. Way et al. (2022, p. 4) note that "Successful technologies tend to follow an 'S-curve' for deployment, starting with a long phase of exponential growth in production that eventually tapers off due to market saturation". They find that, for their four chosen technologies (PV, batteries, wind, electrolysers), experience curves derived from past data (e.g. 1976–1990) actually predict future prices (e.g. to 2020) remarkably well, and they use these curves to project the prices of these key low-carbon technologies into the future.

Way et al. (2022, Figure 5C, p. 9) shows their results for batteries, the future price development of which will be critically important for the success or otherwise of EVs. The main figure shows the cost development of two kinds of battery – Li-ion consumer cell batteries and Li-EV battery packs. In 1995 the cost of the former was around US$5,000/kWh and in 2010 the cost of the latter was around US$1,000/kWh. By 2020 the cost of these had broadly converged at around US$150/kWh. Way et al. (2022, Figure 5C, p. 9) also show various cost projections for EV batteries from Integrated Assessment Models (IAMs) or the IEA, which have consistently underestimated the cost reductions that have in fact taken place.

Way et al. (2022) further estimated for three different scenarios the probabilities of further cost reductions in their four technologies on the basis of the cost reductions that had occurred in the past. The three scenarios envisaged No Transition, with deployment of these technologies on the basis of past rates of deployment, a Slow Transition with increased rates of deployment, and a Fast Transition with much increased rates of deployment. The key message from this exercise was that the cost of these technologies depends on how fast they are deployed, with most cost reduction occurring in the scenario with the highest deployment (Fast Transition).

A note of caution around these results may be introduced related to the discussion of those minerals that are critical for the low-carbon energy technologies, discussed in Chapter 8. If there is a failure to produce enough of these minerals over the requisite timescale, or to develop technologies that are less dependent on them, then their increase in cost might negate the cost reductions that might otherwise occur through a Fast Transition scenario. There were signs in 2022 that this might already be happening. The IEA reported in 2022 that key material costs for solar PV (silicon metal), wind turbines (steel) and lithium batteries (lithium) had all risen substantially, increasing the overall costs of solar PV and wind turbines in 2021 and slowing the cost reduction in batteries ((International Energy Agency, 2022, p. 20).

Of course, achieving large-scale deployment of any technology requires significant investment and it is the scale of the investment required, and where it is likely to come from, that is the subject of the next section.

Investment, finance and business

This chapter quotes extensively from the IEA's *World Energy Investment* reports from both 2021 and 2022, because the latter saw some departures from trend due to the Russia/Ukraine war, and because some of the breakdowns in the earlier version of the report are more appropriate for this book. The 2023 *World Energy Investment* was published shortly before this book went to press, and there is a brief review of the numbers there too, after the discussion of the 2021 and 2022 editions.

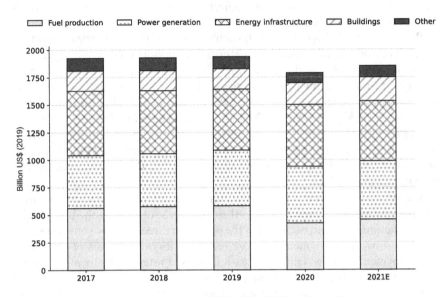

Figure 11.1 Global energy investment, 2017–2021 (2021 estimated).
Note: Energy infrastructure includes midstream and downstream oil and gas infrastructure, electricity networks and batteries.
Source: International Energy Agency, 2021, p. 6.

In the normal course of events the world invests a large amount in energy, just short of US$2 trillion per year in 2017–2019, with a dip because of Covid in 2020, and an estimated rebound in 2021 (Figure 11.1).

This investment is broken down by sector in Figure 11.2. While it is encouraging that 70% of 2021 investment in power generation was estimated to go to renewables (and this share increased to an estimated 80% in 2022), less positive is that fossil fuel (coal and gas) power generation was still estimated to attract US$100 billion, and US$350 billion a year is still being spent upstream (i.e. exploration and production) in oil and gas, when the IEA makes clear that any new oil and gas wells or coal mines are inconsistent with its Net Zero Emissions (NZE) trajectory. The war between Russia and Ukraine in 2022 increased investment in fossil fuels – coal as well as oil and fossil methane gas (International Energy Agency, 2022, p. 16), but 2021 had also seen the approval of 30 GW of new coal-fired power plants (International Energy Agency, 2022, p. 43), which hardly reflects a 'phase-down' of coal – something to which countries committed themselves at the COP26 climate conference at the end of that year.

Total investment in clean energy technologies was expected to be over US$1,400 billion in 2022 (International Energy Agency, 2022, p. 11), with investment growing in both renewable power and energy efficiency, where an energy efficiency investment is defined as the incremental spending on

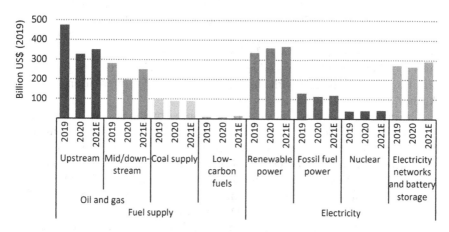

Figure 11.2 Global energy investment by sector, 2017–2021 (2021 estimated).

Source: International Energy Agency, 2021, p. 7.

new energy-efficient equipment or the full cost of refurbishments that reduce energy use. However, the IEA estimates that this will need to double even to meet all the Nationally Determined Contributions (NDCs) and pledges made at COP26 (broadly consistent with the 2°C Paris target), and triple to meet the 1.5°C target (International Energy Agency, 2022, p. 24). The good news, as will be seen, is that such investment levels are perfectly feasible if the investment incentives (i.e. the ratios of risk to return) are right. The bad news is that there is currently a chronic shortage of projects that satisfy this requirement.

The IEA's 2021 conclusion was "Today's investment spending on fuels appears caught between two worlds: neither strong enough to satisfy current fossil fuel consumption trends nor diversified enough to meet tomorrow's clean energy goals" (International Energy Agency, 2021, pp. 11–12). Oil and gas prices started to rise in response to this situation in the latter part of 2021, a trend that was reinforced with a vengeance in 2022, especially in Europe, by the war in Ukraine. At the time of writing, it is still not clear what this will do to energy investment flows. In his 'A Call to Clean Energy' published in December 2022,[3] the IEA Executive Director noted that the high oil and gas prices could stimulate increased investment in renewables, or fossil fuels, but at the time was doing neither, running the risk of "the worst of both worlds" – ongoing high fossil fuel prices and excessive carbon emissions.

3 Article in *Finance and Development*, published by the International Monetary Fund (IMF), December 22, 2022, www.imf.org/en/Publications/fandd/issues/2022/12/a-call-to-clean-energy-fatih-birol. Accessed April 24, 2023.

The scale of the investment challenge is most stark in the electricity sector, in which investments over 2011–2020, at around US$1.5 trillion for the ten years, are not at the required level even to meet the 2026–2030 requirements of the IEA's Stated Policies Scenario (STEPS), let alone its Sustainable Development Scenario (SDS, broadly compatible with the Paris 2°C target) or its NZE scenario (broadly compatible with the Paris 1.5°C target) (International Energy Agency, 2021, p. 22).

The increase in energy efficiency (as measured by the decline in the energy intensity of the world economy[4]) was only 0.8% in 2020, and needs to increase to 4% per year according to the NZE scenario. This too will require a significant increase in investment levels, in both energy efficiency and more efficient end use. The IEA estimates that the US$300 billion total for 2016–2020 energy efficiency investments will need to increase in 2026–2030 to US$700 billion for STEPS, to US$1.1 trillion for SDS, and to over US$1.4 trillion for NZE (International Energy Agency, 2021, p. 49). Of course, the greater the increase in energy efficiency, the lower the investment needs in new energy (electricity) supply. If the investments in energy efficiency projections in these three scenarios do not materialise, then the scale of the investments in new supply will have to be correspondingly greater.

The UNEP 2022 *Emission Gap Report* (United Nations Environment Programme, 2022) confirmed this substantial shortfall in the investment to 2030 required to achieve net-zero emissions in 2050. The total investment required is US$4–6 trillion per year, a factor increase of 3–6 across the global economy (United Nations Environment Programme, 2022, p. XXVI). Table 11.3 shows that the shortfall is across all economic sectors and regions, with AFOLU among the sectors, and lower- and low-income countries, particularly falling short. The investment challenge is especially great for the Middle East, with its current heavy reliance on fossil fuels.

However, a brighter spot on the investment landscape was the investment in batteries for energy storage, which, as discussed in Chapter 5, will be critical for the stability of power systems when they have a high share of renewables. The IEA estimated that battery storage investment doubled in 2022, with increases both at grid scale and 'behind the meter' (e.g. linked to roof-top PV) (International Energy Agency, 2022, p. 54). The IEA's 2023 *World Energy Investment* report (International Energy Agency, 2023) had both good and bad news. On the positive side, estimated 2023 investment in clean energy rose from around US$1.6 trillion in 2022 to more than US$1.7 trillion, with the investment led by solar PV, but also with substantial investments in energy efficiency and battery storage, which are crucial complements to investment in renewable power, as has been seen. The bad news is that over US$1 trillion was invested in fossil fuel supply and unabated power, including US$500 billion in upstream oil and gas. Given that much of this investment will become stranded if the world

---

4 Energy intensity is energy use divided by economic output.

Table 11.3 Mitigation investment needs by sector, type of economy and region to 2030 for net zero by 2050

Sector	Approx. current annual investment 2017–2020 (US$ billion)	Factor increase in required annual mitigation investment to 2030	
		Low	High
Energy efficiency	250	x2	x7
Transport	150	x7	x7
Electricity	300	x2	x5
Agriculture, forestry and other land use (AFOLU)	50	x10	x31
Type of economy			
Developing countries	400	x4 (4% GDP)	x7 (9% GDP)
Developed countries	350	x3 (2% GDP)	x5 (4% GDP)
Region			
East Asia	250	x2	x4
North America	150	x3	x6
Europe	200	x2	x4
South Asia	50	x7	x14
Latin America and Caribbean	50	x4	x8
Japan, Australia and New Zealand	50	x3	x7
Eastern Europe and West Central Asia	<50	x7	x15
Africa	<50	x5	x12
South East Asia and developing Pacific	<50	x6	x12
Middle East	<50	x14	x28

Source: Compiled from data in United Nations Environment Programme, 2022, Figure 7.1, p. 66, itself adapted from IPCC, 2022b, Figure TS.25, p. 134.

gets on track to limit global warming to 1.5°C, this shows that the fossil fuel industry is still betting heavily against this target being achieved. Also negative was the very small proportion of the investment in clean energy that took place in emerging market and developing economies – except for China, which saw a larger increase in estimated investment over 2019–2023 than either the European Union (EU) or the USA. Despite these increases in investment, the gap between actual investment and that required in the IEA's NZE scenario remains large.

The low-carbon energy transition will not take place unless the very large required investments that have been identified in this section are mobilised and committed to the low-carbon technologies that require them – what is sometimes called 'shifting the trillions'. Polzin & Sanders (2020) show that in principle the necessary investment is available in aggregate in the required quantity, but that there is often a mismatch between the types of funding that are available and the kinds of projects, or stage of the TRL journey, that they are able to fund. In addition to this, Hafner et al. (2019, p. 6) identify numerous barriers to investment in activities to reduce emissions, including:

- immaturity or lack of climate change policy frameworks and stable policies;
- policies in favour of 'brown' energy infrastructure (e.g. fossil fuel subsidies);
- constraints on decision making within investor companies;
- returns on renewable infrastructure investments being too low to justify high initial capital investments, leading to limited projects with acceptable risk-return profiles;
- projects lacking an adequate credit rating;
- risk associated with uncertain and unproved technologies;
- lack of transparency of climate-related disclosure and data;
- lack of suitable financial vehicles/financial instruments;
- high transaction costs or fees;
- lack of knowledge/technical advice on green infrastructure investment.

A particular problem in emerging market and developing economies is the cost of capital. Financing costs can account for up to 60% of the total cost of renewables, and the cost of capital (CoC) in emerging market and developing economies can be seven times that in developed economies, which reflects the level of risk perceived by private investors (International Energy Agency, 2022, p. 39).

Table 11.4 shows the difference that this makes to the cost of solar PV projects. Sudan and Democratic Republic of the Congo (DRC), with much more sun than Switzerland, nevertheless generate more costly solar electricity because their CoC is eight and four times higher respectively. This explains why

Table 11.4 Cost of capital (CoC) for solar PV projects in selected countries (the source gives 152 countries), and the effect on the levelised cost of energy (LCOE) in those countries

Country	Cost of capital for solar PV projects (%)	Cost of electricity generated by solar PV (€/MWh)
Sudan	25.2	32
Democratic Republic of the Congo (DRC)	12.1	21
Italy	6.1	15
Peru	4.7	10
South Korea	3.8	12
Switzerland	3.1	13

Source: Egli et al., 2019, Figure 1, p. 2.

Note: LCOE is a standard way of calculating the cost of an energy technology.

many developing economies, despite having much better solar resources, get far less investment in solar PV.

This, and barriers identified by Hafner et al. (2019), comprise a formidable list of problems that will need to be solved if investments at the necessary scale are to flow into low-carbon technologies and infrastructure, especially in emerging market and developing economies. Through the setting up of such financial sector initiatives as the Paris Aligned Asset Owners, the 53 members of which are said to have around US$3.3 trillion under management,[5] or the Glasgow Financial Alliance for Net Zero (GFANZ),[6] with its more than 500 members and reputed US$133 trillion under management or advice,[7] it would appear that the financial sector has the appetite for low-carbon investment.

Climate change represents substantial risk to investors in two distinct ways. The first risk derives from potential impacts from climate change. Any business that depends on resources that may be affected by climate change (e.g. energy, food, travel and tourism) is clearly subject to such risks. The second risk derives from businesses' own effect on the climate (e.g. fossil fuel companies). These are clearly at risk both from the sentiments of investors who care about climate change, and may avoid or engage critically with such companies, affecting their reputations, and from policymakers who seek to mitigate climate change in ways that affect their businesses (e.g. through carbon taxes, emissions trading systems, or, in the limit, banning certain uses of fossil fuels leading to stranded assets). These two sources of risk comprise the 'double materiality' of companies' exposure to climate change.[8]

It was to enable investors to understand these risks that the Financial Stability Board in 2015 set up the Task Force on Climate-related Financial Disclosure (TCFD), which released its first guidance report in 2017,[9] and recommendations as to the implementation of that guidance by both the financial and other climate-exposed sectors (energy, transportation, materials and buildings, agriculture, food and forest products) in 2021.[10] While the 2022 TCFD status report concluded that more companies each year were reporting in line with at least some of the TCFD guidance and recommendations, it was also clear that *"more urgent progress* is needed in improving transparency on the actual and potential impact of climate change on companies" (The Task Force on Climate-related Financial Disclosures, 2022, p. 3).

5 Paris Aligned Asset Owners – Investing for a net zero future. www.parisalignedassetowners. org/. Accessed April 25, 2023.

6 www.gfanzero.com/about/. Accessed April 25, 2023.

7 www.argusmedia.com/en/news/2359675-gfanz-issues-draft-framework-for-net-zero-portfolios. Accessed April 25, 2023.

8 For a longer discussion of this, see www.lse.ac.uk/granthaminstitute/news/double-materiality-what-is-it-and-why-does-it-matter/. Accessed April 10, 2023.

9 https://assets.bbhub.io/company/sites/60/2021/10/FINAL-2017-TCFD-Report.pdf. Accessed April 10, 2023.

10 https://assets.bbhub.io/company/sites/60/2021/07/2021-TCFD-Implementing_Guidance.pdf. Accessed April 10, 2023.

Undoubtedly one of the barriers to businesses taking action on climate (and other environmental) issues is the plethora of reporting standards and methodologies with which they are confronted, and the complexities of calculating and reporting businesses' environmental impacts – measuring Scope 3 emissions (see Chapter 4) from global value chains can be particularly challenging. In order to standardise corporate reporting on climate and broader sustainability issues, in 2021–2022 the International Financial Reporting Standards (IFRS) Foundation set up the International Sustainability Standards Board (ISSB), which in 2022 released for consultation its Exposure Draft of recommendations for climate-related disclosure.[11] In June 2023 it was announced that the IFRS Foundation will take over from the TCFD the monitoring of progress on companies' climate-related disclosures in 2024.[12] It is to be hoped that the ISSB standards will simplify the sustainability reporting landscape and will be readily agreed across both the financial and business communities, so that the information disclosed is material, comprehensive and comparable between companies.

The risk from climate change to the financial system is also being taken seriously by central banks and supervisors, who in 2017 set up the Network for Greening the Financial System (NFGS),[13] which in March 2023 had 123 members of central banks and supervisors. Governments are also legislating for disclosure of climate risks to become mandatory, for example, the EU's 2019 Sustainable Finance Disclosure Regulation (SFDR).[14] Furthermore, in order both to facilitate investments into zero-carbon and other environmentally beneficial areas, and to prevent misleading environmental claims ('greenwashing') being made by the plethora of new supposedly 'sustainable' funds, many countries have now set up a taxonomy to define the criteria according to which such claims can be made. One of the most advanced of these taxonomies is the one recently agreed by the EU, which describes it as an "EU-wide classification system for sustainable activities".[15] The EU Taxonomy, embodied in a Regulation in 2020, grew out of the EU's 2018 Sustainable Finance Action Plan, which, in addition to the Taxonomy, envisaged an EU Regulation on investors' duties and disclosures, and benchmarks for different economic activities to identify those which were most sustainable.

The Taxonomy itself sets out six environmental objectives (climate change mitigation, climate change adaptation, the sustainable use and protection of water and marine resources, the transition to a circular economy, pollution

---

11 www.ifrs.org/content/dam/ifrs/project/climate-related-disclosures/issb-exposure-draft-2022-2-climate-related-disclosures.pdf. Accessed April 10, 2023.

12 www.ifrs.org/news-and-events/news/2023/07/ifrs-foundation-publishes-comparison-of-ifrs-s2-with-the-tcfd-recommendations/. Accessed August 16, 2023.

13 www.ngfs.net/en/about-us/governance/origin-and-purpose. Accessed April 10, 2023.

14 https://eur-lex.europa.eu/legal-content/EN/TXT/?uri=celex%3A32019R2088. Accessed April 10, 2023.

15 https://finance.ec.europa.eu/sustainable-finance/tools-and-standards/eu-taxonomy-sustainable-activities_en. Accessed April 25, 2023.

prevention and control, and the protection and restoration of biodiversity and ecosystems). To be 'sustainable' under the Taxonomy, an investment must make a positive contribution to at least one of these objectives, while doing no significant harm to any of the others.

Both GFANZ and the EU Taxonomy are at the centre of controversy. A 2023 report from a climate research and campaigning group, Reclaim Finance,[16] found that GFANZ members were still investing heavily in fossil fuels, accusing GFANZ of 'greenwashing' and its members of being 'climate arsonists'.[17] However, a number of investors withdrew, or threatened to withdraw, from GFANZ in late 2022, for a variety of reasons, including excessive commitments to decarbonisation, hostility from US Republicans to any decarbonisation commitments from investors, and concerns about potential conflicts with investment regulations.[18] GFANZ itself watered down its membership commitments by dropping its requirement that its members should align themselves with the UN's 'Race to Zero' campaign.[19] The exit from GFANZ continued in 2023, with Munich Re withdrawing from the Net Zero Insurance Alliance (part of GFANZ) on anti-trust concerns (though it remains a member of the Net Zero Asset Owners Alliance, another part of GFANZ),[20] and the ethical bank GLS withdrawing because other GFANZ members were still heavy investors in fossil fuels.[21] It remains to be seen whether GFANZ can weather these legal uncertainties and political storms.[22]

With respect to the EU Taxonomy, a row erupted in 2022 when the European Commission proposed to include in the list of 'sustainable activities' both nuclear power and certain fossil methane gas investments, which opponents again characterised as blatant examples of 'greenwashing'. However, the European Parliament passed the Commission's proposal in July 2022, whereupon the governments of Austria and Luxembourg, as well as Greenpeace, said that they would challenge this decision in the courts.[23]

16 https://reclaimfinance.org/site/en/2023/01/17/throwing-fuel-on-the-fire-gfanz-members-prov ide-billions-in-finance-for-fossil-fuel-expansion/. Accessed April 25, 2023.

17 www.theguardian.com/environment/2023/jan/17/banks-still-investing-heavily-in-fossil-fuels-despite-net-zero-pledges-study. Accessed April 25, 2023.

18 *Financial Times*, December 7, 2022, www.ft.com/content/48c1793c-3e31-4ab4-ab02-fd5e94b64 f6b. Accessed April 25, 2023.

19 www.esgtoday.com/mark-carney-led-gfanz-drops-requirement-for-race-to-zero-commitment/. Accessed August 16, 2023.

20 www.insurtechinsights.com/munich-re-has-announced-its-withdrawal-gfanz-due-to-material-legal-risks/. Accessed August 16, 2023.

21 https://capitalmonitor.ai/sdgs/sdg-13-climate-action/gfanz-why-gls-banks-departure-is-a-seri ous-setback/. Accessed August 16, 2023.

22 www.reuters.com/business/sustainable-business/esg-watch-is-it-curtains-mark-carneys-green-alliance-or-just-teething-problems-2023-04-26/. Accessed August 16, 2023.

23 www.euronews.com/my-europe/2022/07/06/meps-back-controversial-eu-plan-to-label-nuclear-and-gas-investments-as-green. Accessed April 25, 2023.

The purpose of these taxonomies and 'sustainable funds' is to facilitate investment in the businesses that are actually going to have to deliver 'real zero'. There are now numerous organisations of businesses that claim that their members have credible commitments to decarbonisation in line with the temperature targets of the Paris Agreement. For example, the Climate Group claims a membership of over 500 multinational businesses in 175 markets.[24] The Science Based Targets initiative (SBTi) describes itself as "a global body enabling businesses to set ambitious emissions reduction targets in line with the latest climate science" (Science-Based Targets Initiative, 2022, p. 3).

According to the Carbon Trust, "a carbon emissions target is defined as science-based if it is in line with the scale of reductions required to keep [the] global temperature increase below 2°C above pre-industrial temperatures".[25] The targets derive from a complicated methodology that seeks to allocate the carbon budgets for 2°C (or 1.5°C) to individual sectors, and thence to individual companies within those sectors. For 1.5°C this should involve a commitment to net-zero emissions by 2050. Founded in 2015, the SBTi's 2021 Progress Report claims that it is now working with 2,253 companies, of which 1,082 have science-based targets approved by SBTi, and the rest of which have commitments to adopt such targets. These companies have a market capitalisation of US$38 trillion, one third of the total (Science-Based Target Initiative, 2022, p. 6). As of April 2023, SBTi's website claimed it was working with 4,799 companies, of which 1,748 had net-zero commitments.[26] However, despite the increase in commitments to net zero, serious concerns remain about their credibility, with a June 2023 analysis finding that only 4% of net-zero corporate commitments met the 'Starting Line criteria' of the UN Race to Zero campaign.[27]

Despite all the initiatives in the financial sector, such as GFANZ, NFGS and a sustainability taxonomy in different countries, so far the barriers to low-carbon investment, and the continuing financial attraction of fossil fuel investments, seem to be preventing investors from committing their resources to the clean energy transition at anything like the required scale. For investors, it is clear that they do not yet take the risk of 'stranded assets' referred to in Chapter 4 at all seriously. Both the financial sector itself, and the policymakers that set the economic context in which they invest, have a huge task to dismantle these barriers to investment.

In respect of businesses, it is clear that initiatives such as the Climate Group and SBTi are not yet having a material effect on global emissions, which are

24 www.theclimategroup.org/about-us. Accessed April 10, 2023.
25 www.carbontrust.com/news-and-insights/insights/what-exactly-is-a-science-based-target. Accessed April 10, 2023.
26 https://sciencebasedtargets.org/. Accessed April 10, 2023.
27 https://zerotracker.net/insights/net-zero-targets-among-worlds-largest-companies-double-but-credibility-gaps-undermine-progress. Accessed August 16, 2023.

Table 11.5 Global renewable energy employment by technology in 2020

	Jobs (thousands)
Solar photovoltaic	3975
Liquid biofuels	2411
Hydropower	2182
Wind energy	1254
Solar heating/cooling	819
Solid biomass	765
Biogas	339
Geothermal energy	96
Municipal and industrial waste	39
CSP	32
Others	105

Source: International Renewable Energy Agency, 2021, Figure 4, p. 20.

Note: CSP is concentrated solar power; Others include tide, wave and ocean energy, and jobs not broken down by particular technology.

still rising. Whatever businesses are doing to reduce their emissions, it is clear that they still need to do a lot more.

Labour markets, employment and skills

It is not only technologies that require huge investments. There is also the labour force that is required to design, build and maintain them. IRENA estimates that in 2020 the renewable energy sector employed around 12 million people globally (International Renewable Energy Agency, 2021), with the breakdown across renewables technologies as in Table 11.5.

At present China has over a third of these jobs, as shown in Table 11.6. All countries face the challenge, as the world decarbonises, of growing their renewable energy sector and capturing some of the enormous job opportunities that will be provided by these renewable technologies – IRENA estimates that, in a scenario compliant with the Paris 1.5°C target, employment in this sector could grow to 38 million jobs by 2030 and 43 million by 2050. Some jobs (IRENA estimates 7 million) will be lost too, as some sectors, principally those supplying fossil fuels, downsize through the low-carbon transition. The transition employment challenge will be to ensure that there will be suitably qualified people to fill these new jobs. The majority of these jobs will not require high-level certification, although a fifth to a third of them will need STEM (science, technology, engineering, mathematics) professionals, depending on the technology (International Renewable Energy Agency, 2021, Figure 15, p. 66).

Apart from the quantitative challenge of training this number of people to do these jobs, IRENA identifies four main 'misalignments' of the new renewable sector jobs, with those that will be lost: temporal misalignments, where jobs are lost before the new jobs are available; spatial misalignments, where

Table 11.6 Location of renewable energy jobs in selected countries in 2020

	Jobs (thousands)
China	4732
EU	1300
Brazil	1202
United States of America	838
India	726
Germany	297
North Africa	23
Southern Africa	73
Rest of Africa	228

Source: International Renewable Energy Agency, 2021, Figure 9, p. 34.

the jobs lost are in different places to those gained; educational misalignments, where old skills are not required, and new skills need to be learned; and sectoral misalignments, where jobs lost in one sector need to be shifted to a completely different sector. Labour market policies will be crucial in seeking to smooth these misalignments as far as possible, so that people with the right skills are available in the right places at the right time in the right sectors as decarbonisation gathers pace. This will require an appropriate mix for the country concerned of on-the-job training, apprenticeships, formal vocational training and higher education.

The 'energy transition scenario' to 2030 of the International Labour Office gives further insights into the possible sectoral implications of the jobs gained and lost through decarbonisation (Figure 11.3), but also shows a substantial net gain in jobs in a wide range of sectors, with construction, mining, manufacturing, transport showing more than 3 million new jobs each.

Clearly the employment implications of decarbonisation will differ for different countries, depending on the availability of renewable energy sources, their level of development and their current economic and industrial structure. For the UK a modelling study carried out for the Climate Change Committee by Cambridge Econometrics indicated what might be expected in terms of employment in the meeting of the UK's target of net-zero emissions by 2050.

Cambridge Econometrics found that meeting the 2050 net-zero target would lead to a net increase of 300,000 jobs in the UK, job losses in some sectors being outweighed by job gains in others, as shown in Table 11.7. The biggest shift in employment, not surprisingly, is the loss of jobs in mining fossil fuels and refining them, and the gain in renewables jobs in the utilities sector.

A sectoral breakdown such as that in Table 11.7 can hide job needs in particular industries that could, if not met, prove serious bottlenecks in decarbonisation. In the UK two such industries are the retrofit of buildings for energy efficiency and the installation of heat pumps. In respect of building

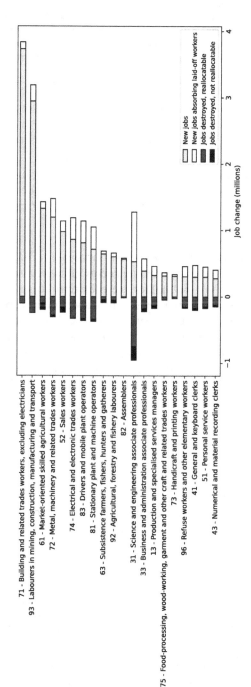

Figure 11.3 Jobs gained and lost in different sectors in the ILO's energy transition scenario to 2030.

Source: International Labour Office, 2019, Figure 6.3, p. 132.

Table 11.7 Change in sectoral employment between a scenario that meets the UK's target of net-zero emissions by 2050 and one based on current policies

(% difference from baseline)	*Output*		*Employment*	
	2030	*2050*	*2030*	*2050*
Agriculture, etc.	4.5	4.3	4.2	2.9
Mining & Refinery	(19.0)	(42.9)	(7.8)	(11.0)
Utilities	20.5	34.6	4.5	35.5
Manufacturing & Construction	7.7	5.5	1.1	0.5
Distribution, Retail, Hotel & Catering	1.3	1.0	1.8	0.9
Transport & Communications	2.4	0.3	2.0	0.1
Services	1.8	1.7	0.2	0.0

Source: Cambridge Econometrics, 2020, Table 3.1, p. 17.

retrofits, the Royal Institution of Chartered Surveyors (RICS) calculated in 2020 that achieving a high level of building energy efficiency would create 5.3–6.7 million jobs, of which 4.7–5.9 million were for retrofitting homes, and the rest retrofitting non-residential buildings.[28] In respect of heat pumps, a 2022 report from Nesta suggested that meeting the UK Government's target of installing 600,000 heat pumps per year by 2028[29] would require 27,000 heat pump installers, a nine-fold increase from the current level of around 3,000.[30] This rate of required jobs and skills increase puts the achievement of targets in these areas in serious doubt, because there is as yet nothing like the training programmes required for heat pump installers, or the required commitment to home energy retrofits for decarbonisation by the middle of the century.

Economic growth

Some people call the very significant investments in energy efficiency and low-carbon electricity supply required for the NZE and related scenarios 'costs', and indeed they are very large expenditures. But they are more correctly termed investments, because they would provide the foundation for delivering energy

28 ww3.rics.org/uk/en/journals/property-journal/retrofit--saving-energy-and-creating-jobs.html. Accessed April 25, 2023. This is calculated on the basis that investment of £1 million in retrofit creates 10–13 jobs, and the total retrofit investment requirement for retrofit is £444 billion. No timescale is given over which the investment would be made, and therefore how many jobs would be created each year.

29 This was one of the targets in the then Prime Minister's 2020 '10-Point Plan for a green industrial revolution', https://assets.publishing.service.gov.uk/government/uploads/system/uploads/attachment_data/file/936567/10_POINT_PLAN_BOOKLET.pdf. Accessed April 25, 2023.

30 https://media.nesta.org.uk/documents/How_to_scale_a_highly_skilled_heat_pump_industry_v4.pdf. Accessed April 25, 2023.

services to 2050 and beyond. And investment is fundamental to delivering economic growth and prosperity in the future.

Economic growth has been one of the defining characteristics of the industrial age. Before the industrial revolution economic output per person increased slowly if at all. Indeed, human societies did not think in these terms, and had no tools or metrics to measure aggregate economic output. GDP, and the related metrics of gross national product (GNP) and net national product (NNP), also called national income (NI),[31] only became global standards in widespread use in the 1950s.

It is now well established in theories of economic growth that it is the product of the three factors that have been discussed in the previous three sections of this chapter: innovation, leading to technical change; investment leading to renewal, increase and improvement of the capital stock; and both the quantity and skills of the workforce. It is therefore to be expected that decarbonisation, which, as has been seen, requires great innovation (in technology, business organisation and institutions), huge investment, and more jobs with enhanced skills, will generate economic growth, and such intuitions are borne out by the available evidence, only a small portion of which will be cited here.

A study which was particularly designed to illustrate the impact of decarbonisation on economic growth is Drummond et al. (2021), some results of which were also published in Ekins et al. (2022), from which the figures below are taken. The decarbonisation pathway of this study has already been shown in Chapter 2. Of more interest here is its impact on economic growth. For this scenario Figure 11.4 shows the relative contribution to global economic growth of population growth, and investment in technology and skills, leading to increases in the capital stock, greater productivity and technical progress. The growth of population peters out later in the century, but the economic change deriving from investment and technical progress persists, albeit at a lower rate.

The result of the global economy growing over the 21st century as in Figure 11.4 is that by 2100 it is five times its size in 2015 (Ekins et al., 2022, Figure 13, p. 12), with the share of investment and public consumption in gross world product (GWP) gradually growing through the century. Given that GDP globally is the sum of investment and consumption, this must mean that private consumption must take a smaller share of world product. This is hardly surprising given the enormous investment that is required by the clean energy transition. It can hardly be imagined that poorer countries will reduce their consumption over this century. Therefore it will be the richer countries that will have to channel a greater proportion of their output into investment for the clean energy transition to come about. From thinking of themselves as 'consumer societies', they should begin to regard themselves as 'investor societies',

---

31 GNP = GDP plus net income from abroad; NNP = NI = GNP minus capital depreciation; gross world product (GWP) is the summation of the world's national GDPs.

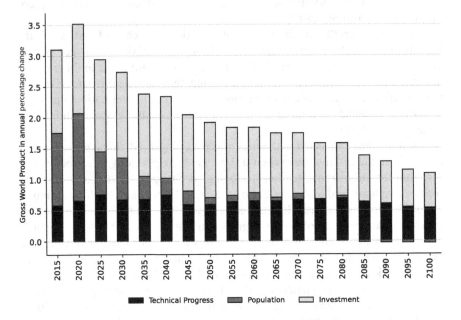

Figure 11.4 Relative contribution to global economic growth to 2100 of investment, population growth and technical progress.

Source: Ekins et al., 2022, Figure 12, p. 11.

holding back somewhat on their consumption so that they can make possible a clean energy future.

In this context, it needs to be remembered that the relation between investment and economic growth is a two-way street. It is certainly true that investment is a prime driver of economic growth, but it is also true that, in the currently existing economy of Western countries, it is the prospect of economic growth that drives further investment, especially from the private sector, which is where most of the investment for the clean energy transition is going to have to come from.

Outcomes from deep decarbonisation that show continuing economic growth are very much the norm in the scientific literature. Figure 11.5 shows the economic trajectories of five of the scenarios from the IPPC 1.5°C scenarios database. The scenarios come from different economic models (REMIND, AIM, MESSAGE, WITCH), with different assumptions about the global socio-economic contexts in which they unfold (shown as S1, S2, S4, S5), and include the Low Energy Demand (LED) scenario that was also encountered in Chapters 2 and 3. For each scenario, except the LED, two lines are shown: the solid line is the baseline, a model run with no decarbonisation beyond current

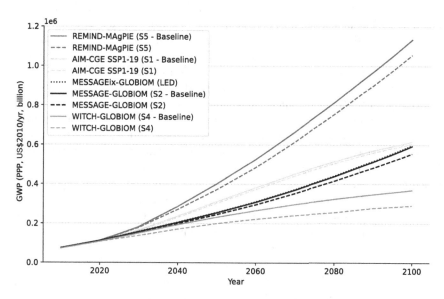

Figure 11.5 Gross world product (GWP) projections from the IPCC scenarios database.

Source: Ekins et al., 2022, Figure 7, p. 6. Data from Rogelj et al., 2018 and Huppmann et al., 2018.

policies; the dashed line is the scenario with carbon emission reduction policies such that the 1.5°C Paris temperature target is met.

Two broad messages are conveyed by Figure 11.5. The first is that GWP increases in all the 1.5°C scenarios. The second is that GWP grows slightly slower with decarbonisation than in the baseline without it, i.e. there is a small net GWP cost associated with decarbonisation. However, this second conclusion is somewhat misleading, because the baselines do not take account of the damages associated with global warming, which the decarbonisation measures in the 1.5°C scenarios are explicitly intended to reduce. Remembering the projected loss of GDP from global warming in Burke et al. (2018) (see Chapter 1), where even 2°C of warming could result in a loss of 5% of GWP by mid-century, and perhaps 20% by 2100, it begins to seem that the cost of meeting the Paris targets is at worst good insurance, given the uncertainties of damages from climate change, and at best a strategy for economic growth, given the conservatism in the projection of cost reductions from low-carbon technologies that was revealed in Way et al. (2022, Figure 5C, p. 9), discussed earlier in this chapter.

Such aggregate results for the world as a whole mask a wide range of different outcomes for different countries. For the UK there are good reasons to think that decarbonisation will lead to net economic benefits, even before counting the benefits of avoided climate damages, and 'co-benefits' (discussed below). This is because, not only would the UK benefit from the economic boost of technical change and greater investment, but it would also reduce its imports of fossil fuels, substituting these with its own wind and solar resources. In its modelling for the Climate Change Committee's assessment of the Sixth Carbon Budget, Cambridge Econometrics found that achieving the UK's 2050 net-zero target would lead to an increase in UK GDP of 2% by 2030 and 3% by 2050 (Cambridge Econometrics, 2020).

It may seem surprising that the huge investments for decarbonisation, that manifest themselves as 'costs' for some businesses and households, should actually increase the prosperity of the economy as a whole. Cambridge Econometrics (2020, p. 14) explains that this is because the direct costs to industry/households are business opportunities and income to companies in other sectors. The overall impact of these costs/investments on the economy as a whole can only therefore be ascertained through the use of economic models that take all these interactions in the economy into account.

In the light of such evidence it is remarkable that the myth persists that decarbonisation in general, and UK decarbonisation in particular, will be unaffordably costly. For example, it is said that the Net Zero Scrutiny Group of UK Conservative MPs believe that attempting to achieve the UK Government's net-zero target for 2050 would be "economically and politically disastrous".[32] Leaving the politics aside (though the YouGov poll in July 2022 showed 67% of those polled were either very or fairly worried about climate change, against 28% who were not[33]), these fears of unaffordability seem to derive either from a lack of understanding of macroeconomics, or from climate scepticism more generally, which it is no longer acceptable to admit to. The findings of the Net Zero Review of Chris Skidmore, also a Conservative MP, robustly refute this economic scepticism. Part I of the Review was entitled 'Net zero is the growth opportunity of the 21st century' (Skidmore, 2023). His 129 recommendations showed in considerable detail how the UK Government, with industry, local government and civil society, could realise the economic benefits of this opportunity.

More surprising than the politically and ideologically charged views of the Net Zero Scrutiny Group are those of some academics and environmentalists who believe that decarbonisation at the required rate cannot be achieved without *reductions* in GDP, called 'degrowth', in rich countries at least. One review of the literature characterised degrowth "as a radical call for a voluntary and equitable downscaling of the economy" (Weiss & Cattaneo, 2017, p. 220). Kallis et al. (2018, pp. 292, 309) consider that degrowth "signifies radical political and economic reorganization leading to drastically reduced resource and energy throughput" which, because this was "physically unlikely"

---

32 https://conservativehome.com/2022/03/31/tory-wars-over-climate-change-the-conservative-environment-network-v-the-net-zero-scrutiny-group/. Accessed April 25, 2023.

33 https://docs.cdn.yougov.com/hdemoi825d/Internal_ClimateChangeTracker_220720_GB_W.pdf. Accessed April 25, 2023.

with continuing economic growth, would require such growth to stop or go into reverse.

The core arguments of this literature are that economic growth has historically been associated with increased energy use and GHG emissions, that there is no evidence that this relationship can be sufficiently changed so that emissions fall at the required rate (i.e. that emissions can be adequately decoupled from economic growth), that some of the technologies shown to be necessary for achieving the 1.5°C target (e.g. CCS, BECCS, negative emission technologies (NETs), direct air carbon capture (DACC)) are either too risky or infeasible, and that therefore the size the global economy will need to shrink if environmental targets are to be met. If in this context the economies of lower- and low-income countries are to continue to grow (as is generally considered desirable in this literature), the economies of rich countries will need to shrink proportionately more.

Ekins et al. (2022), where more references to the degrowth literature may be found, contested this view, and the arguments that follow largely draw on that paper.

Global GHG emissions, which are responsible for anthropogenic global warming, have increased along with growth in the economy since records began, as was seen earlier in this book with, historically, the emissions of the old industrial countries increasing most and first (see Figure 2.3), and global emissions increasing, though at different rates, since 1990 (Figure 1.5).

This trajectory is not surprising, as the major part of GHGs are CO_2 emissions, which come from burning fossil fuels. Fossil fuels have been the predominant source of energy since the industrial revolution, and energy use is at the heart of most economic activity. The question now is, if emissions are reduced at the rate required to achieve the Paris target to limit global warming well below 2°C, while aiming for 1.5°C, can the economy keep on growing? In other words, can emissions be "decoupled" from economic growth to the required extent? As noted above, those advocating degrowth suggest that this is "physically unlikely".

Emissions (Em) from economic activity may be expressed as the product of four terms:

$$Em = P \cdot \frac{A}{P} \cdot \frac{En}{A} \cdot \frac{Em}{En}$$

where P is population, A is affluence, measured by GDP, En is energy use. A/P is therefore GDP per person, En/A reflects the energy per unit of GDP (the energy intensity of the economy), and Em/En is the emission intensity of energy.

Reducing emissions with ongoing growth in population (P) and GDP per person therefore requires more than proportional reductions in the energy intensity of the economy (En/A) and the emission intensity of energy (Em/En). As has been seen in earlier chapters of the book, this can come about in two main ways:

- increasing efficiency in the use of energy and structural changes in the economy whereby consumption of low-energy goods and services replaces consumption of high-energy goods and services (thereby reducing En/A);
- replacing fossil fuels with low- or zero-carbon energy sources (thereby reducing Em/En).

These developments occur in different countries in different ways and to different extents, but in those countries taking emission reduction most seriously, they have been driven by public policy stimulating and reinforcing some market forces and constraining others, as is discussed in some detail in the next chapter.

And these policies have worked. Between 1990 and 2016 the EU economy, for example, grew by more than 50%, while CO_2 emissions from fuel combustion fell by 25% (Figure 11.6).

Moreover, European countries are not the only ones to have reduced CO_2 emissions. Among countries with populations greater than 1 million, 32 of them, including Jamaica, Japan, Singapore and USA outside Europe, reduced their territorial (production) carbon emissions over 2005–2019, while their economies kept growing. Moreover, this emission reduction applied to consumption emissions (production emission plus emissions from producing imports, minus emissions from producing exports) over the same period too (Hausfather, 2021).

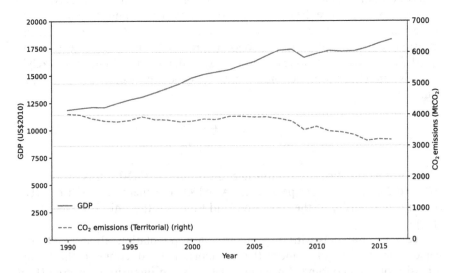

Figure 11.6 Gross domestic product (left axis) and CO_2 emissions (right axis) from fuel combustion in the EU (EU-28, including the UK), 1990–2016.

Source: Ekins et al., 2022, Figure 2, p. 3.

Given that the decarbonisation policies introduced by these countries have not achieved the rates of CO_2 emission reduction to reach the Paris Agreement targets, it is very likely that such achievement will require a considerable intensification of the policies that have generated or reinforced these developments.

As input to the IPCC's Special Report on Global Warming of 1.5°C (Rogelj et al., 2018), 90 decarbonisation scenarios in the modelling literature that were consistent with the 1.5°C target were identified, using different modelling frameworks. As has been extensively discussed in earlier chapters, the modellers chose the values for the parameters in their scenarios in the following areas: developments in economic structure and output; energy demand and efficiency; material demand and efficiency; use of low-carbon energy carriers and technology; availability and use of CCS and NETs; land use and the availability of biomass for energy; choice and implementation of policy measures; and costs of carbon reduction technologies. Many of the scenarios generated were in line with the Paris target of limiting warming to 1.5°C by 2100. *All* the scenarios, a small sample of which are shown in Figure 11.5, showed continuing growth in the economy, as well as meeting the 1.5°C climate target. None of the scenarios came anywhere near declines in economic output from the 2020 level (i.e. they all experienced economic growth). The scenarios therefore all exhibited the necessary "decoupling" of emissions from economic growth to reach the 1.5°C target.

To understand the economic results of these scenarios, it may be noted that economic growth for the world as a whole could be negatively affected by decarbonisation if reducing carbon emissions were to raise the cost of energy or reduce rates of technical progress. With renewable electricity now competitive with that produced from fossil fuels in many countries (International Renewable Energy Agency, 2022), as shown above in some detail, the cost impact on economic growth from the switch to zero-carbon energy sources seems likely to be limited and may even be positive. With regard to technical progress, this is likely to be stimulated by decarbonisation, rather than the reverse, with fossil fuel industries being relatively mature and low-carbon energy generating whole new industries. It is therefore unsurprising that in these scenarios the global economy experienced continuing economic growth as it decarbonised in line with the Paris Agreement 1.5°C target.

One critique in the degrowth literature of the evidence cited above (that deep decarbonisation is consistent with continuing economic growth throughout this century) is that the modelling approach that is adopted in these studies 'assumes' economic growth in the future. In fact, *all* projections of the future make assumptions about future technological and social developments. For example, projections of population growth assume some continuation or departure from the human relations and social norms that produce children and affect health and life expectancy. Economic growth, as was seen earlier in this chapter, arises from capital accumulation (investment), population growth, and technical progress. Such processes have been widely observed in many different countries over the past 200 years. 'Assumptions' of economic growth are therefore based on a perception that, for the rest of this century at least, these processes will continue to produce economic growth. Given historical and current levels of investment and rates of technical change, and the investment and technical change that is required for decarbonisation, such a perception seems not unreasonable.

There are claims by those advocating degrowth that, notwithstanding the reduction in GDP, and therefore their average incomes, people in rich countries could still expect 'a good life' (O'Neill et al., 2018), although it is clear that persuading them of this would be politically challenging. Nor have the policies to bring about such a societal transformation, which might include "making social services growth-independent, reducing working hours, introducing basic incomes and a maximum wage, decelerating life and democratising (economic) decision-making" (Kuhnhenn et al., 2020, p. 9) been worked through in detail, with their economic and social, as well as their environmental implications, made apparent. Keyßer & Lenzen (2021, p. 9) acknowledge: "it is clear that a degrowth transition faces tremendous political barriers".

There is also the issue that, given the twin feedback between economic growth and investment (higher growth is driven by higher investment, but also prospects for growth drive higher investment), if degrowth were to occur, it is most unlikely that the huge investments required for decarbonisation would be forthcoming.

Even more than that, arguments that degrowth is necessary to achieve climate targets are a gift to those who, like the Net Zero Scrutiny Group referred to above, want to derail the decarbonisation agenda altogether. Given current life aspirations of the great majority of people in all countries, if degrowth were necessary to achieve deep decarbonisation, then it is vanishingly unlikely politically that such decarbonisation would be implemented by governments. Luckily, as has been argued here, there is no evidence that this is the case.

The degrowth debate includes consideration of necessary reductions of resource use and environmental impacts apart from GHG emissions, which are not discussed in this book. That said, many of the arguments rehearsed above, about the potential compatibility of environmental improvement with economic growth, seem to apply more widely (Ekins & Zenghelis, 2021).

Co-benefits of climate mitigation

The purpose of climate change mitigation policies is to reduce GHG emissions, and hence to reduce the extent of global warming and climate change. This reduction in emissions and climate change will have many effects on the natural environment, as well as on human societies, including reducing biodiversity and forest loss, ecosystem change, ice melt, ocean warming, and the intensity of extreme weather events, compared to a situation in which climate change proceeds as a result of current emission trajectories. These effects may be called the 'ancillary environmental benefits' of reducing climate change.

However, there may be other benefits to humans and the environment from reducing GHG emissions, that have nothing to do with the reduction in global warming that is the principal aim of emission reduction. Such benefits are here called the 'co-benefits' of climate mitigation, and they are the subject of this section.

Air pollution and health co-benefits

The most important co-benefit of climate mitigation derives from the reduction in local air pollution from the combustion of fossil fuels, as they are increasingly replaced by other energy sources. Premature and avoidable deaths from outdoor air pollution due to the burning of fossil fuels in 2015 was estimated at 3.61 million (Lelieveld et al., 2019, p. 7193), or 65% of all deaths from anthropogenic air pollution. For comparison, the World Health Organization is currently estimating that global deaths from Covid-19 were 3.3 million.[34] Covid-19, of course, was a single pandemic event. The deaths from pollution from fossil fuels happen every year, year-in, year-out.

A more recent estimate of premature mortality from the smallest particles ($PM_{2.5}$) from burning fossil fuels (Vohra et al., 2021) put the figure for 2012 much higher, with a global estimate of 10.2 million premature deaths, with the highest burdens in China (3.91 million per year) and India (2.46 million per year). However, deaths in Europe (1.4 million per year) and the USA (355,000 per year) are also substantial. Of course, many of these deaths could be avoided by pollution control measures, as witnessed by the death rate from all sources of outdoor pollution in China falling by 30–50% over 2013–2018 (Zhai et al., 2019). But these premature deaths from fossil fuel pollution could be removed forever by stopping the burning of fossil fuels completely, through attaining real zero.

The Centre for Research on Energy and Clean Air (CREA) estimated the monetary costs of this level of air pollution from fossil fuels (Myllyvirta, 2020). The 20 countries with the highest absolute costs, and when those are expressed as a share of GDP, are shown in Table 11.8. The total cost of the pollution in 2018, taking account of premature death, the costs of healthcare and of loss of labour participation and productivity was US$1.2 trillion, or 3.3% of GDP. The highest costs per person are in more affluent countries, which tend to have a higher valuation per year of life lost and more expensive healthcare systems. The highest cost in terms of percentage of GDP was in less wealthy countries with high levels of pollution and a relatively high level of chronic diseases (China, and the Central and Eastern European countries in the % GDP part of the table).

---

34 www.who.int/data/stories/the-true-death-toll-of-covid-19-estimating-global-excess-mortality. Accessed April 10, 2023.

Table 11.8 Countries with the highest costs of air pollution from fossil fuels in US$, per capita and as a percentage of GDP

Highest cost per capita by country/region (US$)		*Highest cost as % of GDP by country/region*	
Luxembourg	2600	China Mainland	6.6
United States	1900	Bulgaria	6.0
Switzerland	1900	Hungary	6.0
Austria	1700	Ukraine	5.8
Germany	1700	Serbia	5.8
Netherlands	1200	Belarus	5.4
Denmark	1200	India	5.4
Norway	1100	Romania	5.3
Belgium	1100	Bangladesh	5.1
South Korea	1100	Moldova	5.0
Canada	1100	Poland	4.9
Czech Republic	1000	Slovak Republic	4.8
Italy	1000	Bosnia and Herzegovina	4.6
United Kingdom	1000	Czechia	4.5
Japan	1000	Croatia	4.4
Hungary	1000	Lithuania	4.2
Slovak Republic	900	Russian Federation	4.1
Slovenia	900	North Macedonia	4.1
France	800	Georgia	4.0
Lithuania	800	Montenegro	3.8
Ireland	800	Latvia	3.6
		Germany	3.5
		Kosovo	3.4
		Slovenia	3.4
		South Korea	3.4

Source: Myllyvirta, 2020, p. 6.

The Organisation for Economic Co-operation and Development (OECD) (Roy & Braathen, 2017) estimated the monetary costs of mortality in 41 countries due to outdoor air pollution (from all sources, not just fossil fuels) in 2015 to be US$5.1 trillion, or 6.7% of world GDP at that date. Their breakdown of these costs across different countries, as a percentage of the countries' GDP, showed that air pollution in the Russian Federation, India and Latvia cost those countries the equivalent of more than 10% of their GDP (Roy & Braathen, 2017, Figure 4, p. 23).

A move to clean fuels would also much reduce the deaths from exposure to household air pollution indoors, which the World Health Organization (WHO) estimates was responsible for an estimated 3.2 million deaths per year in 2020, and in 2019 the loss of 86 million healthy life years.[35]

35 www.who.int/news-room/fact-sheets/detail/household-air-pollution-and-health. Accessed April 25, 2023.

Of course, outdoor and indoor air pollution also cause significant illness, short of death. Barwick et al. (2018) calculated that in China in 2015 annual expenditure on illness from air pollution in excess of the WHO standard for $PM_{2.5}$ was US$42 billion, or 7% of China's expenditure on healthcare in that year.

Combustion of fossil fuels is not the only cause of air pollution in urban areas. The wear and tear of roads, brakes and tyres from road traffic are also a major cause of the small particles ('particulate matter', PM_{10} and $PM_{2.5}$, with the numerical subscripts referring to their size) that are particularly damaging to health. In the low-carbon scenario for the UK of Lott et al. (2017, Figure 6, p. 48), the large-scale replacement of internal combustion engines by electric vehicles by 2050 reduced PM_{10} emissions only by around 10% compared to 2010, and $PM_{2.5}$ emissions only by around 30%, due to the growth of traffic over this period. Really deep reductions in the "around 40 000 deaths related to air pollution per year in the UK" (Jennings et al., 2020, p. e426) will require road traffic levels to be reduced as well, through investment in public transport and an increase in active travel (walking and cycling) over short distances. Of course, the latter would bring further health benefits (Avila-Palencia et al., 2018), and less traffic would reduce the economic costs of congestion in urban areas, the cost of which in London in 2019 was estimated at £6.9 billion, with drivers each spending an average of 149 hours 'stuck in traffic'.[36]

The International Monetary Fund (IMF) classifies air pollution from fossil fuel use as a subsidy to that use, in addition to explicit monetary subsidies to the production or use of fossil fuels, and forgone consumption taxes. The IMF's world total of such subsidies in 2020 comes to a mind-boggling US$5.9 trillion, equivalent to 6.8% of world GDP (International Monetary Fund, 2021). Its detailed calculations show that the costs of air pollution and climate change are both considerably larger than the explicit subsidies to fossil fuel producers and consumers and forgone consumption taxes (International Monetary Fund, 2021, Figure ES3, p. 4). On this reading, the co-benefits from reduced air pollution would considerably outweigh the primary benefit of reduced global warming. The IMF further estimates that removing these subsidies in 2025 "would reduce global carbon dioxide emissions by 36% below baseline levels, which is consistent with the 1.5°C target, would raise revenues worth 3.8% of global GDP and prevent 0.9 million deaths from local air pollution each year" (International Monetary Fund, 2021, p. 2).

---

36 www.autoexpress.co.uk/car-news/consumer-news/94871/traffic-jams-costs-in-the-uk. Accessed April 25, 2023.

Energy efficiency and health co-benefits

Chapter 3 laid out the importance of energy efficiency in reducing carbon emissions, with residential buildings offering major opportunities for improvement. In countries with cold winters and an energy-inefficient housing stock, like the UK, increasing the energy efficiency of housing would offer considerable health benefits in addition to emission reduction. The five-year moving average of excess winter deaths in the UK to 2018–2019 (pre-Covid) was 36,162.[37] Jennings et al. (2020, p. e426) report that in 2016 about a third of these deaths were attributable to living in a cold home, and that energy efficiency measures that result in warmer homes could save the UK National Health Service £2.5 billion per year from avoided use of health services. The 2009 report of the UK Chief Medical Officer estimated that £1 invested in energy efficiency would save the NHS £0.42.[38] It is important also to note that, to avoid negative health outcomes from retrofitting homes for energy efficiency, proper ventilation must be ensured.

The cost of heating in cold climates has led, in countries with cold winters and energy-inefficient housing, to the phenomenon of fuel poverty, which leads to a pernicious cycle of impacts, not just on health, but on education, work, and the economy. Figure 11.7 shows how this can be turned into a virtuous cycle by an improvement in the energy efficiency of the housing of the fuel-poor. The vicious cycle shows that poor communities are more likely to live in inefficient homes, which they cannot afford to heat adequately, so that they get ill, with children missing school and adults missing work, imposing costs on the NHS, and reducing household income still further. Improving the energy efficiency of their home initiates a virtuous cycle. It saves them money, so that they can afford more heating, which is good for their health, reducing their use of the NHS, so that children get more education and adults miss fewer days at work. Their productivity and incomes increase. Taxes rise and NHS expenditure is reduced.

Low-meat diets and health co-benefits

Chapter 10 showed that GHG emissions from the food system were about a quarter of the global total, and that 80% of these emissions came from livestock. It showed the importance of low-meat diets to emission reduction. Springmann et al. (2016) estimated both the health benefits and emission reductions of a healthy global diet (HGD), a healthy vegetarian diet (VGT) and a healthy vegan (VGN) diet, "which maintained the regional character of food consumption", against a reference diet, projected to 2050 by the Food

---

37 www.ons.gov.uk/peoplepopulationandcommunity/birthsdeathsandmarriages/deaths/bullet ins/excesswintermortalityinenglandandwales/2020to2021provisionaland2019to2020final. Accessed April 25, 2023.

38 www.sthc.co.uk/Documents/CMO_Report_2009.pdf. Accessed April 25, 2023.

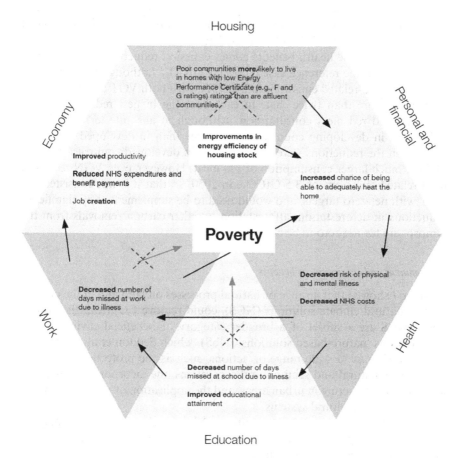

Figure 11.7 Multiple impacts of improvements in the energy efficiency of the housing of fuel-poor households.

Note: The crosses indicate how the energy efficiency intervention breaks links in the cycle of poverty and achieves multiple benefits in respect of personal and financial circumstances, health, education, work, and the economy.

Source: Jennings et al., 2020, Figure 2B, p. e428.

and Agriculture Organization (FAO). HGD comprised, per day, a "minimum of five portions of fruits and vegetables, fewer than 50 g of sugar, a maximum of 43 g of red meat, and an energy content of 2,200–2,300 kcal, depending on the age and sex composition of the population" (Springmann et al., 2016, p. 4147).

Their calculations suggest that HGD could avoid 5 million deaths annually, with VGT and VGN increasing that to more than 7 million and 8 million respectively (Springmann et al., 2016, Figure 1, p. 4148). The largest number of avoided deaths comes from reduced red meat consumption. The change in diets also reduced the cost of illness: by US$735 billion per year for HGD, and by 32% and 45% more than that for VGT and VGN respectively, these reduced costs amounting to 2.3–3.3% of world GDP.

Along with these health benefits comes emission reduction. HGD reduces emissions from the reference case by over 3 $GtCO_2e$ (although this is still 7% more than food-related emissions in 2005/07), with both VGT and VGN reducing them more than twice that. In each case the biggest reductions come from reduced red meat consumption. Although in absolute terms emission reductions in developing countries were larger than in developed countries, per person the reductions were twice as large in developed countries, because of their much larger consumption of red meat. However, even VGN results in food-related emissions of 3.5 $GtCO_2e$ in 2050, so that it is not consistent by itself with net-zero targets, and would need to be supplemented by significant reductions in deforestation, afforestation or other carbon removals from the atmosphere for net zero to be achieved.

Environment and ecosystem co-benefits

Chapter 7 showed that a range of natural processes on land and in the oceans, termed natural climate solutions (NCS), could remove CO_2 from the atmosphere. NCS are a subset of a broader category of beneficial environmental interventions 'nature-based solutions' (NbS), which Seddon et al. (2020, Box 1, p. 2) define as "a wide range of actions, such as the protection and management of natural and semi-natural ecosystems, the incorporation of green and blue infrastructure in urban areas, and the application of ecosystem-based principles to agricultural systems".

Like any environmental interventions NbS involve risks of damage if implemented inappropriately, but Seddon et al. (2020) give many examples of how "well-designed NbS that incorporate diverse native species, avoid damaging biodiverse ecosystems and respect social safeguards offer good opportunities for mitigation with key benefits for local people" (Seddon et al., 2020, p. 4). These include, in different countries: protection from erosion, inland flooding, coastal hazards and sea level rise; moderating urban heat waves and heat island effects; managing storm water and flooding in urban area; sustaining natural resources in drier and more variable climates; buffering communities from climate shocks by enhancing and diversifying ecosystem services; empowering communities; and improving governance and access to resources (Seddon et al., 2020, Table 1, pp. 3–4). The money values of these benefits are not estimated, but they would be significant.

Conclusion

The first conclusion to note about the economic implications of the clean energy transition is that nowhere in the quantitative scientific literature is it suggested that it will result in negative economic growth. On the contrary, the least optimistic of the IPCC scenarios reviewed above (S4 in Figure 11.5) suggests that a 1.5°C scenario will result in the economy growing over 2010–2100 from around US$62 trillion to around US$300 trillion. It should also be noted that the 'baseline' projections of global output (without large-scale emissions reductions) in these scenarios take no account of climate change, because of uncertainties in modelling this. Yet it is clear from the earlier discussion in this chapter of climate risks to the global financial system, and the likely impacts of climate change noted earlier in this book, that it is highly likely that investing to reduce emissions to meet the Paris temperature targets is, for the world as a whole, the most profitable strategy in which it could engage.

These investments in the decades to 2050 will need to come from the rich countries reducing the share of consumption in their GDPs. Coupled with the innovation in new technology this investment will support, the growth in investment is more likely to increase these countries' economic growth rates than reduce them, especially if low-carbon technologies continue to experience the dramatic rates of cost reduction that have been seen in recent years. In fact, such cost reductions, taking the cost of clean energy technologies below those of fossil fuels, could mean that a global clean energy transition has *higher* rates of economic growth than continuing on the current fossil-fuelled path even when climate damages are not taken into account.

Most of the new technologies and investments will be deployed by businesses. Those businesses that grasp the opportunities will be the global businesses of the future, but to attract the investment they need they will have to become more transparent in their reporting of climate (and other environmental) risks. The standardising of reporting to enable them to show their leadership over other companies is well under way. It will increasingly become mandatory.

In addition to potentially higher rates of growth, the clean energy transition could lead to higher employment than sticking with fossil fuels. But many of the new jobs will need different skills to the old, and there is a huge training and re-training challenge that is currently hardly being addressed in many countries.

Finally, it has been seen that the 'co-benefits' of climate change mitigation go much wider than the reduction of climate damages, even if that is their main purpose. In particular, the replacement of fossil fuels with the far more energy-efficient use of clean fuels could bring about major improvements in public health, through the reduction of outdoor air pollution. In addition, there would be very significant health benefits, as well as emission reductions, from moving to healthier diets, in particular in respect of reduced red meat consumption in countries where it is excessive, and environmental co-benefits from using NbS to remove CO_2 from the atmosphere. With their more extensive definition of 'co-benefits', the systematic literature review of Deng et al. (2017) identifies further co-benefits relating to greater efficiency of material resource use, and greater energy and food security.

These co-benefits are not only important in respect of their addition to the quantity of the overall benefits of climate change mitigation. They are also significant because they improve the lives of people in the here and now, whereas the benefits from reducing global warming can often be seen to accrue mainly to future generations. An emphasis on the co-benefits of GHG emission reduction from the mitigation of climate change may serve to make the policies to bring about such emission reduction, reviewed in the next chapter, easier to implement.

References

Avila-Palencia, I., Int Panis, L., Dons, E., Gaupp-Berghausen, M., Raser, E., Götschi, T., Gerike, R., Brand, C., de Nazelle, A., Orjuela, J. P., Anaya-Boig, E., Stigell, E., Kahlmeier, S., Iacorossi, F., & Nieuwenhuijsen, M. J. (2018). The effects of transport mode use on self-perceived health, mental health, and social contact measures: A cross-sectional and longitudinal study. *Environment International, 120*, 199–206. https://doi.org/10.1016/j.envint.2018.08.002

Barwick, P., Li, S., Rao, D. & Zahur, N. (2018). *The morbidity cost of air pollution: Evidence from consumer spending in China (Mimeo)*. Cornell University. https://barwick. economics.cornell.edu/Morbidity%20Cost.pdf. Accessed August 18, 2023.

Burke, M., Davis, W. M., & Diffenbaugh, N. S. (2018). Large potential reduction in economic damages under UN mitigation targets. *Nature, 557*(7706), 549–553. https://doi.org/10.1038/s41586-018-0071-9

Cambridge Econometrics. (2020). *Economic Impact of the Sixth Carbon Budget*. Cambridge Econometrics for the Climate Change Committee, Cambridge. www.theccc.org.uk/publication/economic-impact-of-the-sixth-carbon-budget-cambridge-econometrics/. Accessed May 6, 2023.

Deng, H.-M., Liang, Q.-M., Liu, L.-J., & Anadon, L. D. (2017). Co-benefits of greenhouse gas mitigation: a review and classification by type, mitigation sector, and geography. *Environmental Research Letters, 12*(12), 123001. https://doi.org/10.1088/1748-9326/aa98d2

Drummond, P., Scamman, D., Ekins, P., Paroussos, L., & Keppo, I. (2021). *Growth-Positive Zero-Emission Pathways to 2050*. Sitra, Helsinki. www.sitra.fi/en/publicati ons/growth-positive-zero-emission-pathways-to-2050/

Egli, F., Steffen, B., & Schmidt, T. S. (2019). Bias in energy system models with uniform cost of capital assumption. *Nature Communications, 10*(1), 4588. https://doi.org/10.1038/s41467-019-12468-z

Ekins, P., Drummond, P., Scamman, D., Paroussos, L., & Keppo, I. (2022). The 1.5°C climate and energy scenarios: Impacts on economic growth. *Oxford Open Energy, 1*. https://doi.org/10.1093/OOENERGY/OIAC005

Ekins, P., & Zenghelis, D. (2021). The costs and benefits of environmental sustainability. *Sustainability Science, 16*(3), 949–965. https://doi.org/10.1007/s11625-021-00910-5

Fraunhofer Institute for Solar Energy Systems. (2023). Photovoltaics Report. www.ise.fraunhofer.de/content/dam/ise/de/documents/publications/studies/Photovoltaics-Report.pdf. Accessed August 18, 2023.

Grubb, M., McDowall, W., & Drummond, P. (2017). On order and complexity in innovations systems: Conceptual frameworks for policy mixes in sustainability transitions. *Energy Research & Social Science*, *33*, 21–34. https://doi.org/10.1016/J.ERSS.2017.09.016

Hafner, S., James, O., & Jones, A. (2019). A scoping review of barriers to investment in climate change solutions. *Sustainability*, *11*(11), 3201. https://doi.org/10.3390/su111113201

Hausfather, Z. (2021). *Absolute Decoupling of Economic Growth and Emissions in 32 Countries*. Breakthrough Institute, California. https://thebreakthrough.org/issues/energy/absolute-decoupling-of-economic-growth-and-emissions-in-32-countries. Accessed April 25, 2023.

Huppmann, D., Rogelj, J., Kriegler, E., Krey, V., & Riahi, K. (2018). A new scenario resource for integrated 1.5 °C research. *Nature Climate Change*, *8*(12), 1027–1030. https://doi.org/10.1038/s41558-018-0317-4

International Energy Agency. (2021). *World Energy Investment*. IEA, Paris.

International Energy Agency. (2022). *World Energy Investment 2022*. IEA, Paris.

International Energy Agency. (2023). *World Energy Investment 2023*. IEA, Paris.

International Labour Office. (2019). *Skills for a Green Future: a Global View*. ILO, Geneva. www.ilo.org/wcmsp5/groups/public/---ed_emp/documents/publication/wcms_732214.pdf. Accessed May 6, 2023.

International Monetary Fund. (2021). *Still Not Getting Energy Prices Right: A Global and Country Update of Fossil Fuel Subsidies*. IMF, Washington, DC

International Renewable Energy Agency. (2021). *Renewable Energy and Jobs: Annual Review 2021*. IRENA, Abu Dhabi. www.irena.org/-/media/Files/IRENA/Agency/Publication/2021/Oct/IRENA_RE_Jobs_2021.pdf?rev=98960349dbab4af78777bc49f155d094. Accessed May 6, 2023.

International Renewable Energy Agency. (2022). *Renewable Power Generation Costs in 2021*. IRENA, Abu Dhabi. www.irena.org/publications/2022/Jul/Renewable-Power-Generation-Costs-in-2021. Accessed May 6, 2023.

IPCC. (2022a). *Summary for Policymakers*. P. R. Shukla, J. Skea, A. Reisinger, R. Slade, R. Fradera, M. Pathak, A. Al Khourdajie, M. Belkacemi, R. van Diemen, A. Hasija, G. Lisboa, S. Luz, J. Malley, D. McCollum, S. Some, P. Vyas. In: P. R. Shukla, J. Skea, R. Slade, A. Al Khourdajie, R. van Diemen, D. McCollum, M. Pathak, S. Some, P. Vyas, R. Fradera, M. Belkacemi, A. Hasija, G. Lisboa, S. Luz, & J. Malley (Eds.) *Climate Change 2022: Mitigation of Climate Change. Contribution of Working Group III to the Sixth Assessment Report of the Intergovernmental Panel on Climate Change*. Cambridge University Press, Cambridge, UK and New York, NY, USA. doi: 10.1017/9781009157926.001

IPCC. (2022b). Technical Summary. M. Pathak, R. Slade, P.R. Shukla, J. Skea, R. Pichs-Madruga, D. Ürge-Vorsatz. In: P. R. Shukla, J. Skea, R. Slade, A. Al Khourdajie, R. van Diemen, D. McCollum, M. Pathak, S. Some, P. Vyas, R. Fradera, M. Belkacemi, A. Hasija, G. Lisboa, S. Luz, J. Malley (Eds.) *Climate Change 2022: Mitigation of Climate Change. Contribution of Working Group III to the Sixth Assessment Report of the Intergovernmental Panel on Climate Change*. Cambridge University Press, Cambridge, UK and New York, NY, USA. doi: 10.1017/9781009157926.002

Jennings, N., Fecht, D., & De Matteis, S. (2020). Mapping the co-benefits of climate change action to issues of public concern in the UK: A narrative review. *The Lancet Planetary Health*, *4*(9), e424–e433. www.thelancet.com/journals/lanplh/article/PIIS2542-5196(20)30167-4/fulltext. Accessed May 6, 2023.

Kallis, G., Kostakis, V., Lange, S., Muraca, B., Paulson, S. & Schmelzer, M. (2018). Research on degrowth. *Annual Review of Environment and Resources, 43*, 291–316. https://doi.org/10.1146/annurev-environ-102017-025941

Keyßer, L. T., & Lenzen, M. (2021). 1.5 °C degrowth scenarios suggest the need for new mitigation pathways. *Nature Communications, 12*(1), 2676. https://doi.org/10.1038/s41467-021-22884-9

Kuhnhenn, K., Costa, L., Mahnke, E., Schneider, L., & Lange, S. (2020). A Societal *Transformation Scenario for Staying Below 1.5°C*. Heinrich Boell Foundation, Berlin. www.boell.de/en/2020/12/09/societal-transformation-scenario-staying-below-15degc

Lelieveld, J., Klingmüller, K., Pozzer, A., Burnett, R. T., Haines, A., & Ramanathan, V. (2019). Effects of fossil fuel and total anthropogenic emission removal on public health and climate. *Proceedings of the National Academy of Sciences, 116*(15), 7192–7197. https://doi.org/10.1073/pnas.1819989116

Lott, M. C., Pye, S., & Dodds, P. E. (2017). Quantifying the co-impacts of energy sector decarbonisation on outdoor air pollution in the United Kingdom. *Energy Policy, 101*, 42–51. https://doi.org/10.1016/j.enpol.2016.11.028

Myllyvirta, L. (2020). *Quantifying the Economic Costs of Air Pollution from Fossil Fuels*. Centre for Research on Energy and Clean Air (CREA). https://energyandcleanair.org/wp/wp-content/uploads/2020/02/Cost-of-fossil-fuels-briefing.pdf. Accessed May 6, 2023.

O'Neill, D. W., Fanning, A. L., Lamb, W. F., & Steinberger, J. K. (2018). A good life for all within planetary boundaries. *Nature Sustainability, 1*(2), 88–95. https://doi.org/10.1038/s41893-018-0021-4

Polzin, F., & Sanders, M. (2020). How to finance the transition to low-carbon energy in Europe? *Energy Policy, 147*, 111863. https://doi.org/10.1016/J.ENPOL.2020.111863

Rogelj, J., Popp, A., Calvin, K. v., Luderer, G., Emmerling, J., Gernaat, D., Fujimori, S., Strefler, J., Hasegawa, T., Marangoni, G., Krey, V., Kriegler, E., Riahi, K., van Vuuren, D. P., Doelman, J., Drouet, L., Edmonds, J., Fricko, O., Harmsen, M., ... Tavoni, M. (2018). Scenarios towards limiting global mean temperature increase below 1.5 °C. *Nature Climate Change, 8*(4), 325–332. https://doi.org/10.1038/s41558-018-0091-3

Roy, R., & Braathen, N.-A. (2017). *The Rising Cost of Ambient Air Pollution thus far in the 21st Century: Results from the BRIICS and the OECD Countries*. OECD, Paris. www.oecd-ilibrary.org/environment/the-rising-cost-of-ambient-air-pollution-thus-far-in-the-21st-century_d1b2b844-en. Accessed May 6, 2023.

Science-Based Target Initiative. (2022). *Science-Based Net Zero: Scaling urgent corporate climate action worldwide. Progress Report 2021*. https://sciencebasedtargets.org/reports/sbti-progress-report-2021

Seddon, N., Chausson, A., Berry, P., Girardin, C. A. J., Smith, A., & Turner, B. (2020). Understanding the value and limits of nature-based solutions to climate change and other global challenges. *Philosophical Transactions of the Royal Society B: Biological Sciences, 375*(1794), 20190120. https://doi.org/10.1098/rstb.2019.0120

Skidmore, C. (2023). *Mission Zero: The Independent Review of Net Zero*, His Majesty's Government, London. https://assets.publishing.service.gov.uk/government/uploads/system/uploads/attachment_data/file/1128689/mission-zero-independent-review.pdf. Accessed April 25, 2023.

Springmann, M., Godfray, H. C. J., Rayner, M., & Scarborough, P. (2016). Analysis and valuation of the health and climate change cobenefits of dietary change. *Proceedings of the National Academy of Sciences, 113*(15), 4146–4151. https://doi.org/10.1073/pnas.1523119113

The Task Force on Climate-related Financial Disclosures. (2022). *Task Force on Climate-related Financial Disclosures: 2022 Status Report.* https://assets.bbhub.io/company/sites/60/2022/10/2022-TCFD-Status-Report.pdf

United Nations Environment Programme. (2022). *Emissions Gap Report 2022: The Closing Window – Climate crisis call for transformation of societies.* UNEP, Nairobi.

Vohra, K., Vodonos, A., Schwartz, J., Marais, E. A., Sulprizio, M. P., & Mickley, L. J. (2021). Global mortality from outdoor fine particle pollution generated by fossil fuel combustion: Results from GEOS-Chem. *Environmental Research, 195,* 110754. https://doi.org/10.1016/j.envres.2021.110754

Way, R., Ives, M. C., Mealy, P., & Farmer, J. D. (2022). Empirically grounded technology forecasts and the energy transition. *Joule.* https://doi.org/10.1016/j.joule.2022.08.009

Weiss, M. & Cattaneo, C. (2017). Degrowth – Taking stock and reviewing an emerging academic paradigm. *Ecological Economics, 137,* 220–230. https://doi.org/10.1016/j.ecolecon.2017.01.014

Zhai, S., Jacob, D. J., Wang, X., Shen, L., Li, K., Zhang, Y., Gui, K., Zhao, T., & Liao, H. (2019). Fine particulate matter ($PM_{2.5}$) trends in China, 2013–2018: separating contributions from anthropogenic emissions and meteorology. *Atmospheric Chemistry and Physics, 19*(16), 11031–11041. https://doi.org/10.5194/acp-19-11031-2019

12 Policy and delivery

Summary

Strong climate policy will need to be adopted if the world is to have any hope of meeting the Paris temperature targets. There has been huge innovation in such policy over the last 30 years, and there is now considerable knowledge about what policies are available to reduce emissions, how effective they are, and how they can be implemented in ways that are socially acceptable. Many of the necessary policies have already been introduced in earlier chapters, but this chapter organises the policies according to various categories of instruments and other dimensions of policy: legislation and litigation; economic instruments; regulation; voluntary agreements; information; innovation instruments; circular economy policies; policy mixes; policy evaluation; and subnational (local government, city-level and community) policy. There is also a section on human behaviour change, including personal carbon calculators and personal carbon trading, and another on the institutions for policy delivery. The chapter ends with the key policy timeline identified by the International Energy Agency (IEA) for decarbonisation to net-zero emissions globally, by 2050.

Introduction

Strong government policy at local, national, European Union (EU) and global levels is essential to deliver the total decarbonisation of the global economy that is now required. The framework for such policy now exists.

The aim of all policy, including climate policy, is to change the behaviour of individuals, businesses or other organisations. Policy does this in the short to medium term by changing the incentives and penalties that these actors face as they go about their daily lives or pursue their organisational objectives.

There has now been very great experimentation with and experience of climate policy, and many of the kinds of policies that have been introduced will be mentioned in this chapter, in addition to those mentioned in earlier chapters.

DOI: 10.4324/9781003438007-13

However, the essence of policy is that, beyond broad generalisations, the devil is in the detail. Climate policy, as with other policy, needs to be designed and implemented with the specific national, sub-national, institutional and cultural context in mind. For this reason, most of the policies discussed here will be from the UK and EU, with which I am most familiar. Many of these policies will be relevant to other jurisdictions. But their successful implementation elsewhere will need to be mindful of and sensitive to the specific conditions in these different places. Policies can only rarely be transferred from one place to another without substantial modification for the new context.

At the global level, the UN Framework Convention on Climate Change (UNFCCC) provides the underpinning treaty and context through which countries, all of which are now signatories, can take part in the annual COP (Conference of the Parties) meetings, make their own commitments through their Nationally Determined Contributions (NDCs) and monitor those of other countries. The only problem with this framework is that, as has been seen in Chapter 2, the NDC commitments to date of the great majority of countries are well short of what is required to meet the Paris targets, and the level of policy stringency necessary to deliver even these NDCs is absent in practically all countries.

Notwithstanding the importance of this global framework, most climate policy is made at the national level, although it is clear that much of it needs actually to be delivered at local level, through sub-national regions, municipalities, city governments or whatever other levels of local government countries may have.

The last 30 years has seen huge experimentation with climate policies in many countries, with varying degrees of success. This chapter will first set out a general categorisation of the kinds of policies that have been tried, and then give examples for different sectors of where they have been implemented with some of the results and lessons learned, adding to the examples of climate policies that have already been given in previous chapters. This policy review will be illustrative only. A full description of the field would require several books to itself.

Climate, justice and development

Practically every aspect of climate change has a profound justice dimension: the reduction of greenhouse gas (GHG) emissions, the 'energy transition' away from fossil fuels to low-carbon energy sources, the adaptation to the changes that are taking place and will take place due to climate change, and who will pay for the losses and damages caused by climate change – all these issues raise fundamental questions about responsibility, fairness and compensation.

The scale of the justice issues at stake was brought into sharp relief by two papers, one of which estimated the climate damages for each region from unmitigated global warming, while the other calculated the monetary responsibility of the different countries that were causing them.

Burke et al. (2015, Figure 4, p. 238) showed that even quite extensive global warming may be expected to benefit northern regions, especially Canada and Russia, but also northern and middle Europe. The global loss of gross domestic product (GDP) by 2100 is in excess of 25%, but this is very unequally distributed among the different regions of the world. The rich countries of Europe and North America experience significant gains and insignificant losses respectively, while all the poorer, already hot, regions of the world (South and Southeast Asia and sub-Saharan Africa) experience GDP losses by 2100 of 75% or more.

The second paper (Callahan & Mankin, 2022), which uses the same basic methodology as Burke et al. (2015), calculates the damages caused by each country's emissions to every other country, and comes to the conclusion that five major emitters (the United States, China, Russia, Brazil, and India) "have collectively caused USD 6 trillion in income losses from warming since 1990, comparable to 11% of annual global gross domestic product; many other countries are responsible for billions in losses". But the distribution of the impacts is highly unequal, bringing benefits to high-income, high-emitting countries and substantial damages to low-income, low-emitting countries (Callahan & Mankin, 2022, p. 1). While continuing uncertainties in the calculation of such numbers mean that they are unlikely to be considered robust enough to be used in legal estimates of 'loss and damage', for which a fund was agreed in principle at COP27 in 2022, they give an idea of the sums of money that may be involved in the implementation of 'loss and damage' payments.

The 1992 UNFCCC acknowledges that countries' responses to climate change should be "in accordance with their common but differentiated responsibilities and respective capabilities and their social and economic conditions" (United Nations, 1992, p. 1), and this language was broadly repeated in the 2015 Paris Agreement, which reiterated the commitment to "the principle of equity and common but differentiated responsibilities and respective capabilities, in the light of different national circumstances" (United Nations, 2015, p. 1). In order to give effect to this principle the UNFCCC divided countries into three groups: Annex I, broadly more 'developed' countries, including the countries of Europe, North America, the Russian Federation, Turkey, Japan, Australia, and New Zealand; Annex II, comprising the Annex I countries minus the countries of the former Soviet Union and Turkey; and non-Annex I countries, the rest of the world, including so-called 'developing' countries.

Annex I countries had the responsibility to report on and reduce their own emissions; Annex II countries had the responsibility to help developing countries to develop in a low-emission way through financial assistance and the transfer of technology, and to adapt to such climate change as would take place. The Kyoto Protocol in 1997 set out the required emission reductions for Annex I parties. The Copenhagen Accord at COP15 in 2009 put a figure on the financial commitments required from developed countries to aid developing countries: US$30 billion over 2010–2012, and US$100 billion per year by 2020.

The US$100 billion target was not met in 2020, and is supposed to be met in 2023.

The 27th Conference of the Parties to the Climate Convention (COP27) in Egypt in 2022 agreed to set up a fund for 'loss and damage' suffered by developing countries due to climate change,[1] but all the details for this are still to be worked out: how much money the fund will contain, who will pay into it, who will be entitled to receive payments from it, and the criteria for such payments. In its agreement to establish such a fund, the EU made clear that this was conditional on contributors to the fund being broadened beyond Annex II countries, in particular to include China, now the world's biggest emitter and much richer than it was in 1992.[2] It is likely that negotiations on this and other points relating to the new fund will be complex and hard-fought.

The "ultimate Objective" of the UNFCCC is "to achieve, in accordance with the relevant provisions of the Convention, stabilization of greenhouse gas concentrations in the atmosphere at a level that would prevent dangerous anthropogenic interference with the climate system". However, the "relevant provisions" of the UNFCCC hem this ultimate objective around with a number of very taxing conditions and further ambitions. For example, the UNFCCC acknowledges that States have:

> the sovereign right to exploit their own resources pursuant to their own environmental and developmental policies [as well as] the responsibility to ensure that activities within their jurisdiction or control do not cause damage to the environment of other States or of areas beyond the limits of national jurisdiction

[tropical forests, that are clearly both resources subject to the sovereign right of exploitation but that also play an important role in climate stability, are probably the most egregious example of potential conflict between these principles]

> responses to climate change should be coordinated with social and economic development in an integrated manner with a view to avoiding adverse impacts on the latter, taking into full account *the legitimate priority needs of developing countries for the achievement of sustained economic growth and the eradication of poverty.*

(emphasis added)

> all countries, especially developing countries, need access to resources required to achieve sustainable social and economic development and

---

1 https://unfccc.int/news/cop27-reaches-breakthrough-agreement-on-new-loss-and-damage-fund-for-vulnerable-countries. Accessed April 23, 2023.
2 www.euractiv.com/section/climate-environment/news/eu-opens-the-door-to-a-loss-and-dam age-facility-if-china-contributes/. Accessed April 23, 2023.

that, in order for developing countries to progress towards that goal, their energy consumption will need to grow taking into account the possibilities for achieving greater energy efficiency and for controlling greenhouse gas emissions in general, including through the application of new technologies on terms which make such an application *economically and socially beneficial.*

> (emphasis added)

specific needs and special circumstances [of developing countries] should be given full consideration, [including in respect of] actions related to funding, insurance and the transfer of technology, to meet the specific needs and concerns of developing country Parties arising from the adverse effects of climate change and/or the impact of the implementation of response measures, [while taking] full account of the specific needs and special situations of the least developed countries in their actions with regard to funding and transfer of technology

[all quotes from the text of the UNFCCC, United Nations, 1992, various pages]

For its part the 2015 Paris Agreement (United Nations, 2015) went even further than the UNFCCC in articulating conditions and constraints on climate action: emphasising "the intrinsic relationship that climate change actions, responses and impacts have with equitable access to sustainable development and eradication of poverty" [in the context of the agreement earlier in 2015 of the 17 Sustainable Development Goals (SDGs), of which No.1 is the eradication of poverty]; recognising "the fundamental priority of safeguarding food security and ending hunger" [SDG 2 is ending hunger]; stipulating that

Parties should, when taking action to address climate change, respect, promote and consider their respective obligations on human rights, the right to health, the rights of indigenous peoples, local communities, migrants, children, persons with disabilities and people in vulnerable situations and the right to development, as well as gender equality, empowerment of women and intergenerational equity;

taking into account the "imperatives of a just transition of the workforce and the creation of decent work and quality jobs in accordance with nationally defined development priorities"; and noting "the importance for some of the concept of 'climate justice', when taking action to address climate change" (United Nations, 2015, various pages).

These concerns, commitments and obligations related to climate justice have been elaborated in a voluminous literature that seeks to articulate the numerous dimensions of climate justice and the 'just transition', which include at least the following set of issues:

- the distribution between countries of the remaining carbon budget to stay within any particular rise in global average temperature;
- rich countries (people) paying poor countries (who may or may not then give the money to poor people) for 'loss and damage' from climate change;
- rich countries (people) paying poor countries for, or in some other way enabling, the implementation of low-carbon technologies in poor countries;
- rich countries (people) paying poor countries (people) to help them adapt to the climate change that cannot be avoided;
- countries undergoing transition helping workers in high-carbon industries (who may not be poor) to adapt or retrain following the loss of their jobs;
- established fossil fuel producers receiving compensation because large quantities of such fuels will have to stay in the ground for climate stability;
- countries with unexploited fossil fuel reserves, who would like to become fossil fuel producers, receiving compensation for forgoing their exploitation;
- countries wanting to reduce emissions being sued for compensation by energy companies for not allowing them to develop their fossil resources (e.g. through the Energy Charter Treaty, see below).

None of these issues has a straightforward resolution. While most people will doubtless declare themselves in favour of 'justice' and 'fairness' in principle, in practice they have very different ideas as to what these concepts entail and how they should be implemented. Some of the issues above are likely to be deeply contested in principle (e.g. the final point in the list above), and the detail of all of them will be deeply contested in practice (as with who pays, and how much, for 'loss and damage').

The academic literature on this topic is now extensive and complex. The sections that follow outline how justice issues have been addressed in three areas of 'climate justice': the sharing of carbon budgets and, more broadly, the efforts of climate action; the transition of the energy system from fossil fuels to low-carbon energy sources; and the country distribution of remaining fossil fuel extraction consistent with the carbon budgets.

Sharing the effort of climate action

An early review of the ways to 'share effort' in responding to the issues enumerated above was produced by Höhne et al. (2014). This synthesis of the issues, which found its way into the Fifth Assessment Report of the IPCC in 2015, articulated them in respect of five principles: responsibility, equality, capability, need and cost-effectiveness (Höhne et al., 2014, Figure 1, p. 125).

Responsibility seeks to take into account different countries' contribution to climate change, and is usually interpreted as cumulative GHG emissions from some date. It is also based on one of the core principles of environmental policy, the 'polluter pays principle'. The *equality* criterion seeks to express the position that all humans have an equal right to the remaining pollution absorption capacity of the atmosphere (i.e. the carbon budget relating to a particular temperature target), and is usually implemented through allocation of equal emissions per person, either now or at some future date. The quest for *cost-effectiveness* is justified by the fact that if emissions targets can be achieved as cheaply as possible globally, then in principle countries that benefit from such an allocation (i.e. those who make least effort because their emission reductions are relatively expensive) can compensate those who make most effort. Such an approach is often implemented in climate-energy-economy models through a global carbon tax or trading system, because of the relative simplicity of this approach, but its practical implementation depends on there being an institutional arrangement both to effect payments between countries and to ensure that such payments actually take place. At present, no such institutions, or the political will to create them, exist. *Capability* acknowledges the fact that countries have very different 'abilities to pay' for climate action, as reflected in the position in the UNFCCC that the countries with most ability should make most effort. *Need* is based on the twin facts that some countries (e.g. small-island developing states) are more vulnerable and find it harder to adapt to climate change, and that developing countries in general and least developed countries in particular have relatively large populations with unfulfilled basic needs. In the language of the Paris Agreement, the 'right to development' of these countries should be given 'full consideration' in the sharing of the efforts of climate action.

Höhne et al. (2014) allocated the remaining carbon budgets for different temperature targets according to these different criteria in three different ways: 'equal cumulative per capita emissions' sought to reflect the principles of equality and responsibility; a second allocation was based on the principles of 'responsibility, capability and need'; while a third approach sought to reflect to some extent all the principles over the course of the transition to net-zero emissions. Not surprisingly, when these different allocation criteria are applied to different countries and regions, the divisions of the remaining carbon budgets for different temperature targets are very different.

The approach of Höhne et al. (2014) to these issues has been roundly criticised by, among others, Dooley et al. (2021), on the grounds that the range of issues considered in this and other earlier papers is much too limited, and that too often papers on this topic failed to articulate transparently the ethical positions and assumptions on which they were based, or even tried to present themselves as 'value neutral'. Other issues which Dooley et al. (2021) list as having been explored in the climate justice literature include "environmental justice and transitional justice, ecological debt, intergenerational equity, survival emissions, progressivity, prioritarianism [priority should be given to the worse off] and egalitarianism" (Dooley et al., 2021, p. 303).

Williges et al. (2022) adapted the approach of Höhne et al. (2014) to seek to respond to the critique of Dooley et al. (2021), with four adjustments to allocations of the total carbon budget based either on equal per capita emissions from 2017 to 2050, or on existing per capita emissions in 2017 that converge to equality in 2050. The four adjustments are:

- redistribution to developing countries of a notional amount required for them to meet the basic needs of all their population;
- computing the carbon budget from 1995, with the actual emissions from countries since then deducted from their share of the budget (leaving high-emission countries with less budget for the future), to take account of historical emissions;
- redistributing the carbon budget further to account for the benefits received particularly by developed countries from emissions before 1995, which enabled them to build up their infrastructure; and
- adjusting the budgets in recognition of the fact that very fast emission reduction is either very costly or completely infeasible, so that the emission reduction rate for all countries is constrained not to exceed 7% per year.

With these adjustments to an initial equal per capita allocation, and a carbon budget to remain well below a 2°C temperature increase, North America, Europe, Australia and Japan have more or less exhausted their carbon budgets by 2020. Given that all these countries are still emitting substantial GHGs, the conclusion can only be that the world will not stay well below 2°C under this interpretation of climate justice unless they 'compensate' for these emissions by paying countries with significantly larger remaining carbon budgets (e.g. in Africa) to emit substantially below their allocation.

A 'just' energy transition

The shift from fossil fuels to zero-carbon energy sources will inevitably involve changes with profound social and economic implications, in terms of both running down the production of fossil fuels and establishing the new energy sources. Applications of the idea of justice to this transition have identified a number of different kinds of justice, the principles of which will need to be met for the transition overall to be considered 'just'.

From their review of the literature Carley & Konisky (2020) identify the following distinct kinds of justice relevant to the energy transition: *distributional justice* refers to how the costs and benefits of the transition are shared across populations; *procedural justice* seeks to ensure that the voices of those affected by the transition are adequately considered in its implementation; *recognition justice* seeks to take account of past and present inequalities; and *restorative justice* has the objective of correcting for past injustices of different kinds. Carley and Konisky (2020)'s examples of relevant issues to be considered across these dimensions include: the treatment of workers displaced from the fossil fuel industry and the communities which may be based on this industry; potential increases in energy prices, which may impact most heavily on the poorest households; inequities, based on gender or ethnicity, of access to opportunities created by new energy technologies; and the participation or otherwise of, especially, indigenous communities in decisions about the siting of clean energy technologies.

Carley & Konisky (2020, p. 573) discuss five different kinds of public interventions that have been introduced to address these dimensions of justice: "workforce and economic diversification programmes; energy assistance and weatherization [increasing energy efficiency in buildings]; expansion of energy technology access; collective action initiatives; and new business development". They analyse how such programmes have been implemented, mainly in the US, and find significant room for improvement in terms of the way these interventions address the justice considerations they identify earlier in their paper.

In a European context, Sovacool et al. (2019) analyse through a justice lens the way four low-carbon energy interventions have been implemented: "nuclear power in France, smart meters in Great Britain, electric vehicles in Norway, and solar energy in Germany". They identify no fewer than 120 "distinct energy injustices" in terms of the effects of these initiatives on vulnerable groups, energy prices, unemployment, taxpayers, and future generations.

Homberg & McQuistan (2019) apply many of the same ideas, with a focus on technology, in a developing country context. Distributive, compensatory, transitional and procedural justice all suggest that communities in developing countries, and especially vulnerable and disadvantaged communities, should be given access to clean energy technologies which they can use and which provide a basis for further innovation. This may require capacity building, in terms of both the skills in the communities and the institutions (markets, public authorities) that both facilitate and regulate the use of the technologies. The technology and capacity building will require finance from richer countries, both of a compensatory kind, to reflect their greater responsibility for climate change, and to reflect their greater ability to pay for its mitigation.

The production of fossil fuels

Constraints on the increase in global temperatures that human societies are prepared to tolerate not only limit the available carbon budget from the consumption of fossil fuels, they also limit the quantity of fossil fuels that can still be produced. An earlier section discussed ways that have been proposed to divide up between countries the carbon budget from fossil fuel consumption, in line with the concept of 'climate justice'. This section discusses how that concept might be applied to the issue of which countries either get to extract the fossil fuels consistent with different carbon budgets, thereby benefiting from the associated income and employment, or leave them underground. As with most issues related to climate justice, there is no easy answer to this question.

Most climate-energy-economic models answer the question by extracting the cheapest fossil fuels consistent with energy demands and remaining carbon budgets. These mainly come from established oil and gas producers, which have the most easily accessible resources and the infrastructure with which to extract them. This means that many developing countries that have unexploited fossil fuel resources, for example in Africa, never get to produce them. It might be regarded as unjust that richer countries that have already benefited greatly from fossil fuel production continue to do so, while poorer countries are denied these benefits.

Pye et al. (2020) explored this issue through the construction of two scenarios, one of which made the production of fossil fuels in very highly developed regions (e.g. North America, Western Europe, Australia) much more expensive, and the other which did the same for those regions that have already benefited most from fossil fuel production (e.g. Middle East, North America). Both scenarios thereby privileged fossil fuel production from developing countries without a history of it. The results were instructive, and not entirely expected.

New fossil fuel production in developing countries did occur in these scenarios, but not by as much as might have been expected, given that it takes time to develop large-scale new fossil fuel infrastructure (e.g. pipelines, ports), so that it was 2040 before new developing country production was ready to come on stream at scale, by which time the demand for fossil fuels was already much reduced because of climate constraints. Moreover, for these new developing country producers to become competitive with the large existing producers, very large cost penalties needed to be applied to this latter group. This in turn significantly increases the cost of fossil fuels to importing developing countries, who are made worse off at the same time as some producing developing countries are made better off. Greater justice for the producer developing countries has led to injustices for importing developing countries.

There is also the consideration that it is highly likely that much of the fossil fuel from developing countries will be produced by large multi-national oil and gas companies that are not based there (e.g. France's Total, which is heavily involved in the infrastructure development to extract Mozambique's large gas resources and has already come in for considerable criticism about the way it has treated existing communities in the area concerned[3]). This means that much of the income from this new fossil fuel production will not benefit the country in which it takes place, but will go to expert expatriate workers and foreign shareholders. Finally, given that fossil fuel resources are already more expensive than renewables in many countries, it is possible that demand for them will fall off faster than projected in the model, so that new developing country producers will be left indebted for the construction of infrastructure to produce fossil fuels that are in excess global supply, reducing their prices, and therefore the benefit to the developing country producers, still further.

---

3 https://thewire.in/world/oil-and-gas-multinational-total-is-making-a-mess-in-mozambique. Accessed April 23, 2023.

All these considerations mean that, despite their undoubted best intentions, statements from those like Mary Robinson that African countries should be encouraged to develop their fossil fuel resources[4] may not actually be promoting an action that is in those countries' best interests.

Climate justice in practice

There is no settled interpretation of climate justice, or of a 'just transition', that leads to clear conclusions about how these concepts should be implemented. Some of the most egregious injustices are obvious, and it is acknowledged in the agreements under the UNFCCC that they should be addressed: it is clearly unjust that countries that have contributed least to climate change should suffer most from it, and justice demands that they receive both some compensation for the loss and damage they have incurred, and help to adapt to the further climate change that is now unavoidable. It is also clearly just that highly developed countries, which have benefited most from past fossil fuel use, and produced most GHG emissions, and which have developed the technologies to deliver zero-emission energy, should aid developing countries to use these technologies for their sustainable development.

Whether, how and to what extent these demands of justice are met in practice is inevitably subject to political negotiation and decision, as the tortuous processes in the COP meetings make clear.

Developed countries need to be aware that, despite climate threats, the priority in developing countries is still overwhelmingly *development*, as conventionally conceived, and that unless they are helped to achieve this with low emissions they will find no shortage of willing sellers of fossil fuels to put them on a high-emission development path, and climate catastrophe at some time in the future.

A relevant historical analogy is the Marshall Plan at the end of the Second World War. Just as it was in the medium- and long-term self-interest of the United States to help Europe rebuild its shattered economies after 1945, so it is now in the medium- and long-term self-interest of the developed world to ensure that the developing countries of Africa, Asia and Latin America and the Caribbean can and do meet their aspirations for a better life without the use of fossil fuels.

The implementation of any such 'Climate and Development Plan' will be politically fraught. Great power competition will ensure that richer countries are unwilling to feel that they are impoverishing themselves through such a plan to the advantage of their competitors, as witness the EU insistence that the new loss-and-damage fund contributors include China and potentially other non-Annex II countries.

---

4 www.theguardian.com/environment/2022/jun/07/let-africa-exploit-natural-gas-reserves-mary-robinson. Accessed April 24, 2023.

What seems essential is that, in the inevitable and protracted debates about climate justice and the just transition, sight is not lost of the biggest justice objective of all: to reduce GHG emissions to zero by mid-century to avoid a climate catastrophe that will overwhelm any politically feasible contributions to a loss-and-damage fund, any possibility of adaptation, and any prospects of economic development, with, of course, the most vulnerable and poorest people and countries suffering first and worst.

Considerable aid from developed to developing countries will be required to achieve this overarching objective of emissions reduction. Within countries, too, it may be found to be politically necessary, as well as just, to ease the transition away from fossil fuels to zero-carbon energy and low-GHG agriculture through well-targeted public interventions.

But to insist that a transformation of the scale required to get to zero emissions in 30 years can be achieved without losers and some injustices runs the real risk that the transformation itself will not be achieved. The world is not currently just, and it is inconceivable that the energy transition in itself will make it so. Doubtless it is desirable that, where there is broad agreement about principles of climate justice and the just transition, moves to a zero-emission economy at least do not increase injustices, and preferably reduce them. But it is necessary to be aware that aspects of justice and decarbonisation may not always be mutually supportive, and that to insist on the former in all its dimensions may delay or prevent achievement of the latter. This would result in the greatest possible climate injustice to those who can least cope with it.

National climate policy

Government responses to climate change are formulated in terms of laws and policies, sometimes with an overarching strategic legal framework for climate policy, within which individual policy instruments are implemented. There are five basic kinds of environmental policy instruments, of which instruments for climate policy are a subset. In the overview of climate policy and associated instruments that follows, many of the example policies come from the UK and Europe, but in principle they are applicable to any country. However, as noted above, the way they are implemented will vary enormously according to different countries' contexts, cultures and institutional/government capacity.

In general, climate policy seems to work best when the instruments are combined into a portfolio or 'policy mix' which is sufficiently flexible to respond to different sectors and conditions in different ways and can be adjusted according to experience with their operation and effectiveness. Like many categorisations of complex situations, the categories overlap somewhat, but it is their effectiveness rather than their category that is important.

Legislation and litigation

Increasingly climate policy is being put into strategic legislative frameworks. These are intended to set the overall context within which individual, specific instruments can be developed and implemented. They may include targets, and dates by which they should be achieved, and make provision for progress towards the targets to be measured and monitored. In parallel to this, climate policy, and the responses of companies (especially fossil fuel companies) to the imperative for climate change, have been subjected to legal challenge.

Market/incentive-based (also called economic) instruments

These policies seek to put an explicit price on carbon emissions, or give some other explicit financial incentive for their reduction. They include carbon taxes and carbon emissions trading, subsidies for clean energy (and the removal of subsidies to fossil fuels), procurement of cleaner products, and liability for climate damages.

Regulation instruments

These seek to define legal standards in relation to technologies or greenhouse gas emissions. They can also include the imposition of obligations, for example in respect of the installation of energy efficiency measures or shares of renewable energy in companies' generation portfolios.

Self-regulation through voluntary (also called negotiated) agreements

These are agreements between governments and businesses to increase efficiency or reduce emissions. Businesses may agree to enter into these agreements in order to avoid the introduction of other policies, which they might think they would find more burdensome.

Information/education-based instruments

As their name implies, this involves the provision of information, either on a mandatory or voluntary basis. Examples that have already been encountered in this book are labels for the energy efficiency of appliances or homes, and smart meters that give readings of energy use in real time.

Innovation instruments

These encourage the development and deployment of new, low-carbon technologies through the Technology Readiness Levels (TRL) journey described in Chapter 11, as well as new business models, institutions, and financial mechanisms that can facilitate decarbonisation. The instruments include research and development (R&D) budgets, capital allowances or tax credits for investment in clean technologies, public–private partnerships for demonstrations, and long-term contracts for deploying clean technologies or purchasing their outputs.

Policy mixes

Conventional policy thinking has it that efficient policy making requires the implementation of one, and only one, policy instrument per policy objective. In climate policy it has become increasingly recognised that effective policy requires a mix of policy instruments, with regulation, information, and market-based and innovation instruments reinforcing each other in different ways across different sectors and emitters.

Each of these aspects of climate policy, with examples, are now explored in more detail.

Legislation and litigation

Climate legislation

As at April 2023 the Climate Change Laws of the World database[5] recorded 3150 pieces of climate legislation, with every country represented. These laws consisted of everything from strategic legislative frameworks, to laws on GHG emissions, energy, climate adaptation and executive orders. They included a wide range of the policies discussed below, applied to all relevant sectors (Eskander et al., 2021).

A prime example of a strategic legislative framework is the UK Climate Change Act, passed in 2008.[6] The Act sets a statutory target for GHG emissions in 2050, which has been progressively tightened, and in 2019 was set to 'net-zero' emissions. It also provides for five-yearly carbon budgets to be adopted on the way to 2050, and set up the independent Climate Change Committee, the duties of which are to recommend the carbon budgets, suggest measures through which they might be achieved and report annually on progress towards them. The publications of the Climate Change Committee (CCC) have become an indispensable resource for anyone interested in emissions reduction in the UK, and have been extensively cited in this book.

The budget-setting process is illustrated in Figure 12.1, showing the CCC's Balanced Net Zero Pathway to 2050, with the interim carbon budgets. However, having a target is not the same as meeting it. The CCC found "either significant risks or a policy gap for 38% of the required emissions reduction to meet the Sixth Carbon Budget" (Climate Change Committee, 2022, p. 22).

Climate litigation

The Climate Change Laws of the World database at April 2023 also contained 2242 cases of climate litigation. Most of these were filed in the United States.

---

5 https://climate-laws.org/. Accessed April 24, 2023.
6 www.legislation.gov.uk/ukpga/2008/27/contents. Accessed April 24, 2023.

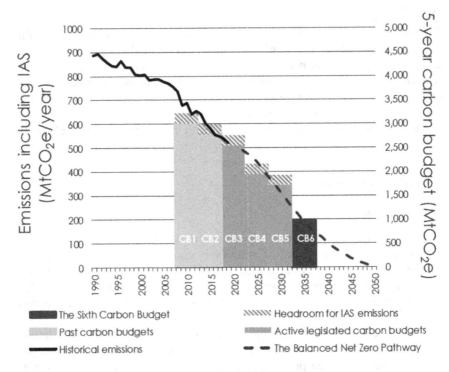

Figure 12.1 UK carbon budgets to net zero in 2050.

Note: IAS = International aviation and shipping, which were included in the recommendation for the Sixth Carbon Budget.

Source: Climate Change Committee, 2020, Figure 1, p. 14.

Most of the non-US cases were filed against governments (Eskander et al., 2021). As an example, in the Netherlands in 2015, a non-governmental organisation, the Urgenda Foundation, and 886 Dutch individuals, sued the Dutch Government for not doing enough to reduce emissions.[7] Urgenda won the case, but the Dutch Government appealed against the ruling all the way up to the Supreme Court. In December 2019 the Supreme Court upheld the judgement against the Dutch Government, ruling that emissions had to be reduced by at least 25% below the 1990 level by the end of 2020. As another example, in 2021 Friends of the Earth Netherlands, other non-governmental organisations and 17,379 individuals sued the Shell oil company for not reducing its emissions in line with the reductions required to meet the Paris temperature target. The court ruled against Shell and ordered it to reduce its emissions by 35% by 2030.

7 www.urgenda.nl/en/themas/climate-case/. Accessed April 24, 2023.

This was the first court ruling against a company to reduce their emissions in line with the Paris Agreement targets, and was notable in that it applied to the whole of Shell's global business, and not just to that in the Netherlands.[8] Shell filed an appeal against the ruling in March 2022,[9] which as of writing is still waiting to be heard.

Litigation is also likely to play an increasing part in discussions about whether high-emitting countries, or fossil fuel-producing companies, should pay compensation for 'loss and damage' to countries or individuals who suffer from the impacts of climate change. The science of attribution of human-caused climate change to damages from extreme weather events is proceeding apace, and this attribution can be then apportioned to countries according to the quantity of their emissions over a given period. As has been seen above, the first scientific paper to attempt to do this found that "the top five emitters (the United States, China, Russia, Brazil, and India) have collectively caused USD 6 trillion in income losses from warming since 1990" (Callahan & Mankin, 2022, p. 1). However, the use of such estimates to support litigation is likely to be much complicated by both the continuing uncertainty in the estimates and the fact that some countries (e.g. those in northern latitudes) are estimated to gain significantly from global warming, in its initial phases at least, as discussed in Chapter 1.

Climate-related litigation can work against the climate too, as is shown by cases under the Energy Charter Treaty. This treaty was signed in 1994 and came into force in 1998, and has 53 states as Contracting Parties. While it is said to be intended "to promote energy security through the operation of more open and competitive energy markets, while respecting the principles of sustainable development and sovereignty over energy resources",[10] it has become controversial through its provisions to protect foreign energy investments. In February 2022, the *Financial Times* reported that five energy companies were suing four European countries under the Energy Charter Treaty for almost €4 billion compensation for restricting their fossil fuel operations.[11] In August 2022 one of these cases resulted in the British oil company Rockhopper winning €190 million in compensation from the Italian Government for blocking a planned fossil fuel project off Italy's Adriatic coast.[12] Some EU countries (e.g. France, Poland, Spain, Italy and the Netherlands) have all either withdrawn from the Treaty or expressed an intention to do so. The European Commission proposed a 'modernisation' of the Treaty, but this was not adequately supported

8 www.clientearth.org/latest/latest-updates/opinions/six-reasons-the-shell-ruling-made-history-for-climate-litigation/. Accessed April 24, 2023.

9 www.reuters.com/business/sustainable-business/shell-filed-appeal-against-landmark-dutch-climate-ruling-2022-03-29/. Accessed April 24, 2023.

10 www.energycharter.org/process/energy-charter-treaty-1994/energy-charter-treaty/. Accessed April 24, 2023.

11 *Financial Times*, February 21 2022. www.ft.com/content/b02ae9da-feae-4120-9db9-fa6341f661ab. Accessed April 24, 2023.

12 www.climatechangenews.com/2022/08/24/british-company-forces-italy-to-pay-e190m-for-offshore-oil-ban/. Accessed April 24, 2023.

by its Member States, so the EU is now proposing that all EU countries withdraw from the Treaty in order to maintain common investment conditions across the EU.[13] However, a provision in the Treaty protects investments for up to 20 years following withdrawal, so the Treaty may continue to have a chilling effect on climate policy for some time to come. The problem is not confined to Europe. A 2022 article in the journal *Science* estimated that "Global action on climate change could generate upward of USD 340 billion in legal claims from oil and gas investors" (Tienhaara et al., 2022, p. 703). Another estimate puts the potential costs of compensation even higher (Saheb, 2020).[14] Given that governments have not yet found the US$100 billion that they promised in 2009 for climate action in developing countries by 2020, the prospect of having to pay more than three times that to fossil fuel companies cannot but chill their desire to move away from fossil fuels.

Market-based (economic) instruments

Chapter 1 made clear that the emission of CO_2 and other GHGs into the atmosphere imposes substantial costs on people currently alive and on future generations. Chapter 2 showed that some people, firms and economies emit very much more GHGs than others. The most elementary considerations of both economic efficiency, as well as social justice, lead to the conclusion that those who emit the damaging gases should pay for the damage they cause, and those who emit the most should pay the most. The fact that this does not happen led Nicholas Stern in his famous review of the Economics of Climate Change to refer to climate change as "market failure on the greatest scale the world has seen" (Stern, 2007, p. 27).

The economist's standard remedy for market failures of this kind is to put a price on the relevant pollutant. Instead of people and organisations being allowed to emit them for free, they would then be required to pay. It is one of the basic insights of economics, that seems to hold in the great majority of cases, that if people have to pay more for a certain activity, they will do less of it.

The price that is set may seek to reflect the damage being caused by the pollution, or it may seek to reduce pollution to the socially desired level, or its level may be intended to be high enough to lead to investment that will provide the good or service without the pollution. In all cases the quantity of pollution would be reduced.

Many countries have now introduced carbon pricing, which may take the form of a carbon tax (often levied on a fossil fuel in relation to the carbon

13 https://energy.ec.europa.eu/news/european-commission-proposes-coordinated-eu-withdrawal-energy-charter-treaty-2023-07-07_en. Accessed August 17, 2023.
14 www.openexp.eu/sites/default/files/publication/files/modernisation_of_the_energy_charter_treaty_a_global_tragedy_at_a_high_cost_for_taxpayers-final.pdf. Accessed April 24, 2023.

emissions when it is burned), or an Emissions Trading System (ETS), both of which are discussed further below. These two kinds of instruments in 2022 had been introduced by 46 countries and some sub-national jurisdictions, including Canada, the EU, China, and several US states, covering just over 23% of global emissions.[15]

The World Bank publishes an annual State and Trends of Carbon Pricing (World Bank, 2022), which stated that in 2021 carbon pricing raised a total of US$84 billion in revenue. There is an enormous range of carbon prices in different countries, from less than US$1/tCO$_2$ in Poland to US$137/tCO$_2$ in Uruguay (imposed in 2021), levied on a wide range of fuels and other sources of GHG emissions (World Bank, 2022, Figure 6, p. 26). The World Bank recommended 'corridor' of carbon prices in order to meet the Paris temperature targets is US$50–100/tCO$_2$e by 2030, but only New Zealand, Switzerland, Finland, the EU's Emissions Trading System (EU ETS, see below), Norway, UK, Liechtenstein, Sweden and Uruguay currently have a carbon price within or above this corridor, as seen in Table 12.1.

Carbon taxes

As noted above, carbon taxes are normally levied on the fuels or other substances that emit GHGs, based on their carbon content or global warming potential. The policy maker levying the tax sets the level of the tax (the price of carbon), and the quantity of carbon emitted declines. The amount of the decline depends on how important the activity is to the emitter, and how easy it is for them to find a substitute for that activity that still enables them to achieve their objectives. For example, a tax on carbon emissions from vehicles in a city with a good public transport system may be expected to bring about a greater proportionate emission reduction than a carbon tax in a rural area with very limited alternative transport opportunities. Given that richer households tend to use more energy than poorer households, a carbon tax will always raise enough revenue to be able to compensate poorer households for the extra cost, if that is desired. A design of carbon tax that has won considerable support from economists in the USA[16] and Europe[17] is the 'carbon fee and dividend approach', whereby the revenues from a carbon tax are distributed to individuals on a per person basis, ensuring that those who use less carbon-based energy than average (mainly poorer individuals) actually end up better off. The same arguments would apply to the removal of subsidies from fossil fuels, if the saved public revenues were redistributed on an equal person basis, as noted below. Disappointingly, Mildenberger et al. (2022) found that in Canada and Switzerland, where a 'carbon fee and dividend' has been introduced, this has

---

15 https://carbonpricingdashboard.worldbank.org/. Accessed April 24, 2023.
16 www.econstatement.org/. Accessed April 10, 2023.
17 www.eaere.org/statement/. Accessed April 10, 2023.

Table 12.1 Carbon pricing in particular jurisdictions

Carbon price (US$/tCO₂e)	Jurisdictions	
	Carbon tax	Emissions trading
<10	Poland, Ukraine, Japan, Singapore, Mexico, Chile, Colombia, Argentina, Norway (reduced rate for Liquified Petroleum Gas and natural gas in the greenhouse industry)	Massachusetts, Shenzhen, Kazakhstan, Fujian, Tianjin, Tokyo, Chongqing, Beijing, Hubei, China, Shanghai
10–20	South Africa, Tamaulipas, Spain, Latvia, Slovenia, Iceland (F-gases)	Guangdong, RGGI, Korea
20–30	United Kingdom, Prince Edward Island, Portugal, Denmark, Luxembourg (all other fuels)	British Columbia
30–40	Northwest Territories, Iceland (fossil fuels), Ireland (other fossil fuels)	California, Québec, Germany
40–50	Canada, British Columbia, New Brunswick, Newfoundland and Labrador, Luxembourg (gasoline), Ireland (transport fuels), Netherlands, France	Canada, Alberta, New Brunswick, Newfoundland and Labrador, Saskatchewan
50–60	Finland (other fossil fuels)	New Zealand
60–70		Switzerland
70–80		
80–90	Finland (transport fuels), Norway (general tax rate)	European Union
90–100		United Kingdom
>100	Liechtenstein, Sweden, Uruguay	

Source: World Bank, 2022, Figure 6, p. 26.

Notes:
The World Bank's suggested 2030 'price corridor' is US$50-100.
RGGI is the Regional Greenhouse Gas Initiative, an emissions trading system involving states in the north-east USA (Connecticut, Delaware, Maine, Maryland, Massachusetts, New Hampshire, New Jersey, New York, Pennsylvania, Rhode Island, Vermont and Virginia). See www.rggi.org/sites/default/files/Uploads/Fact%20Sheets/RGGI_101_Factsheet.pdf

had a limited impact in increasing public support for carbon pricing in those countries.

Emission trading systems (ETSs)

In an ETS an authority issues a number of carbon emission allowances corresponding to the desired level of emissions, and emitters are then required to acquire these, and are able to trade them with each other. At the end of the year, each emitter has to submit back to the authority the number of allowances corresponding to its emissions in that year. If the authority wishes to reduce emissions over time, the number of allowances issued each year declines. The price of the allowances is set by the carbon market, and depends on the balance between the supply of emissions and the demand for them, and on the cost of reducing emissions from the activities covered by the ETS.

There are numerous complexities in how both carbon taxes and ETSs operate in real life, but in essence the carbon tax sets the price of the emissions, and the quantity of emissions adjusts depending on how easy it is to reduce them; the ETS sets the quantity of the emissions and the price is set by the market, again depending on how easy it is to reduce them.

The largest of the ETSs is the EU ETS, which covers around 40% of the emissions in the EU – those from energy-intensive industries and intra-EU aviation, though there are plans to extend it to shipping, road transport and heating for buildings as well.[18] The EU ETS was established in 2008, and from 2009–2018 the carbon allowance price was rather low (below €20/tCO_2), but then the European Commission instituted a Market Stability Reserve to remove emission allowances from the market when there was an excess, and this had the desired effect of pushing the prices up. They reached €80/tCO_2 in mid-2022, though this had fallen back to €65/tCO_2 by September that year. The carbon allowance price went above €100/tCO_2 in mid-February, but was trading at €94/tCO_2 on April 21, 2023,[19] right at the top of the World Bank's carbon price 'corridor'.

Following the UK's withdrawal from the EU, the UK developed its own ETS, which went live on January 1, 2021. The price trajectory of UK emission allowances broadly followed that of the EU since then but fell significantly below it in mid-2023 following the issuance of more allowances by the UK Government.

The carbon price in the EU ETS is very much higher than the carbon prices in most other countries that have implemented them. Many countries have no carbon price at all. In fact, the situation is worse than that. Many countries give substantial financial *subsidies* to the production or consumption of fossil fuels, which amounted to US$700 billion in 2021, about the same level as it was in 2010.[20] A subsidy amounts to a *negative* price on carbon, and adds a further economic inefficiency on top of Stern's 'greatest' market failure. In fact, if the positive carbon prices and negative carbon prices are added together, weighted by the amount of carbon emissions to which they apply, the net global carbon price is actually negative. Romanello et al. (2022, Figure 14, p. 1642) found that a number of countries actually subsidise the production of carbon emissions by US$50/t$CO_2$ or more.

A 2021 report from the International Institute for Sustainable Development (IISD) found that reforming the subsidies in 32 of the countries with the highest

18 https://ec.europa.eu/commission/presscorner/detail/en/qanda_21_3542. Accessed April 24, 2023.
19 https://tradingeconomics.com/commodity/carbon. Accessed April 23, 2023.
20 www.oecd.org/fossil-fuels/. Accessed April 24, 2023.

subsidies would reduce GHG emissions by 5.46 $GtCO_2$, and cumulatively save governments around US$3 trillion by 2030.[21] It is sometimes argued that the subsidies are necessary to reduce the cost of energy for low-income people. This is simply wrong. Most energy is consumed by higher-income people and they are therefore the main beneficiaries of the subsidy. Removing it would give the government substantial revenues, far more than it would need to fully compensate low-income people for the removal of the subsidy, if it so wished.

Feed-in tariffs

Feed-in tariffs (FITs) were discussed in Chapter 5. They are contracts, normally between governments and electricity generators, ranging from households to large-scale suppliers, which guarantee a certain price for any electricity generated over a period of time. Feed-in premiums (FIPs) are similar to FITs but guarantee to pay a certain premium over the market price. This price guarantee allows individuals and companies to invest the relatively large sums required to install the renewable technology, in the knowledge that they will be paid a certain amount when the electricity starts to be generated. In 2004 33 countries had implemented a FIT, and by 2016 this number had risen to 83.[22] FITs, along with Renewable Portfolio Standards (discussed in the next section) have been largely responsible for the astonishing growth in renewable capacity seen over the last 15 years. However, the main problem with FITs is knowing where to set the fixed price: too low and they don't incentivise deployment, too high and a rush for deployment makes them unaffordable. This is the reason that the UK moved from fixed-price FITs to auction-based FITs based on Contracts for Difference, as described in Chapter 5.

Regulation

Regulations define legal standards in relation to technologies, environmental emissions, or some other characteristic of a product or process that affects its energy use or emissions. Common regulations in relation to climate change stipulate the energy efficiency of buildings, vehicles or energy-using equipment. Regulations may also impose obligations on businesses to use, or not use, certain technologies, or undertake measures to improve the energy efficiency of their customers.

A huge range of regulations have been applied in many countries to reduce GHG emissions. Very often these focus on increasing the energy efficiency of vehicles, buildings, and appliances, as discussed in some detail in Chapter 3.

21 www.iisd.org/articles/press-release/fossil-fuel-subsidy-reform-could-reduce-co2-emissions-equivalent-those-1000. Accessed April 24, 2023.
22 www.statista.com/statistics/830641/countries-with-feed-in-tariff-adoptions-of-renewable-energy/. Accessed April 24, 2023.

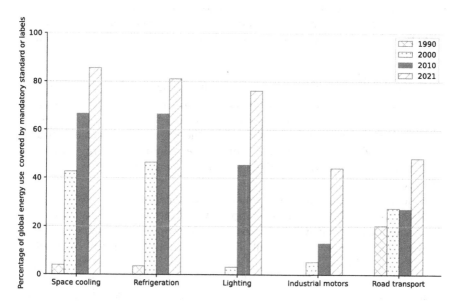

Figure 12.2 Percentage of global energy use covered by mandatory standards or labels, 1990–2021.

Note: Space cooling, refrigeration and lighting cover residential use only.

Source: International Energy Agency, 2021b, p. 33.

Figure 12.2 shows that the application of regulations stipulating efficiency standards for energy use technologies has grown dramatically since 1990, and now cover 80%, or nearly 80%, of appliances for three main residential uses of electricity (space cooling/air conditioning, refrigeration and lighting). In addition, the energy efficiency of buildings is widely regulated: "54 countries had mandatory codes on the national level for both residential and non-residential buildings" (International Energy Agency, 2021b, p. 37).

Some other examples of regulations related to climate and energy are:

- In 2005 in the UK, it became mandatory to replace old gas boilers with much more efficient condensing gas boilers, which were widely installed in other countries, but still relatively rare in the UK. In 2019 54.8% of all dwellings in England had condensing gas boilers, up from 1.7% in 2003.[23] The UK Government is planning to take the same approach with new (fossil methane) gas boilers (banning them from 2035) in its push for net zero by 2050.

---

23 www.statista.com/statistics/292259/boiler-types-in-dwellings-in-england/. Accessed April 24, 2023.

- Another example of the banning or phasing-out of carbon-emitting technology is Norway, which will ban the sale of new petrol and diesel cars in 2025,[24] with many other countries, including the UK, following suit from 2030.[25]
- Building regulations (or codes or mandates) in many countries require that new buildings meet certain standards for energy efficiency, which typically get stricter over time. However, for such regulations to be effective, it is imperative that they are enforced, which is not always the case when governments have cut back funding for the relevant enforcement bodies. There is evidence that enforcement of the Minimum Energy Efficiency Standards introduced in England in 2019 leaves a considerable amount to be desired.[26]
- A number of countries have implemented renewable portfolio standards (RPS, also called renewable obligations), which require electricity companies to generate from renewable sources a certain proportion, usually increasing over time, of the power they sell. This has been a popular way, often allied with tax credits, to require generators in the US to increase their supply of renewable power. The US Berkeley Lab reports that around half of all growth in US renewable electricity generation and capacity since 2000 was associated with state RPS requirements, although in 2019 that percentage had declined to 23% of all US renewable capacity additions.[27]
- A number of European countries have instituted Energy Efficiency Obligations, which require energy suppliers to implement, in their customers' homes, energy efficiency measures that are estimated to save a defined amount of energy or carbon. In Chapter 3 it was seen that such obligations have delivered large energy savings at relatively low costs and have leveraged investments in energy efficiency.[28]
- Both renewables generators and energy efficiency installers qualify in some countries to receive green or white certificates respectively, to acknowledge the amount of energy they have generated or saved. These certificates can sometimes be sold (traded) to other organisations that have obligations, rather like carbon allowances.

24 https://electrek.co/2021/09/23/norway-bans-gas-cars-in-2025-but-trends-point-toward-100-ev-sales-as-early-as-april/. Accessed April 24, 2023.

25 https://en.wikipedia.org/wiki/Phase-out_of_fossil_fuel_vehicles. Accessed April 24, 2023.

26 https://assets.publishing.service.gov.uk/government/uploads/system/uploads/attachment_data/file/825485/enforcing-enhancement-energy-efficiency-regulations-English-private-rented-sector.pdf. Accessed April 24, 2023.

27 https://emp.lbl.gov/publications/us-renewables-portfolio-standards-3. Accessed April 23, 2023.

28 https://climatepolicyinfohub.eu/energy-efficiency-policy-instruments-european-union.html. Accessed April 24, 2023.

Voluntary agreements

Voluntary (also called negotiated) agreements between governments and industry have a long history, but became especially prominent in Europe in environmental policy in the 1990s and early 2000s, being hailed as a 'new model of cooperative environmental management' (de Jongh & Captain, 1999). They were defined by the European Commission in 1996 as "agreements between industry and public authorities on the achievement of environmental objectives" (Commission of the European Communities, 1996, p. 5), who then considered that they "can promote a pro-active attitude on the part of industry, can provide cost-effective, tailor-made solutions and allow for a quicker and smoother achievement of objectives". (ibid., p. 3). They were implemented by many countries, being most numerous in Austria, Germany, Denmark and the Netherlands (Wurzel et al., 2013, p. 108).

These agreements may be truly voluntary, with no sanctions for non-compliance, or legally binding. Businesses are prepared to enter into them either because there are financial incentives (e.g. the tax reduction firms get if they meet their energy efficiency targets under the Climate Change Agreements in the UK), or because they consider them preferable to the regulations or taxes that are likely to be imposed in the absence of an agreement.

To be environmentally effective, such agreements clearly need to go beyond the regulations then in force. For governments the hope is that a negotiated agreement will be easier to implement and get wider stakeholder buy-in than a new regulation. For businesses there may be a hope that an agreement can forestall new regulation, and provide an opportunity to influence the nature of the agreement and any target that might be set as part of it.

The Netherlands is widely regarded as among the leaders of this type of environmental policy (Bressers et al., 2009). While the survey by Bressers et al. (2009) reveals a generally positive assessment of a large subset of Dutch environmental agreements, it is clear that the context and process of the agreement plays a large part in both its cost and environmental effectiveness, and fully assessing their effects is in any case made more complex by the fact that they were often implemented in conjunction with other instruments (e.g. subsidies, information, and technology development programmes).

In respect of decarbonisation, the major use of negotiated agreements has been to seek to improve energy efficiency in industry by setting targets to be achieved in the future. The doubts as to whether the agreements actually do result in businesses achieving greater energy efficiency improvements than they would have in the absence of the agreement are revealed by two studies of the UK's target-setting Climate Change Agreements (CCAs), which came to opposite conclusions on this matter. Ekins & Etheridge (2006) concluded that they did achieve greater energy efficiency than would have been achieved under 'business as usual'; Martin et al. (2014) found that they did not.

A study of energy efficiency agreements in Denmark, Sweden, Germany, France and the Netherlands produced findings that "give no hint that voluntary agreements are a cheap solution", and that their environmental effectiveness "strongly depends on the performance of the broader policy mix in which the agreement scheme is embedded" (Krarup & Ramesohl, 2002, p. 115). Although the use of voluntary or negotiated agreements in climate policy is now less prominent than it was, one area in which they are still an option for industry is in relation to the EU's Ecodesign Directive, which seeks to set minimum energy performance standards for some electrical equipment. In this context, making the regulatory alternative to voluntary agreements explicit, the European Commission has written: "Self-regulation may achieve the ecodesign policy objectives more quickly or at lesser expense than mandatory requirements. Therefore, industry sectors may propose voluntary agreements as alternatives to potential ecodesign regulations".[29] Such agreements have been concluded for set top boxes, imaging equipment and games consoles.[30] However, environmental groups have been very critical of the effectiveness of this approach.[31]

The voluntary nature of the agreements and the increasing stringency of emission reduction commitments, often embedded in law, means that voluntary agreements have a much lower profile now than they did at the turn of the century. Governments want greater assurance now that their targets and other policies will achieve the decarbonisation that they have committed themselves to. And that has led to greater reliance on regulation and economic instruments.

Information instruments

Information plays an absolutely critical role in relation to knowledge about the cause and nature of climate change, and about what can be done about it by government, businesses, community organisations and individuals. With regard to climate change itself, huge damage has been done by the systematic *mis*information about climate change that has been disseminated by the fossil fuel industry, both to deny that the climate is changing because of the combustion of their products,[32] and to lobby against climate policy that could reduce emissions.[33] The effectiveness of these campaigns is most clearly shown in the United States, the most technologically advanced country in the world, and the

29 https://energy.ec.europa.eu/system/files/2016-12/list_eco-design-voluntary_agreements_0.pdf. Accessed April 24, 2023.

30 https://commission.europa.eu/energy-climate-change-environment/standards-tools-and-lab els/products-labelling-rules-and-requirements/energy-label-and-ecodesign/energy-efficient-products/recognised-voluntary-agreements-under-ecodesign-legislation_en. Accessed April 24, 2023.

31 https://ecostandard.org/news_events/voluntary-agreements-get-a-cold-shower/. Accessed April 24, 2023.

32 www.theguardian.com/environment/2021/nov/18/the-forgotten-oil-ads-that-told-us-climate-change-was-nothing. Accessed April 24, 2023.

33 https://news.harvard.edu/gazette/story/2021/09/oil-companies-discourage-climate-action-study-says/. Accessed April 24, 2023.

second largest emitter of GHGs. Here a majority of the elected representatives of one of the two great political parties, the Republicans, and of the people who vote for them, continue to deny the very existence of human-caused climate change despite the wildfires, droughts, hurricanes and floods that afflict that country with increasing frequency and intensity, and incontrovertible scientific evidence of the role of human-caused climate change in these events.

Against the billions of dollars of fossil fuel industry disinformation, even governments convinced by the climate science find it difficult to persuade their citizens of the seriousness of the situation and to introduce policies that are commensurate with its risks. Given the evidence of these risks and of life-threatening climate-related events that are seen regularly on television screens, it is remarkable that still only 41% of the global population sees climate change as a 'very serious threat' to their country (although a further 28% say that they are 'somewhat concerned' about it).[34] So far this level of concern has not been sufficient to compel governments, through the annual COP meetings, to make commitments to the necessary levels of emission reduction, and then put the policies in place to deliver them back home.

Awareness of and concern about the impacts of climate change among the general population has been much increased in recent years by pictures of its increasingly dramatic impacts (floods, wildfires, etc.). However, awareness of its causes, and of ways of reducing human contributions to it, is still generally low. Policies of information and education seek to increase such awareness, and to lead to actions that will reduce emissions.

The main areas in which such measures, which often comprise labelling, have been put in place are the energy efficiency of buildings, the energy efficiency of vehicles and electrical equipment, and the environmental impacts of timber and food (especially meat) production.

In the UK, building energy efficiency labels are called Energy Performance Certificates (EPCs) in residential buildings, as discussed in Chapter 9, and Display Energy Certificates (DECs) in government and commercial buildings. They are very similar in appearance and information conveyed. Energy Performance Certificates were mandated by the EU's Energy Performance in Buildings Directive, which was passed in 2002. Grading buildings from A to G according to their energy efficiency, they are required for all residential buildings that are rented, sold or newly built. DECs are required for public buildings over a certain size that are frequently visited by the public.

For vehicles, new car fuel consumption and CO_2 emissions data need to be shown whenever a new car is sold. In the UK a wide range of other data (e.g. fuel cost, tax band) is also shown in the standardised label that has been produced in consultation with the industry. The grading A–G is similar in concept and look to that for buildings.

---

34 https://wrp.lrfoundation.org.uk/2019-world-risk-poll/the-majority-of-people-around-the-world-are-concerned-about-climate-change/. Accessed April 24, 2023.

Energy labels are also required for a wide variety of new appliances, and again they give other useful information, depending on the appliance. The labels were revised in 2020 to take account of the fact that the energy efficiency of appliances has improved rapidly over the years, such that previously A-rated appliances were no longer best-in-class, and the practice of reflecting this in A+, A++ and A+++ ratings was proving confusing to consumers. Now, an A-rating will be best-in-class when purchased new, but the same machine could be given a different rating when sold in the future if more energy-efficient appliances have appeared in the meantime.

Smart meters for electricity and, where relevant, for fossil methane gas, are now widely installed in industrial countries. They give consumers real-time information about their energy use, and how much it is costing them, and in due course will allow time-of-use pricing, where non-critical energy use can be scheduled for off-peak times when energy is available at lower costs. This functionality is especially useful for electricity systems with a high proportion of intermittent renewables. The issue was discussed in Chapter 5.

Non-governmental organisations have introduced a wide range of other labels to reflect supposedly superior environmental performance. Relevant to climate change are the Forestry Stewardship Council and Soil Association, which have strict criteria for timber and food production respectively, which are intended to reflect the higher environmental sustainability of their production methods compared to industry averages. Because of the complexity of the issues involved, and the difficulties in measuring relevant outcomes, there is inevitably debate as to the extent to which they actually do so.

While the information instruments that have been put in place (e.g. labels), for example in respect of appliances (Chapter 3), vehicles and buildings (Chapter 9), have had a significant effect on the emissions associated with some purchased goods, information about the GHG-intensity of other high-emitting goods and services (e.g. meat and flights) is conspicuous by its low profile or absence.

Beyond labelling, there is clearly a role for comprehensive energy information and advice for householders as to how they can make their homes more energy-efficient, and reduce their energy bills, while maintaining adequate ventilation (to prevent damp and mould) and safeguarding the fabric of the building. The nature of this information and advice will obviously depend on the country and context. In the UK such information has been delivered for many years by the Energy Saving Trust,[35] which also gives advice on generating renewable energy at home and buying energy-efficient products.

For some time it was imagined that the simple provision of information about climate change, and what to do about it, would lead to the significant changes in behaviour that are required to address it. It is now widely held that this 'information deficit model' (Suldovsky, 2017) is unrealistic, because it fails

---

35 https://energysavingtrust.org.uk/energy-at-home/. Accessed April 24, 2023.

to recognise constraints in human nature (e.g. the force of habit, aversion to the sacrifice of valued activities), which act as a barrier to behaviour change, creating a 'gap' between pro-environmental attitudes and actual behaviours (Park & Lin, 2020). It also does not take account of external constraints, such as upfront cost or lack of public transport, that prevent people from installing energy efficiency measures or driving less.

Innovation instruments

There has been enormous innovation, to reduce carbon and other GHG emissions, in both the supply and use of energy in the last few decades, as has been discussed in earlier chapters. This has not come about by accident, but is the result of consistent funded effort by governments, industry, universities and other research and development institutions of all kinds.

In a market economy, innovation occurs naturally as businesses seek to develop new products and processes for competitive advantage. The prospect of profit is the driving force. Because carbon emissions are often not priced, the incentive for businesses and households to save carbon and make other environmental improvements through 'green' innovation is often weak. It will not happen at anything like the pace required to address climate change without strong policy support.

Innovation is often thought to consist of research and development, and these are indeed important. But they are only part of the picture. Innovation takes place, as shown in Figure 12.3, through a whole chain of activities, from the generation of the initial ideas, through to the deployment at scale of new processes and products, that correspond to the journey of a technology through the Technology Readiness Levels (TRLs) discussed in Chapter 11.

Innovation policy is the broadest category of policy and refers not just to support for research and development (R&D), though this is obviously important, but to any policy that will help to develop the many technologies described at the beginning of Chapter 11, as well as those that will pull these technologies through the TRLs shown in Table 11.1. And the policies need to stimulate innovation not only in technology, but also in all the other areas that are crucially important for the scale-up and diffusion of technologies, including business organisation, business models and supply chains, consumer preferences and behaviours, financing, regulation and standard-setting in markets, and institutions of research and governance and infrastructure (Grubb et al., 2017).

Green innovation, as shown in Figure 12.3, involves different actors and networks, who are incentivised through the provision of resources and infrastructure, and appropriate regulations and other policy instruments, to push new ideas through R&D into niche markets. If successful there, further public support may sometimes be necessary to enable them to diffuse more widely through society through market demand. Public funding is most important in the early stages of the chain, and green public procurement (the purchase by

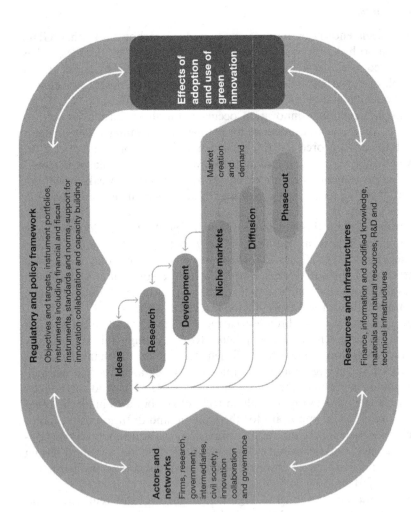

Figure 12.3 The innovation chain.

Source: Miedzinski et al., 2020, Figure 1, p. 31. www.ucl.ac.uk/bartlett/sustainable/sites/bartlett/files/the_commissio
ns_final_report.pdf. Accessed April 23, 2023.

the government of early-stage low-carbon products) can be important in kick-starting markets. As markets develop and demand increases, private, commercial funding becomes dominant.

Figure 12.3 shows that a wide range of policy instruments have been used to drive the innovation process. Particularly important in Europe in respect of renewable energy have been feed-in tariffs or premiums (FITs/FIPs). This approach to incentivising innovation and new technologies has recently been taken to new levels by the US Inflation Reduction Act (IRA), which envisages making available tax credits and other subsidies amounting to US$370 billion to reduce the cost of clean energy, provided that it encourages economic activity in the US.[36] While in principle, given the urgency of the climate crisis, such public support for acceleration of the clean energy transition is justified, there are obvious implications for the global trading system, which has sought to limit such subsidies, arising from the openly nationalistic way in which it is being implemented. Worried that support on this scale would encourage European clean energy firms to relocate to the other side of the Atlantic, the European Commission has responded with its own Green Deal Industrial Plan.[37] This has led to fears in Europe that the larger more industrial Member States, and large companies, will benefit most from these arrangements, undermining development in smaller Member States and reducing competition and ultimately competitiveness in the EU.[38]

Circular economy policies

Chapter 8 showed that some 50% of current GHG emissions are associated with the extraction and processing of materials. If materials could be used far more efficiently, such that they stay in the economy for far longer before being disposed of, then the need for the extraction and processing of new materials would fall and so would the emissions of GHGs associated with these processes. This is part of the thinking behind advocacy for a more 'circular' economy, rather than the currently dominant 'linear' (take-make-dispose) use of materials that is prevalent today.

Figure 12.4 lists ten strategies for reducing the quantity of materials required in society, and thereby reducing the associated GHG emissions (and other environmental impacts of extraction and processing). R0, R1 and R2 involve decisions about the manufacture and first use of a product that reduce the associated material use. R3–R7 inclusive involve extending the life of a product. The extent to which these options are possible depends to a large extent on the original design of the product. R8 and R9 involve recovering the

---

36 www.csis.org/analysis/electric-debate-local-content-requirements-and-trade-considerations. Accessed April 10, 2023.

37 https://ec.europa.eu/commission/presscorner/detail/en/ip_23_510. Accessed April 10, 2023.

38 www.euractiv.com/section/competition/opinion/eu-response-to-inflation-reduction-act-must-not-damage-competitive-markets/. Accessed April 10, 2023.

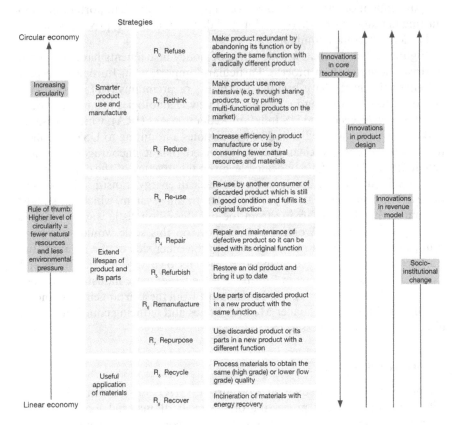

Figure 12.4 Strategies for increasing the circular use of materials.

Source: Potting et al., 2016, Figure 1, p. 5.

materials from the product, or using the energy from its incineration for electricity or heating.

The Organisation for Economic Co-operation and Development (OECD) has found that there are substantial obstacles to the implementation of strategies R0–R9. Table 12.2 outlines the nature of these obstacles. The most important major obstacles are: reaching a critical scale for circular activities, the lack of financial resources, the regulatory framework and the existence of cultural barriers.

In a paper commissioned by the OECD, Ekins et al. (2019) go into more detail about the nature of these barriers, as shown in Table 12.3.

In Figure 12.4 only R0 and R1 are driven only by consumer decisions. The extent to which R2–R9 are implemented depends on whether they are required or facilitated by public policy or whether they are economically profitable for the companies or other economic actors concerned.

Table 12.2 Percentage of survey respondents identifying major and important obstacles to the implementation of a circular economy (where 'major' represents a greater obstacle than 'important')

	Major obstacle (%)	*Important obstacle (%)*
Cultural barriers	29	44
Regulatory framework	29	44
Financial resources	32	38
Holistic vision	21	41
Adequate information	12	50
Incoherent regulation	24	35
Financial risk	18	38
Awareness	12	44
Critical scale	38	15
Human resources	21	29
Private sector engagement	6	35
Political will	15	21
Technological solutions	9	26

Source: Organisation for Economic Co-operation and Development, 2019, p. 6 www.oecd.org/cfe/regionaldevelopment/Circular-economy-brochure.pdf. Accessed April 23, 2023.

One approach that has been successful in a number of countries is 'industrial symbiosis', which involves establishing a network of businesses to share information about resources and wastes which enables them to be used, or reused, in a mutually beneficial way. This business network may involve businesses in close geographical proximity, as in 'industrial eco-parks', which are common in some Asian and European countries, and which have been defined as "industrial zones that promote collaborations between businesses and with local communities, generating environmental, social and economic benefits" (Aggeri, 2021, p. 60).

Alternatively, the business network may be facilitated through a third party, either through an electronic platform or through physical workshops, or both. The first national network of this kind was the UK National Industrial Symbiosis Network (NISP), which from 2005 to 2010 was supported by the UK's Department for Environment, Food and Rural Affairs (Defra) through cumulative investment of £27.7 million. Table 12.4 shows the environmental and economic benefits from NISP. In terms of the environmental effectiveness of the public investment, it can be seen that the environmental benefits were very cheaply delivered: CO_2 reduction at £0.51/tCO_2, reduction in the use of virgin materials and water at £0.32/t, hazardous waste reduction at £8/t – these are extraordinarily cost-effective environmental improvements. The programme also leveraged nearly five times as much private investment as Defra put in over the five years, both created and saved jobs, generated extra sales and reduced costs at a rate of £1 for every £0.10 of public expenditure. It also returned three times as much money to the Treasury (through the jobs created and saved) as the government had provided. This was a truly outstanding

Table 12.3 Barriers to moves towards a circular economy (CE)

Barrier cluster	Examples of barriers
Governmental issues	Ineffective, insufficient or unsupportive policies; lack of performance indicators; unclear vision
Economic issues	Weak incentives, lack of internalisation of external costs; high upfront costs and insufficient short-term benefits prevent investment; resource-efficient options can be more expensive
Technological issues	Product complexity inhibits separation of materials making recycling harder; challenges monitoring product quality throughout the life cycle, and maintaining product quality with recovered or remanufactured materials; lack of accurate information in tracking material composition of products to enable recycling and manufacturing
Knowledge and skills issues	Lack of public information and awareness to support participation in reuse/recycle/remanufacturing; lack of necessary skills in workforce; consumer awareness about refurbished or remanufactured products – perception that quality is lower
Management issues	Lack of interest in or leadership on circular economy within firms at management level; higher priority given to other supply chain issues; organisational structures within firms inhibit implementation of CE practices
Circular economy framework issues	Lack of successful business models; complexity of transnational supply chains, including for waste management; tendency to focus on recycling when other CE practices might be more beneficial
Culture and social issues	Lack of good relationships in supply chain; linear technologies and practices deeply rooted; negative customer perceptions of remanufactured products; 'thrill' of newness
Market issues	Challenges to operating take-back systems with multiple companies involved, and legal problems for service providers retaining the sold product; lack of standards and variable quality of refurbished products; lack of consumer acceptance of 'service' rather than ownership models; remanufacturing requires experience and knowledge

Source: Adapted from Govindan & Hasanagic (2018) in Ekins et al. 2019, Table 6.2, pp. 42–43. www.oecd.org/cfe/regionaldevelopment/Ekins-2019-Circular-Economy-What-Why-How-Where. pdf. Accessed April 23, 2023.

programme, and it was frankly extraordinary that in 2010 the then Labour Government decided not to continue it. The company which facilitated NISP, International Synergies, as of March 2021 had applied the NISP model in 40 countries.[39]

Returning to Figure 12.4, the fact that the core 'extend lifespan' options, R3–R7, are very much the exception rather than the rule in the current economy shows that, in the main, they are neither mandated by governments, nor are

39 www.international-synergies.com/about-us/our-story/. Accessed April 23, 2023.

Table 12.4 Environmental and economic benefits of the UK National Industrial Symbiosis Programme, 2005–2009

	Actual 5-year total[1]	*Cumulative[2] over 5 years*	*Value for money (Public investment/ unit output)[3]*
Environmental benefits			
Landfill diverted (Mt)	7.0	12.6	0.44 (£/t)
CO_2 reduction (Mt)	6.0	10.8	0.51 (£/t)
Virgin materials saved (Mt)	9.7	17.5	0.32 (£/t)
Hazardous materials reduced (Mt)	0.36	0.7	7.9 (£/t)
Water saved (Mt)	9.6	17.2	0.32 (£/t)
Economic benefits			
Extra sales (£ m)	176	317	0.087 (£/£)
Costs saved (£ m)	156	281	0.099 (£/£)
Extra government revenue (£ m)		89	0.31 (£/£)
			Fiscal multiplier: 3.2 (£/£)
Private investment (£ m)	131		
Jobs created	3683		
Jobs saved	5087		

Source: Author calculation from data in Laybourn & Morrissey, 2009. https://assets.plottcdn.co.uk/international-synergies/wp-content/uploads/2021/07/20121910/Appendix-D-Pathway-Report.pdf. Accessed April 23, 2023.

Notes:
[1] Total over 5 years, computed by simply summing the results for each year (independently verified)
[2] Total over 5 years assuming NISP contribution to savings of only 60%, but persistence of savings to subsequent years, declining by 20% per year
[3] Public investment of £27.7 million over five years. For environmental categories, this is assumed to be split equally between five categories (i.e. £5.5 million per category), divided by results in Cumulative column; for economic categories, the full public investment figure (i.e. £27.7 million) is used as the numerator.

they profitable economically. Put another way, the incentives for businesses to adopt these processes are not strong enough. If policy makers wish to move towards a 'circular economy', then they will need to address and remedy these facts.

In its 2020 report[40] on incentive policies, the European Circular Economy Stakeholder Platform (ECESP) considered that there were just three possible ways in which policy makers could address the current lack of incentives for a circular economy. These were:

- Demand creation, especially through public procurement being directed to products and processes that make more efficient and environmentally preferable use of resources.

---

40 https://circulareconomy.europa.eu/platform/en/about/cg-activities-documents/ecesp-leadership-group-economic-incentive-policies-reflections-summer-2020. Accessed April 24, 2023.

- Pricing policies that increase (reduce) the price of less (more) circular products and processes. Such policies could include: ensuring that producers (rather than taxpayers) pay the cost of the disposal of their products; taxes on virgin materials but not on recycled materials; lower VAT rates for products produced by R3–R8 processes; and CO_2 pricing.
- Regulations mandating more circular products and processes (e.g. through required recycled content); labelling procedures showing the recycled content in products; and a quality assurance scheme for recycled materials.

Of particular importance in encouraging greater circularity (i.e. uptake of strategies R2–R8 in Figure 12.4) are policies that change the *design* of products in such a way that they address the barriers to these strategies. This was at least part of the intention of the environmental policy approach of Extended Producer Responsibility (EPR), which "aims to make producers responsible for the environmental impacts of their products throughout the product chain, from design to the post-consumer phase" (Organisation for Economic Co-operation and Development, 2016, p. 13). By 2016 there were over 400 such schemes in operation, mainly in Europe and North America, focusing mainly on small electronic goods (35%), tyres (18%), packaging (17%), and vehicles and batteries (12%). The policy measures implemented by these schemes were mainly 'take-back' measures (70%, where producers take back products at the end of their lives, or pay for third parties to do so), advance disposal fees (17%, where producers charge for end-of-life arrangements at the time of purchase of the product) and deposit-refund schemes (11%, where a payment at purchase is refunded to the consumer when the item is returned).

While these EPR schemes have been successful in increasing the proportion of relevant products that is recycled (R8), there is little evidence that they have affected product design, which is the key to implementation of the 'extended lifespan' circular strategies (R3–R7). It has been suggested that a far greater influence on product design would be exerted if producers retained the ownership of the products they put on the market, and only sold their lifetime services, being required to take back their products at the end of their lives, thereby greatly extending the range of products subject to a take-back obligation. Such a scheme could remove from consumers and local authorities the responsibility and cost of the disposal of unwanted products; reduce illegal disposal of materials ('fly-tipping'); incentivise producers to produce longer-lasting and repairable products, maintained through service models and other business models; and reduce the environmental impacts of ever-increasing material use (Domenech et al., 2019).

In an intriguing twist to 'take-back' policy, Jenkins et al. (2023) have suggested that it should be applied to fossil fuel producers, i.e. they should be forced to build the necessary carbon capture and storage (CCS) facilities and pay to capture and store the amount of CO_2 corresponding to the amount of fossil fuel that they put on the market. While this is an interesting and intellectually elegant proposal both to kick-start at scale the CCS industry that

Table 12.5 Characterisation of environmental policy instruments, many of which are used in climate policy in 'policy mixes'

Primary type	Primary purpose		
	Technology push	*Demand pull*	*Systemic*
Economic instruments	RD&D grants and loans, tax incentives, state equity assistance	Subsidies, feed-in tariffs, trading systems, taxes, levies, deposit-refund systems, public procurement, export credit guarantees	Tax and subsidy reforms, infrastructure provision, cooperative RD&D grants
Regulation	Patent law, intellectual property rights	Technology/performance standards, prohibition of products/practices, application constraints	Market design, grid access guarantee, priority feed-in, environmental liability law
Information	Professional training and qualification, entrepreneurship training, scientific workshops	Training on new technologies, rating and labelling programmes, public information campaigns	Education system, thematic meetings, public debates, cooperative RD&D programmes, clusters

Source: Rogge & Reichardt, 2016, Table 2, p. 1624.

Note: RD&D stands for research, development and demonstration, where the last is a step between development and niche markets, not shown in Figure 12.3.

pretty well all 1.5°C scenarios suggest will be necessary for the target to be met, and implement the 'polluter pays principle', it seems most unlikely that policy makers, who cannot even agree to phase out fossil fuel subsidies, will mandate its implementation any time soon.

Policy mixes

One of the most frequent characteristics of climate policy is that it uses different policy instruments at the same time, in 'policy mixes'. Table 12.5 shows three of the categories of policy instruments described above, and against each lists a number of specific instruments that have been used in 'policy mixes' for climate mitigation. These instruments are further distinguished between 'technology push', 'demand pull' and 'systemic' instruments, where the first of these seeks to develop new technologies and products, the second seeks to pull them through into markets, and the third acts to transform the contexts, or systems, within which these instruments operate.

At the micro-level (individuals and small companies), regulatory standards, and engagement and dissemination campaigns that ensure that people are aware of these, tend to be most effective. These also tend to be most effective for products that are already on the market. At the meso-level, larger companies and economic sectors are most responsive to pricing instruments, market structures, planning and regulation. Such policies work throughout the value chain. At the macro-, whole-economy level, and early in the innovation and technology development journey, strategic investment in R&D and infrastructure, demand-pull policies like government procurement, and industrial development policy, are required (IPCC, 2022a, Figure 1.6, pp. 1–50).

Policy evaluation

Once implemented, it is obviously important to be able to evaluate how effective policy instruments have been across a range of criteria. Peñasco et al. (2021) evaluated ten policy instruments across seven dimensions of effectiveness through a review of 211 scientific articles and reports. The policy instruments were:

- building codes and standards;
- renewable energy obligations (or renewable portfolio standards, RPS);
- government procurement;
- public R&D funding;
- feed-in tariffs or premiums (FITs/FIPs);
- energy auctions;
- energy taxes and tax exemptions;
- GHG emissions allowance trading schemes;
- tradable green certificates; and
- white certificates (or energy efficiency standards).

The seven effectiveness dimensions were: environmental, technological, cost-related, innovation, competitiveness, distribution and other social outcomes.

While the paper shows that this literature is by no means unanimous about the effectiveness of different instruments across different dimensions, it raises many important points about the potential impacts, advantages and disadvantages of different instruments. The paper is supplemented by an online tool[41] that enables many of the relevant issues to be explored in more detail than in the paper.

The IPCC assessed the effectiveness of different policies in different sectors, as shown in Table 12.6, according to whether they directly affect emissions, or the immediate drivers of emissions, or the take-up of low-carbon technology. All the different kinds of policies discussed above are included in the table, and many of them have also been implemented as constituents of 'policy mixes' to

---

41 https://dpet.innopaths.eu/#/. Accessed April 23, 2023.

Table 12.6 Policies implemented in different sectors

Sector	Effects on emissions	Effects on immediate drivers	Effects on low-carbon technology
Energy supply	Carbon pricing, emissions standards, and technology support have led to declining emissions associated with the supply of energy	Carbon pricing and technology support have led to improvements in the efficiency of energy conversion	A variety of market-based instruments, especially technology support policies, have led to high diffusion rates and cost reductions for renewable energy technologies
Agriculture, forestry and other land use (AFOLU)	Regulation of land-use rights and practices have led to falling aggregate AFOLU-sector emissions	Regulation of land-use rights and practices, payments for ecosystem service, and offsets have led to decreasing rates of deforestation	
Buildings	Regulatory standards have led to reduced emissions from new buildings	Regulatory standards, financial support for building renovation and market-based instruments have led to improvements in building and building system efficiencies	Technology support and regulatory standards have led to adoption of low-carbon heating systems and high efficiency appliances
Transport	Vehicle standards, land-use planning, and carbon pricing have led to avoided emissions in ground transportation	Vehicle standards, carbon pricing, and support for electrification have led to automobile efficiency improvements	Technology support and emissions standards have increased diffusion rates and cost reductions for electric vehicles
Industry		Carbon pricing has led to efficiency improvements in industrial facilities	

Source: Adapted from IPCC, 2022b, Box TS.13, Table 1 (upper panel), p. TS-118.

regulate for energy efficiency, incentivise energy efficiency and new renewables, and stimulate innovation in low-carbon options across the board.

Policy implementation at sub-national levels

Local government

The climate policies described above are, by and large, instituted and implemented by national governments. But the actions to which they are intended to lead will all be carried out by local governments, businesses and households locally. All climate action is therefore local action.

Many cities, local authorities and municipal governments have now adopted climate strategies and policies, within their areas of remit and competence. For example, the C40 group of 97 cities round the world describes itself as "A global network of mayors taking urgent action to confront the climate crisis";[42] 13 of the 97 cities are in China; UK has just one member, London. Because cities in different countries have very different powers and capacities, the main purpose of the network is to share ideas between members, which may or may not then get translated into their local context.

In London the Mayor's principal powers in the field of climate change relate to buildings (improving their energy efficiency) and transport (curbing car use in favour of public transport, cycling and walking). Some relevant city-level decarbonisation policies and proposals have been discussed in the relevant sections of Chapter 9, but, in the UK, they are crucially dependent on the central UK Government providing adequate funding for their implementation, which is not by any means assured at a time (2023) when there is a Conservative central government and Labour Mayor.

Other local governments in the UK, and the devolved national governments of Scotland, Wales and Northern Ireland, and of course local governments in other countries, have their own strategies and policies. One of the interesting features of these strategies is that they seek to project a very positive vision of a low-carbon town or city that is very different from the road- and car-dominated urban visions of the 1960s and 1970s. The medium-sized city of Nottingham provides a good example. Its vision for 2028 is set out in the six rows in Table 12.7, stressing the benefits of safety, health, opportunity, inclusion, affordability and quality of life which they intend to offer through low-carbon living.

Tingey & Webb (2020) surveyed UK local authority (LA) energy activity, finding that their policies and strategies were at very different stages of development and implementation. While around half of UK LAs (205 out of 408) had declared 'climate emergencies', only 13% were categorised as 'energy leaders', with "a programme of energy initiatives with multiple investments in energy

---

42 www.c40.org/. Accessed April 24, 2023.

Table 12.7 Vision for Nottingham, UK, in 2028

A safer city	By becoming the UK's first carbon neutral city, Nottingham will be helping safeguard the future of our children and future generations. Through adaptation and resilience measures we will ensure the City and its residents are protected.
A healthy city	Nottingham will be one of the healthiest places to live with clean air, green open spaces and locally produced healthy food. New networks of safe cycling routes and high quality vehicle free public spaces will make it easier for people to get regular exercise.
A city of opportunities	The Nottingham economy will be built on new sustainable technologies creating high quality employment for our citizens and a worldwide reputation for innovation and excellence. New infrastructure developments will not only create jobs directly, but make Nottingham one of the best places for businesses to thrive.
An inclusive city	Nottingham will become one of the most equitable cities with new training opportunities to help Nottingham people benefit from the low-carbon economy. We will continue to ensure the benefits from economic growth are felt by our citizens and lift people out of poverty.
A city that takes care of its residents	Nottingham will become one of the cheapest places to live and work in with low household fuel bills, affordable low-carbon public transport, and high quality public services.
A city where everyone is able to reach their full potential	Nottingham will be one of the happiest places to live. Good quality homes, high employment, attractive public spaces and biodiverse ecosystems will improve the overall wellbeing of citizens and communities.

Source: Nottingham City Council, 2020, p. 8.

projects" (Tingey & Webb, 2020, p. 4). Next came 26% of LAs which were 'running hard' with "some investment in energy-related activities" and plans for more; followed by 47% still on the 'starting blocks', with commitments and action plans, but little evidence of activity; 18% were 'yet to join the race'.

The LAs had implemented a wide range of different energy technologies, many of which have been discussed in Chapters 3, 8, and 9, with the front-running technologies being combined heat and power (CHP), building improvements and heat networks, and others including a diversity of renewable technologies (solar PV and solar thermal, biomass boilers, onshore wind, landfill gas capture, anaerobic digestion, hydropower, energy-from-waste (incineration) plant, LED street lighting, transport, hydrogen fuel cells and marine infrastructure) (Tingey & Webb, 2020, Figure 2, p. 6).

One of the 'energy leader' LAs was Nottingham. Their action plan that seeks to deliver the vision in Table 12.7 comprises five areas, led, as in London, by transport and buildings, but also including energy generation, waste and water, and consumption (Table 12.8).

Table 12.8 Action areas in the Nottingham Carbon Neutral Action Plan

Transport	In 2017, nearly a third of Nottingham's total CO_2 emissions came from transport of which, nearly all come as a result of road transport from cars, vans, lorries and buses. Action is needed to reduce car journeys, increase cycling and walking, improve public transport and more low-emission vehicles. Through this, we can achieve better air quality, mobility and health for citizens.
The built environment	In the coming decade, Nottingham will have to improve the efficiency of all buildings to reduce the demand for energy. We will need to heat our buildings with low-carbon and/or renewable heating, change our behaviours towards energy reduction and increase the adoption of energy efficiency technologies in both commercial and domestic buildings.
Energy generation	Nottingham has been active in helping to decarbonise electricity. For instance, by March 2019, over 6,200 solar photovoltaic (PV) installations had been deployed across the city. We will need to take action to significantly expand local low-carbon sources of energy with the capacity to store energy within the city and to be recognised as a test bed for new energy generating technologies.
Waste and water	In 2018–19, 113,000 tonnes of household waste was produced by Nottingham, of which 26.5% was reused, recycled or composted, 64.5% was sent for energy recovery; and 7.0% was sent for landfill. Actions are needed to reduce the volume of all waste and eliminate it from landfill, increase the reuse and recycling of waste, and use the rest for energy. Water use must be managed effectively.
Consumption	The goods we buy have many emissions in their manufacturing and transport, often from around the world. Food and drink in particular has a big impact on wider and imported emissions, the wider environment and our use of finite resources. Areas for reducing emissions include reducing meat and dairy, reducing transport miles, and sourcing from less energy-intensive forms of production.

Source: Nottingham City Council, 2020, p. 15.

However, there have been pitfalls and hiccoughs along the way in implementing these ambitions. In respect of energy generation, Nottingham City Council set up its own energy company, Robin Hood Energy, in 2015, hoping to offer cheaper energy to its low-income residents. It supplied 125,000 customers at its peak, but went out of business in 2020, losing millions of pounds of public money.[43] While this venture was not directly related to decarbonisation ambitions, it exemplifies the risks of local authorities venturing outside those areas where they have evident competence.

43 www.bbc.co.uk/news/uk-england-nottinghamshire-54056695

Community energy

Local authorities and cities are not the only local actors who have played a role in local energy initiatives aimed at decarbonisation. The last 20 years have also seen a great expansion in community organisations active in this area.

The EU has acknowledged this role in legislation by recognising in its Clean Energy Package[44] both 'renewable energy communities' and 'citizen energy communities', where 'energy communities' are defined as "new types of non-commercial entities that, although they engage in an economic activity, their primary purpose is to provide environmental, economic or social community benefits rather than prioritise profit making" (Uihlein & Caramizaru, 2020, p. 4). The specific acknowledgement of renewable energy communities reflects the fact that renewable energy, which, as has been seen in Chapter 5, is likely to play a major role in future low-carbon energy systems, can be implemented at a relatively small scale at a local level, by community organisations.

Different countries have recognised these energy communities in their own legislation in widely differing ways, including energy cooperatives, limited partnerships, community interest companies, trusts and foundations, housing associations, non-profit customer-owned enterprises, public–private partnerships and a public utility company. Most common are energy cooperatives, of which there were an estimated 3,500 in mainly north-western Europe in 2020 (Uihlein & Caramizaru, 2020, p. 4). Table 12.9 shows the number of energy communities in those European countries where they are most prominent. It can be seen that Germany has most of them by quite a large margin. It is estimated that citizens, through these community organisations, owned 42% of the 100 GW of renewable energy capacity in Germany in 2016.[45]

Uihlein & Caramizaru (2020) reports case studies of 24 of the most significant energy communities in Europe, and Table 12.10 shows the very wide range of energy-related activities in which they engage. Nearly all (20 out of 24) are involved in electricity generation, over half (14/24) supply electricity to their members, 10 also engage in energy efficiency activities and 9 play a role in distributing electricity to their customers through networks. (These different functions in relation to electricity have been described in more detail in Chapter 5.) An increasing area of activity is providing mobility services (e.g. car sharing) related to electric vehicles.

One of the case studies is of Ecopower, a cooperative in Belgium that in 2015 was generating around 73 GWh per year from 20 wind turbines, 1 CHP plant, and 322 solar PV installations on the roofs of schools and private houses; was supplying around 41,000 customers with 94 GWh of electricity;

---

44 https://op. europa.eu/en/publication-detail/-/publication/7fa59d21-c7ff-11eb-a925-01aa75ed71a1
45 www.unendlich-viel-energie.de/media-library/charts-and-data/infographic-dossier-renewable-energy-in-the-hands-of-the-people

Table 12.9 Number of energy communities in nine European countries

Country	Number of energy communities
Germany	1750
Denmark	700
Netherlands	500
United Kingdom	431
Sweden	200
France	70
Belgium	34
Poland	34
Spain	33

Source: Uihlein & Caramizaru, 2020, Figure 1, p. 5.

Table 12.10 Activity areas of 24 significant energy communities in Europe

Activity area	Number of energy communities
Generation	20
Supply activity	14
Energy efficiency	10
Distribution activity	9
Electro mobility	8
Consumption and sharing	6
Flexibility & storage	4
Financial services	1

Source: Uihlein & Caramizaru, 2020, Figure 2, p. 13.

and through energy efficiency activities reduced the average energy consumption of its 48,000 members by around 40%.[46]

In Great Britain, community energy has separate coordinating organisations in England, Scotland and Wales. Community Energy England defines community energy as "the delivery of community-led renewable energy, energy demand reduction and energy supply projects, whether wholly owned and/or controlled by communities or through partnership with commercial or public sector partners".[47]

Table 12.11 comes from the Community Energy: State of the Sector Report, 2021, produced by Community Energy England,[48] but including groups in the

46 http://citynvest.eu/sites/default/files/library-documents/Model%2025_%20Cooperative%20 Cases_ECOPOWER.pdf. Accessed April 24, 2023.
47 https://communityenergyengland.org/pages/what-is-community-energy. Accessed April 24, 2023.
48 https://communityenergyengland.org/pages/state-of-the-sector. Accessed April 24, 2023. But for the 2021 State of the Sector report, see UK-SOTS-2021.pdf (regensw.wpenginepowered. com). Accessed April 24, 2023.

Table 12.11 Statistics relating to UK community energy in 2020

	Number of organisations
Total	424
England	290
Northern Ireland	2
Scotland	72
Wales	60
Working in energy efficiency	132
Working in low-carbon heat	53
Working in low-carbon transport	76
Provided COVID-19 relief	83

	Number of people
People engaged on energy and climate change	358,000
Staff (full-time equivalent)	431
New job roles created	84
Volunteers	3,096

	Financial benefits (£ m)
Reduced energy bills	2.9
Total raised from community shares	30.2
Community benefit expenditure	3.13
Diverted from community funds for COVID-19 relief	0.2004

	Power capacity (MW)
Total installed renewable electricity	319
New in 2020	8.2

Source: Community Energy: State of the Sector Report, 2021, https://regensw.wpenginepowered. com/wp-content/uploads/UK-SOTS-2021.pdf. Accessed April 24, 2023.

other nations of the UK. It gives a comprehensive snapshot of community energy activity in the UK, with 424 organisations, 132 of them working on energy efficiency, 53 on low-carbon heat, and 76 on low-carbon transport; 431 people are employed, and over 350,000 people are involved. £2.9 million has been saved on energy bills, and £3.13 million delivered in community benefit. The social objectives of these organisations was well illustrated by the fact that 83 of them provided over £200,000 for Covid relief. Finally, while the 319 MW of community energy-generating capacity is an impressive achievement, it is less than half a percentage point of the UK's total 2020 generating capacity of around 76 GW.

One of the reasons for this relatively small generating capacity compared to, for example, Germany, is that policy support for the sector from the government in England has never been large, and is now minimal. Feed-in tariffs were discontinued in 2019, as was the Urban Community Energy Fund. With the £10 million Rural Community Energy Fund, started in 2019, now drawing to a close, community energy in England could soon find itself with no public support.

In its inquiry into community energy in 2021, the House of Commons Environmental Audit Committee recommended[49] further public support for community energy and recognition of its important role in the government's then-forthcoming Net Zero Strategy.[50] However, this strategy gave only the briefest mention of community energy and committed the government only to further dialogue.

The fact that the economics of community energy remains challenging is ironic given the enormous cost reductions in renewable electricity generation (as seen in Chapter 5), and the very high 2022 electricity prices in the UK, that are forecast to persist into 2023 and perhaps beyond. This situation is largely due to how the electricity market is structured in the UK, which was explored in some detail in Chapter 5.

In 2021, 18 new community energy organisations were formed, taking the number to 495 across the UK; organisations working on low-carbon transport increased to 90; an extra 7.5 MW of new renewable capacity was installed; and an extra £21.5 million was raised for new investment in the UK.[51]

Human behaviour

By definition, all the GHG emissions we are concerned about in this book that are contributing to human-caused climate change are the result of human decisions and human behaviour, either as individuals, or as members of households, businesses or other organisations. All policy aimed at reducing emissions therefore needs to change the behaviours of people, as individuals or in organisations, so that they either modify their emission-causing activities, or develop and adopt new low-emission technologies, or both.

Humans are complicated. They have multiple and widely differing aspirations, both for themselves as individuals, and for their families. Their many roles include members of communities, participants in organisations (e.g. employees, community organisations), consumers, and citizens who can influence policies.

In addition to consumer action, behaviour of relevance for climate action thus encompasses the adoption of low-carbon and climate-resilient technologies (e.g. installing insulation); support for large-scale low-carbon infrastructures (e.g., windfarms); political action to support or demand

---

49 https://committees.parliament.uk/publications/5718/documents/56323/default/. Accessed April 24, 2023.

50 www.gov.uk/government/publications/net-zero-strategy. Accessed April 24, 2023.

51 Community Energy: State of the Sector Report, 2022. https://communityenergyengland.org/files/document/626/1655376945_CommunityEnergyStateoftheSectorUKReport2022.pdf. Accessed April 24, 2023.

climate change measures (e.g., voting and protesting); participation in policy formulation (e.g., through citizen juries) and grassroots activities (e.g., community energy or transport initiatives); and engaging in climate change conversations and interactions with others that raise awareness, enable and normalise low-carbon lifestyles.

(Whitmarsh et al., 2021, p. 76)

In their actions in these roles, humans have many different motivations and influences. They have their values and worldviews, but their actions are also deeply influenced by prevailing social norms, institutions and practices, their cultural contexts, by their neighbours and by those with whom they socialise and compare themselves. They have will and agency, but at the same time are creatures of habit. They are also constrained in their actions by numerous factors, most obviously income (e.g. ability to invest in home insulation, to buy an electric car), but also infrastructure (e.g. availability of public transport, or safe cycle lanes) and information (having the time, energy and intellectual capacity required to access and make sense of information about emission reduction, which can be both complex and technical). In organisations they may have little or no power to change those activities responsible for most of the organisation's emissions, or if they have the power they may be constrained by technical complexities or commercial considerations.

Other important issues are the timescales over which emissions reductions will be achieved, the investments that may be required to achieve them, and whether the actions are one-off or need to be habitual. Purchasing a new fuel-efficient car or fridge freezer, or increasing the energy efficiency of a home, are one-off actions, which most people perform relatively rarely, and which allow lifestyles to be maintained with relatively little change while emissions fall significantly. But they require resources which less well-off people may struggle to find. In contrast, switching a single journey from a car to public transport or a bicycle, or turning the heating thermostat down for one day, although requiring relatively little or no expenditure, may save relatively few emissions. But if this behaviour becomes habitual, the effect can be significant.

People also differ enormously in the emissions resulting from their consumption. Chapter 2 showed the large differences in average per capita emissions across countries, but Table 12.12 shows that the differences are actually between the consumption emissions of the rich (in all countries) and those of the poor.

Table 12.12 shows that the shares of emissions of each income group have changed remarkably little between 1990 and 2015, although the absolute level of CO_2 emissions rose from around 23 $GtCO_2$ to around 36 $GtCO_2$ over this period. The consumption of the richest 10% of the global population (average income in 2021 around US\$122,000[52]) is, and has been for many years,

---

52 www.businessinsider.com/how-much-wealthy-middle-class-poor-make-income-per-year-2021-12?r=US&IR=T. Accessed April 24, 2023.

Table 12.12 Shares of global emissions between three income groups (10% richest, 50% poorest and 40% medium income) of the global population, 1990 and 2015

	Share of global emissions (%)	
	1990	*2015*
Income group		
Richest 10%	50	49
Middle 40%	41	44
Poorest 50%	8	7

Source: Castano Garcia et al., 2021, Figure 3, p. 5.

responsible for 50% of total global CO_2 emissions, while that of the poorest 50% (average income around US$4,000) is less than 10%. It is therefore the behaviour of less than 1 billion people that needs to show the greatest changes if climate change is to be addressed.

The picture is very similar with regard to high-consumption lifestyles within countries. Flying is one of the most carbon-intensive activities. In the UK, more than three quarters of those on incomes less than £25,000 flew internationally once or less in 2019, with well over half flying not at all. In contrast, of those on incomes of more than £50,000, less than a quarter did not fly at all, while over a quarter took four or more flights (Castano Garcia et al., 2021, Figure 4, p. 8).

Thus a person's social and economic context is crucial if they are to be able to effect emission-reducing behaviour change. As Nielsen et al. (2021, p. 132) put it:

> limiting climate change requires interventions at multiple levels and time scales: technology change and policy change are necessary, but do not obviate the importance of individual and household behavior, especially where these have the potential to push forward such systemic change; likewise, individual responses to climate change are necessary but must be supported and enabled by policy and structural change.

Personal carbon emissions

For individuals who want to reduce their CO_2 and other GHG emissions, the first task has to be to measure them, and thereby to understand which activities produce the most emissions.

There are a number of calculators online that enable people to do this, of varying degrees of complexity. The one I have used most is Carbon Independent,[53] because it is relatively simple to fill in the required data (with

---

53 www.carbonindependent.org/. Accessed April 24, 2023.

Table 12.13 Calculation of a personal carbon footprint, 2020

Your CO$_2$ and other greenhouse gas emissions (tonnes CO$_2$ equivalent)
Based on a household size of 2

	Per household	*Per person*
Electricity (2127 kWh)	0.49	0.25
Gas (2320 kWh)	0.47	0.24
Food		1.44
Health, education, etc.		1.1
Train (520 miles)		0.05
Miscellaneous		1.19
Your total		4.26

No or minimal recorded emissions from the following categories: oil, coal, wood, bottled gas, car 1, car 2, car 3, car 4, bus, flights

Source: Author calculation, Carbon Independent. www.carbonindependent.org/. Accessed April 24, 2023.

Notes:
This 4.26 t compares (according to the Carbon Independent website which produced this comparison at the time of the calculation) with a world average of 4.4 t, a UK average of 14.1 t, and GHG emissions in USA (17.6 t), China (6.2 t), India (1.8 t) and Mozambique (0.3 t).
The figure for the UK includes adjustments for greenhouse gases other than CO$_2$. The figures for the other countries do not, as these are not so readily available.

the data to hand it takes about 10 minutes), but there are detailed explanations of how the calculations are derived for those who want to understand this. The results of my entry for 2020 are shown in Table 12.13.

The lifestyle that delivers these numbers is broadly as follows. I live in a small flat with my wife in London, and we seek to minimise our use of hot water and central heating. We eat meat no more than three or four times a week, we have no car and travel outside London by train. 2020, of course, was a lockdown year, with minimal international travel. In 2019 a single return flight to Italy added 1.25 tonnes to my carbon footprint. A train journey instead would have reduced that by 80–90%. That is a resolution for the future. But my carbon footprint, though well below the average, is still some distance from 'real zero' and is therefore work in progress.

The health, education and miscellaneous categories in the calculation are national consumption averages and include imports. The miscellaneous category includes consumer goods like cars, electronics and furniture. There is no individual entry for these categories, which for me, as can be seen, amount to more than one half of my carbon footprint. Some of this consumption is age-related. The education footprint is significant when children are at school or university. The health footprint gets larger as people get older. I would welcome a more detailed calculator that would show the carbon cost of buying a new car or kitchen, or extending a house, or buying furniture. But a moment's thought shows that such a calculator would be impossibly complicated, given the complexity of consumer goods and the variety of supply chains, components and processes that go into creating them. As those who try the Carbon Independent calculator will see, it tries to capture such issues through the question "What is your miscellaneous spending?", and allows for four answers around an average. But this is obviously a crude approximation that is unlikely accurately to reflect individual realities.

That said, a carbon calculator such as Carbon Independent provides pretty clear guidance on some carbon-emitting decisions. For the great majority of richer people, reducing emissions in the short term means using less energy in the home, driving less, flying less, eating less meat and buying less stuff. Over time, new technologies may allow these behaviours to become much less carbon-intensive, as was seen in earlier chapters, with electric vehicles replacing internal combustion engines, heat pumps replacing gas boilers, high-speed trains replacing short- and mid-distance flights, and plant-based foods providing an indistinguishable substitute for meat. But, for the rest of this decade at least, to achieve the 50% reduction in global emissions that the emission scenarios of Chapter 2 showed was required by the 1.5°C temperature targets, there will be no getting away from the need for rich people to cut back on some of the consumption categories that they value.

This is one of the big challenges for policy makers. Before they decide which of the policy instruments considered earlier in this chapter to implement, those who wish to introduce policies that will reduce carbon emissions need to take into account three main factors (Nielsen et al., 2021, pp. 133ff.).

First, there is the technical potential of their policy: if implemented, will it reduce emissions and, if so, by how much? In addition to considering the consumption categories of individuals – their home energy use, their diets and transport modes – policymakers must think about emissions from production processes, which individual consumers can hardly influence. Steel, aluminium, other metals, cement, chemicals, ceramics, pulp and paper – these are the big emitting sectors of industry, and policy makers will need to help these industries develop low- or zero-carbon processes to substitute for current technologies.

Second, there is what psychologists call 'behavioural plasticity'. How easy is it for people or organisations to change their behaviour in a low-carbon direction? What are the barriers to these changes? How might these barriers be reduced or removed by policy?

And, third, what is the feasibility, in both political and administrative terms, of the policy they are considering? Political feasibility is, arguably, variable. Circumstances change, and policies once thought infeasible may become feasible. With climate change, that may happen as more people become aware of and concerned about climate change. Administrative feasibility may change with technology, but constraints are likely to remain. For example, it is well established that good policy will try to compensate low-income households for any energy price increase due to the imposition of carbon pricing. But actually finding the households, and crafting a policy that will target them effectively, is extremely difficult, as the UK Government found with the energy price increases that occurred in late 2021 and early 2022.

The effectiveness of policy depends on the product of all these factors. The best policy will combine high technical potential with high behavioural plasticity with high feasibility. Unfortunately such policies are hard to come by, and most involve some trade-off between these and other factors.

Personal carbon trading

It has been seen that achieving the climate targets of the Paris Agreement requires net global emissions of greenhouse gases to be strictly limited. The level of emissions that can still be emitted consistent with a given temperature target is called the 'carbon budget'. In its 1.5°C report in 2018 the IPCC estimated a "remaining carbon budget of 580 $GtCO_2$ for a 50% probability of limiting warming to 1.5°C, and 420 $GtCO_2$ for a 66% probability" (IPCC, 2018, C.1.3, p. 14) (see also Table 2.2).

Analogously to the Emissions Trading Systems (ETSs) described above that have been set up for and which give (or sell) emission allowances to companies, there have been proposals to institute annual 'personal carbon allowances' (PCAs) for individuals, which would be used up as individuals used carbon-emitting fuels for their travel or home heating. Those who used more than their PCAs in one year would need to buy more from those that used less. This would give a financial incentive to reduce emissions, and benefit lower-income households, as higher-income households tend to use more energy and emit more carbon, as has been seen. The number of annual PCAs would be gradually reduced in line with emission reduction targets.

It is thought that PCAs would reduce emissions through the processes set out in Figure 12.5. Numerous practical and institutional issues, many of which are explored in Fuso Nerini et al. (2021), would need to be addressed and resolved before such a proposal could be implemented in practice, but the proposal has the advantage of reflecting for the first time in an individual's or household's use of energy the absolute nature of carbon budgets, and providing an incentive to reduce emissions that would be relatively beneficial to poorer households. However, implementing such a scheme for a population in excess of 50 million people would not be cheap – a UK Government study in 2008 estimated its running cost at £1–2 billion per year (Defra, 2008, p. 3).[54]

Policy delivery

It must by now be apparent that the transformations of the energy, transport, industry and agriculture sectors, and the built environment, that will be

---

54 Synthesis report of the findings from Defra's pre-feasibility study into personal carbon trading. www.teqs.net/Synthesis.pdf. Accessed April 24, 2023.

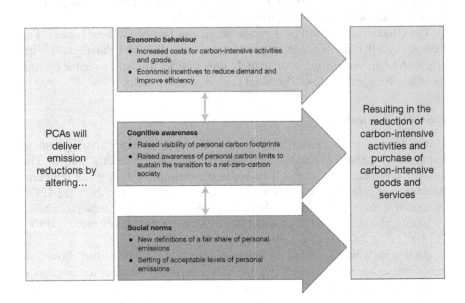

Figure 12.5 Potential impacts of PCAs on the use by individuals of carbon-intensive fuels and activities.

Source: Fuso Nerini et al., 2021, Figure 1, p. 1026.

required to reduce GHG emissions to zero requires an unprecedented (outside wartime) mobilisation of resources. And this mobilisation of resources will require an institutional architecture that is able to join up government departments that traditionally operate in silos.

Different countries will wish to organise this coordination of government bodies in different ways, that suit their existing structures of government, and the following proposals are put forward for the UK alone. But all countries, if they are to be successful in reducing GHG emissions to zero by mid-century will need analogous, and far-reaching, arrangements for policy delivery. Figure 12.6 indicates the kind of institutional architecture that is likely to be necessary.

In the UK many government departments need to be involved in the formulation and delivery of decarbonisation policy, and it is essential that these departments are closely coordinated in their activities in this area. It is suggested that this coordination should be through a Climate Change Emergency Committee, analogous to the existing COBRA Committee, a cross-departmental committee convened to respond to national emergencies, with a clear focus on net zero in less than three decades. The Climate Change Emergency Committee would be chaired by the Prime Minister. It would

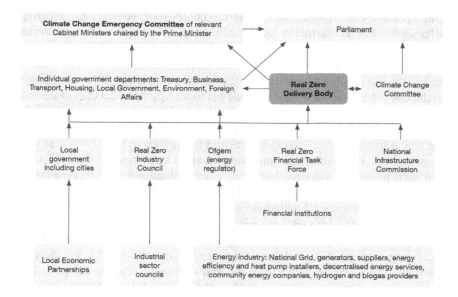

Figure 12.6 Institutional architecture to deliver real zero.

Source: Author, adapted from Allan et al., 2020, p. 14.

include the First Ministers of Scotland, Wales and Northern Ireland, who would ensure that their national strategies were aligned with overall net-zero targets for the UK as a whole.

Some of the departmental imperatives for decarbonisation by the Westminster Government are clear: the UK Department for Energy Security and Net Zero (DESNZ)[55] would lead on ensuring that power generation is zero-carbon by 2030 or soon thereafter, and that there would be enough zero-carbon electricity after 2030 to help the decarbonisation of transport and heat. DESNZ, with the Department for Business and Trade, would also need to challenge and support key energy-intensive industrial sectors (steel, non-ferrous metals, cement, ceramics, chemicals, pulp and paper) to produce first low-carbon and then zero-carbon products, and guarantee to purchase them at a profitable price when they did. The Treasury needs to design and implement an escalating carbon price. The Department for Transport must build on its recent decarbonisation consultation document[56] to ensure that only electric

55 The names of these departments are correct at the time of going to press, but they change frequently.

56 https://assets.publishing.service.gov.uk/government/uploads/system/uploads/attachment_d ata/file/876251/decarbonising-transport-setting-the-challenge.pdf. Accessed April 23, 2023.

cars can be purchased after 2030, that there is an accessible nationwide charging system by then, and that no internal combustion vehicle can be driven after 2040. The Department for Levelling Up, Housing and Communities needs to put in place a massive, well-funded scheme to bring UK housing up to the standards of energy efficiency that are already widespread in Scandinavia, and at the same time embark on a programme to convert by 2050 every home in the UK from fossil methane (natural) gas to heat pumps, fuel cells or district heating (the role of hydrogen gas as a heating fuel is likely to be very limited). Defra (Department for Environment, Food and Rural Affairs) must strengthen the new Agriculture Act[57] to plant trees and to use soils to store carbon as well as grow food. The Foreign, Commonwealth and Development Office (FCDO) will need to oversee the distribution of climate finance to low-income countries. And the Cabinet Office will need to use the Behavioural Insights Team to recommend how low-carbon behaviours can be encouraged across all areas of activity associated with GHG emissions.

The energy industry is obviously critical to the zero-carbon transition. An increasing number of households will be generators as well as consumers of electricity. They will provide balancing and storage services to the grid. The electricity distribution grid will need both to absorb local generation and deliver power for heating and transport as well as appliances. Community energy companies will need clear and efficient conditions for access to the grid that gives them adequate return on their investments. They will all need to work effectively with, and be overseen by, the regulator, Ofgem. Without a clear institutional overview of the whole energy system, and foresight of and advance planning for the huge changes in both the supply and demand side of electricity and other forms of energy, there is a real danger of bottlenecks, dysfunctionality and breakdown.

Industrial sector councils will need to feed into a Real-Zero Industry Council, which, with central government, will need to forge and follow through on the UK Government's *Powering up Britain* report and related strategy documents.[58] Local Economic Partnerships and local government, including cities and their mayoral offices, will need to coordinate their activities so that they can learn from one another. The financial sector will need to ensure that the reporting arrangements, transparency and incentives are in place for the necessary zero-carbon investments to be made. The National Infrastructure Commission will advise on the adequacy of infrastructure, and the need for new infrastructure, for real zero. All these will need to have clear lines of communication with

57 www.gov.uk/government/news/agriculture-bill-to-boost-environment-and-food-production. Accessed April 23, 2023.
58 https://assets.publishing.service.gov.uk/government/uploads/system/uploads/attachment_data/file/1147340/powering-up-britain-joint-overview.pdf. Accessed April 10, 2023.

the relevant government departments and, through them, the Climate Change Emergency Committee.

However, these departmental and various stakeholder and institutional arrangements, while essential, will not be effective unless they are integrated into a coherent Real-Zero Delivery Plan, jointly produced with the Climate Change Committee, that is coordinated across government with clear lines of communication and joint delivery with business, the financial sector and civil society stakeholders. And this Real-Zero Delivery Plan would need to be implemented by a **Real-Zero Delivery Body**, with accountability both to the Climate Change Emergency Committee and to Parliament.

Of course, as noted above, these detailed institutional arrangements will not be appropriate for other countries. But there are certain policy objectives and industrial and economic changes that are essential for all countries if the goal of zero carbon emissions by mid-century is to be achieved. These changes are set out by the International Energy Agency in its Net Zero Emissions (NZE) Pathway, and shown in Figure 12.7, along with the global NZE trajectory to net zero by 2050, in the form of milestones to be reached by a certain date.

Emissions source	*Milestones*						
	2021	*2025*	*2030*	*2035*	*2040*	*2045*	*2050*
Buildings		No new sales of fossil fuel boilers	Universal energy access All new buildings are zero-carbon-ready	Most appliances and cooling systems sold are best-in-class	50% of existing buildings retrofitted to zero-carbon-ready levels	50% of heating demand met by heat pumps	More than 85% of buildings are zero-carbon-ready
Transport			60% of global car sales are electric	50% of heavy truck sales are electric No new ICE car sales	50% of fuels used in aviation are low-emissions		
Industry			Most new clean technologies in heavy industry demonstrated at scale	All industrial electric motor sales are best-in-class	Around 90% of existing capacity in heavy industries reaches end of investment cycle		More than 90% of heavy industrial production is low-emissions

Emissions source	Milestones						
	2021	*2025*	*2030*	*2035*	*2040*	*2045*	*2050*
Electricity and heat	No new unabated coal plants approved for development		1020 GW annual solar and wind additions Phase-out of unabated coal in advanced economies	Overall net-zero emissions electricity in advanced economies	Net-zero emissions electricity globally Phase-out of all unabated coal and oil power plants		Almost 70% of electricity generation globally from solar PV and wind
Other	No new oil and gas fields approved for development; no new coal mines or mine extensions		150 Mt low-carbon hydrogen 850 GW electrolysers	4 GtCO$_2$ captured		435 Mt low-carbon hydrogen 3000 GW electrolysers	7.6 GtCO$_2$ captured

Figure 12.7a

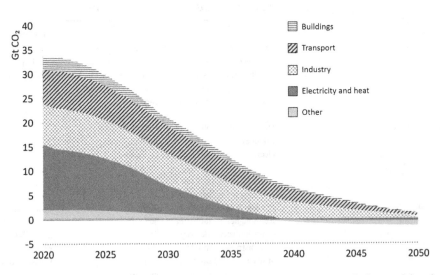

Figure 12.7b Key milestones on the road to net zero by 2050 and the associated emissions pathway.

Source: International Energy Agency, 2021a, p. 20.

The NZE Pathway requires no new unabated coal plants or oil and gas fields from 2021, no new fossil fuel boilers from 2025, and no new sales of internal combustion engines (ICEs) by 2035. The other required changes to industry, transport, buildings, and electricity and heat are clearly mapped in the table above Figure 12.7 across the next three decades, together with the requisite growth of hydrogen and carbon removals from the atmosphere.

It will be noticed that the world has already failed to meet the early milestones, and therefore fallen behind this trajectory. The 1.5°C target will only be achieved if it can catch up and achieve some of the other targets early. There is currently little sign that the global political will exists for this to happen.

Conclusion

The world has never been better placed for policy initiatives to tackle climate change. The huge innovations in climate policy in the last three decades mean that, by and large, policy makers know what to do to reduce emissions, and how to do it effectively.

The most important priority is for developed and developing countries to collaborate on a strategy that will enable the latter to move definitively onto a path of low-carbon development. The developed countries will need to provide the lion's share of the resources to implement this strategy. This is first and foremost an imperative for climate justice, given the responsibility of developed countries (plus China) for the highest cumulative emissions, and the fact that climate change will hit the poorest hardest. But it will also be essential for the human contribution to worsening climate change to be stopped. If support for low-carbon development is not forthcoming from the rich countries, then developing countries will pursue high-carbon development.

For national policy, there is no 'one size fits all'. Individual policies, often in policy mixes, will need to be crafted carefully to fit the relevant social, cultural and economic context, with provisions for policy evaluation built in from the start to assess policy effectiveness. For businesses, financial incentives such as carbon prices will be particularly important. There are clear limits to what businesses can, or should be expected, to do against their commercial interests. For individuals, behaviour is shaped more by social and cultural contexts and expectations. It is very clear that rich people in both richer and poorer countries are the big emitters. Therefore policy, to be effective, will need to address them and their lifestyles. At the same time, to be socially acceptable, policy will need to ensure that it does not impoverish people who are already relatively poor. Special care will need to be taken with the design of policy that, for example, raises energy prices to ensure that it is not regressive, although, as the example of 'carbon fee and dividend' shows, this may not be enough to win public acceptability for the policy.

Ultimately, to be effective climate policy will need to change the behaviour of businesses and individuals. So far climate policy has concentrated on developing and deploying technology, as shown by the US Inflation Reduction

Act, rather than trying to change the carbon-intensive lifestyles of individuals. It remains to be seen whether this technology-only approach, leaving consumer aspirations and lifestyles largely unchanged, will be adequate to keep emissions within the now very tight carbon budgets.

References

Aggeri, F. (2021). Industrial eco-parks as drivers of the circular economy. *Field Actions Science Reports [Online], Special Issue 23*. http://journals.openedition.org/factsrepo rts/6642. Accessed April 23, 2023.

Allan, J., Donovan, C., Ekins, P., Gambhir, A., Hepburn, C., Robins, N., Reay, D., Shuckburgh, E., & Zenghelis, D. (2020). *A Net-Zero Emissions Economic Recovery from COVID-19*. Smith School Working Paper 20-01, Oxford University, Oxford. www.smithschool.ox.ac.uk/sites/default/files/2022-04/A-net-zero-emissions-econo mic-recovery-from-COVID-19.pdf. Accessed April 23, 2023.

Bressers, H., Bruijn, T. de, & Lulofs, K. (2009). Environmental negotiated agreements in the Netherlands. *Environmental Politics, 18*(1), 58–77. https://doi.org/10.1080/09644010802624819

Burke, M., Hsiang, S. M., & Miguel, E. (2015). Global non-linear effect of temperature on economic production. *Nature, 527*(7577), 235–239. https://doi.org/10.1038/natu re15725

Callahan, C. W., & Mankin, J. S. (2022). National attribution of historical climate damages. *Climatic Change, 172*(3–4), 40. https://doi.org/10.1007/s10584-022-03387-y

Carley, S., & Konisky, D. M. (2020). The justice and equity implications of the clean energy transition. *Nature Energy, 5*(8), 569–577. https://doi.org/10.1038/s41560-020-0641-6

Castano Garcia, A., Ambrose, A., Hawkins, A., & Parkes, S. (2021). High consumption, an unsustainable habit that needs more attention. *Energy Research & Social Science, 80*, 102241. https://doi.org/10.1016/j.erss.2021.102241

Climate Change Committee. (2020). *The Sixth Carbon Budget: The UK's path to net zero*. CCC, London.

Climate Change Committee. (2022). *Progress in Reducing Emissions: 2022 Report to Parliament*. CCC, London.

Commission of the European Communities. (1996). *Communication from the Commission to the Council and the European Parliament on Voluntary Environmental Agreements*. COM(96) 561 Final. European Commission. Brussels. https://eur-lex.europa.eu/legal-content/EN/TXT/PDF/?uri=CELEX:51996DC0 561&from=DE

Defra (Department for Environment, Food and Rural Affairs). (2008). *An Analysis of the Technical Feasibility and Potential Cost of a Personal Carbon Trading Scheme*. Defra, London. https://webarchive.nationalarchives.gov.uk/ukgwa/20110405153319/http:/www.decc.gov.uk/en/content/cms/what_we_do/change_energy/tackling_clima/ind_com_action/personal/personal.aspx

de Jongh, P., & Captain, S. (1999). *Our Common Journey: A Pioneering Approach to Cooperative Environmental Management*. Zed Books, London.

Domenech, T., Ekins, P., van Ewijk, S., Spano, C., Miedzinski, M., Kloss, B., Petit, M., Stuchtey, M., & Tomei, J. (2019). *Making Materials Work for Life – Introducing Producer Ownership*. UCL and Systemiq, London. www.systemiq.earth/resource-category/materials/

Dooley, K., Holz, C., Kartha, S., Klinsky, S., Roberts, J. T., Shue, H., Winkler, H., Athanasiou, T., Caney, S., Cripps, E., Dubash, N. K., Hall, G., Harris, P. G., Lahn, B., Moellendorf, D., Müller, B., Sagar, A., & Singer, P. (2021). Ethical choices behind quantifications of fair contributions under the Paris Agreement. *Nature Climate Change*, *11*(4), 300–305. https://doi.org/10.1038/s41558-021-01015-8

Ekins, P., Domenech, T., Drummond, P., Bleischwitz, R., Hughes, N., & Lotti, L. (2019). *The Circular Economy: What, Why, How and Where.* Background Paper for an OECD/EC Workshop on 5 July 2019 within the Workshop Series "Managing Environmental and Energy Transitions for Regions and Cities". OECD, Paris.

Ekins, P., & Etheridge, B. (2006). The environmental and economic impacts of the UK climate change agreements. *Energy Policy*, *34*(15), 2071–2086. https://doi.org/10.1016/j.enpol.2005.01.008

Eskander, S., Fankhauser, S., & Setzer, J. (2021). Global lessons from climate change legislation and litigation. *Environmental and Energy Policy and the Economy*, *2*, 44–82. https://doi.org/10.1086/711306

Fuso Nerini, F., Fawcett, T., Parag, Y., & Ekins, P. (2021). Personal carbon allowances revisited. *Nature Sustainability*, *4*(12), 1025–1031. https://doi.org/10.1038/S41893-021-00756-W

Govindan, K., & Hasanagic, M. (2018). A systematic review on drivers, barriers, and practices towards circular economy: A supply chain perspective. *International Journal of Production Research*, *56*(1–2), 278–311. https://doi.org/10.1080/00207543.2017.1402141

Grubb, M., McDowall, W., & Drummond, P. (2017). On order and complexity in innovations systems: Conceptual frameworks for policy mixes in sustainability transitions. *Energy Research & Social Science*, *33*, 21–34. https://doi.org/10.1016/J.ERSS.2017.09.016

Höhne, N., den Elzen, M., & Escalante, D. (2014). Regional GHG reduction targets based on effort sharing: A comparison of studies. *Climate Policy*, *14*(1), 122–147. https://doi.org/10.1080/14693062.2014.849452

Homberg, M. van den, & McQuistan, C. (2019). Technology for climate justice: A reporting framework for loss and damage as part of key global agreements. In: Mechler, R., Bouwer, L., Schinko, T., Surminski, S., Linnerooth-Bayer, J. (Eds.). *Loss and Damage from Climate Change: Climate Risk Management, Policy and Governance* (pp. 513–545). Springer, Cham. https://doi.org/10.1007/978-3-319-72026-5_22

International Energy Agency. (2021a). *Net Zero by 2050*. IEA, Paris.

International Energy Agency. (2021b). *Energy Efficiency, 2021*. IEA, Paris. https://iea.blob.core.windows.net/assets/9c30109f-38a7-4a0b-b159-47f00d65e5be/EnergyEfficiency2021.pdf

IPCC. (2018). Summary for Policy Makers. In: *Global warming of 1.5°C. An IPCC Special Report on the impacts of global warming of 1.5°C above pre-industrial levels and related global greenhouse gas emission pathways, in the context of strengthening the global response to the threat of climate change, sustainable development, and efforts to eradicate poverty.* V. Masson-Delmotte, P. Zhai, H. O. Pörtner, D. Roberts, J. Skea, P. R. Shukla, A. Pirani, W. Moufouma-Okia, C. Péan, R. Pidcock, S. Connors, J. B. R. Matthews, Y. Chen, X. Zhou, M. I. Gomis, E. Lonnoy, T. Maycock, M. Tignor, T. Waterfield (Eds.). Cambridge University Press, Cambridge, UK and New York, NY, USA, pp. 93–174. https://doi.org/10.1017/9781009157940.004

IPCC. (2022a). Summary for Policymakers. P. R. Shukla, J. Skea, A. Reisinger, R. Slade, R. Fradera, M. Pathak, A. Al Khourdajie, M. Belkacemi, R. van Diemen, A.

Hasija, G. Lisboa, S. Luz, J. Malley, D. McCollum, S. Some, P. Vyas. In: P. R. Shukla, J. Skea, R. Slade, A. Al Khourdajie, R. van Diemen, D. McCollum, M. Pathak, S. Some, P. Vyas, R. Fradera, M. Belkacemi, A. Hasija, G. Lisboa, S. Luz, J. Malley (Eds.) *Climate Change 2022: Mitigation of Climate Change. Contribution of Working Group III to the Sixth Assessment Report of the Intergovernmental Panel on Climate Change.* Cambridge University Press, Cambridge, UK and New York, NY, USA. doi: 10.1017/9781009157926.001

IPCC. (2022b). *Climate Change 2022: Mitigation of Climate Change. Contribution of Working Group III to the Sixth Assessment Report of the Intergovernmental Panel on Climate Change.* P. R. Shukla, J. Skea, R. Slade, A. Al Khourdajie, R. van Diemen, D. McCollum, M. Pathak, S. Some, P. Vyas, R. Fradera, M. Belkacemi, A. Hasija, G. Lisboa, S. Luz, J. Malley (Eds.). Cambridge University Press, Cambridge, UK and New York, NY, USA. doi: 10.1017/9781009157926.001

Jenkins, S., Kuijper, M., Helferty, H., Girardin, C., & Allen, M. (2023). Extended producer responsibility for fossil fuels. *Environmental Research Letters, 18*(1), 011005. https://doi.org/10.1088/1748-9326/aca4e8

Krarup, S., & Ramesohl, S. (2002). Voluntary agreements on energy efficiency in industry – not a golden key, but another contribution to improve climate policy mixes. *Journal of Cleaner Production, 10*(2), 109–120. https://doi.org/10.1016/S0959-6526(01)00032-4

Laybourn, P., & Morrissey, M. (2009). *National Industrial Symbiosis Programme, Pathway to a Low-Carbon Economy.* International Synergies, Birmingham, UK. https://assets.plottcdn.co.uk/international-synergies/wp-content/uploads/2021/07/20121910/Appendix-D-Pathway-Report.pdf

Martin, R., de Preux, L. B., & Wagner, U. J. (2014). The impact of a carbon tax on manufacturing: Evidence from microdata. *Journal of Public Economics, 117*, 1–14. https://doi.org/10.1016/j.jpubeco.2014.04.016

Miedzinski, M., Dibb, G., McDowall, W., & Ekins, P. (2020). *Innovation for a Green Recovery: Business and Government in Partnership.* Report of the Green Innovation Policy Commission, UCL, London.

Mildenberger, M., Lachapelle, E., Harrison, K., & Stadelmann-Steffen, I. (2022). Limited impacts of carbon tax rebate programmes on public support for carbon pricing. *Nature Climate Change, 12*(2), 141–147. https://doi.org/10.1038/s41558-021-01268-3

Nielsen, K. S., Clayton, S., Stern, P. C., Dietz, T., Capstick, S., & Whitmarsh, L. (2021). How psychology can help limit climate change. *American Psychologist, 76*(1), 130–144. https://doi.org/10.1037/amp0000624

Nottingham City Council. (2020). *Carbon Neutral Nottingham: 2020–2028 Action Plan.* www.nottinghamcity.gov.uk/media/2619917/2028-carbon-neutral-action-plan-v2-160620.pdf

Organisation for Economic Co-operation and Development. (2016). *Extended Producer Responsibility: Updated Guidance for Efficient Waste Management.* OECD, Paris. https://doi.org/10.1787/9789264256385-en

Organisation for Economic Co-operation and Development. (2019). *The Circular Economy in Cities and Regions.* OECD, Paris.

Park, H. J., & Lin, L. M. (2020). Exploring attitude–behavior gap in sustainable consumption: Comparison of recycled and upcycled fashion products. *Journal of Business Research, 117*, 623–628. https://doi.org/10.1016/j.jbusres.2018.08.025

Peñasco, C., Anadón, L. D., & Verdolini, E. (2021). Systematic review of the outcomes and trade-offs of ten types of decarbonization policy instruments. *Nature Climate Change*, *11*(3), 257–265. https://doi.org/10.1038/s41558-020-00971-x

Potting, J., Hekkert, M., Worrell, E., & Hanemaaijer, A. (2016). *Circular Economy: Measuring Innovation in Product Chains*. Policy Report. PBL Netherlands Environmental Assessment Agency, The Hague. www.pbl.nl/sites/default/files/downloads/pbl-2016-circular-economy-measuring-innovation-in-product-chains-2544.pdf

Pye, S., Bradley, S., Hughes, N., Price, J., Welsby, D., & Ekins, P. (2020). An equitable redistribution of unburnable carbon. *Nature Communications*, *11*(1), 3968. https://doi.org/10.1038/s41467-020-17679-3

Rogge, K. S., & Reichardt, K. (2016). Policy mixes for sustainability transitions: An extended concept and framework for analysis. *Research Policy*, *45*(8), 1620–1635. https://doi.org/10.1016/j.respol.2016.04.004

Romanello, M., Di Napoli, C., Drummond, P., Green, C., Kennard, H., Lampard, P., Scamman, D., Arnell, N., Ayeb-Karlsson, S., Ford, L. B., Belesova, K., Bowen, K., Cai, W., Callaghan, M., Campbell-Lendrum, D., Chambers, J., van Daalen, K. R., Dalin, C., Dasandi, N., … Costello, A. (2022). The 2022 report of the Lancet Countdown on health and climate change: Health at the mercy of fossil fuels. *The Lancet*, *400*(10363), 1619–1654. https://doi.org/10.1016/S0140-6736(22)01540-9

Saheb, Y. (2020). Modernisation of the Energy Charter Treaty: A global tragedy at a high cost for taxpayers. *OpenExp*, Paris. www.pbl.nl/sites/default/files/downloads/pbl-2016-circular-economy-measuring-innovation-in-product-chains-2544.pdf

Sovacool, B. K., Martiskainen, M., Hook, A., & Baker, L. (2019). Decarbonization and its discontents: A critical energy justice perspective on four low-carbon transitions. *Climatic Change*, *155*(4), 581–619. https://doi.org/10.1007/s10584-019-02521-7

Stern, N. (2007). *The Economics of Climate Change*. Cambridge University Press, Cambridge/New York. https://doi.org/10.1017/CBO9780511817434

Suldovsky, B. (2017). The information deficit model and climate change communication. In *Oxford Research Encyclopedia of Climate Science*. Oxford University Press. https://doi.org/10.1093/acrefore/9780190228620.013.301

Tienhaara, K., Thrasher, R., Simmons, B. A., & Gallagher, K. P. (2022). Investor-state disputes threaten the global green energy transition. *Science*, *376*(6594), 701–703. https://doi.org/10.1126/science.abo4637

Tingey, M., & Webb, J. (2020). Governance institutions and prospects for local energy innovation: Laggards and leaders among UK local authorities. *Energy Policy*, *138*, 111211. https://doi.org/10.1016/j.enpol.2019.111211

Uihlein, A., & Caramizaru, A. (2020). *Energy Communities: An Overview of Energy and Social Innovation*. European Commission, Joint Research Centre, Publications Office.

United Nations. (1992). *United Nations Framework Convention on Climate Change*. FCCC/INFORMAL/84 GE.05-62220 (E) 200705. United Nations, New York. https://unfccc.int/resource/docs/convkp/conveng.pdf

United Nations. (2015). *Paris Agreement*. https://unfccc.int/files/meetings/paris_nov_2015/application/pdf/paris_agreement_english_.pdf

Whitmarsh, L., Poortinga, W., & Capstick, S. (2021). Behaviour change to address climate change. *Current Opinion in Psychology*, *42*, 76–81. https://doi.org/10.1016/j.copsyc.2021.04.002

Williges, K., Meyer, L. H., Steininger, K. W., & Kirchengast, G. (2022). Fairness critically conditions the carbon budget allocation across countries. *Global Environmental Change*, 74, 102481. https://doi.org/10.1016/j.gloenvcha.2022.102481

World Bank. (2022). *State and Trends of Carbon Pricing*. World Bank, Washington, DC.

Wurzel, R., Zito, A., & Jordan, A. (2013). *Environmental Governance in Europe: A Comparative Analysis of New Environmental Policy Instruments*. Edward Elgar, Cheltenham.

13 Conclusions

Summary

This chapter concludes the book by summarising and synthesising the various messages from previous chapters. The topics covered are: the climate science and the global response to it; emission trajectories to 2050; energy efficiency; fossil fuels; the major sources of electricity, nuclear and renewables, and the bioenergy and hydrogen that may be used to supplement them; carbon capture, use and storage, and carbon dioxide removal, including carbon offsets, and geoengineering; the main energy demand sectors of buildings, transport and industry; agriculture and land use; technology, innovation, finance and investment; economic growth, employment and the health co-benefits of mitigation; and policy for 'real zero'.

The climate science

The emissions of greenhouse gases (GHGs) since the industrial revolution have driven atmospheric concentrations of these gases to levels that are well outside all human experience. These elevated concentrations have increased average global temperatures to 1.1–1.2°C above pre-industrial temperatures, with the last eight years (2015–2022) the warmest on record. There is strong evidence that this global heating is having increasingly pronounced effects on the climate and weather in different parts of the world, in terms of enhanced sea level rise, and more severe and frequent floods, droughts, heat waves and storms. Increasingly climate scientists are able to identify the human, as opposed to the natural, contribution to these phenomena, and to attribute probabilities to the extent to which the human contribution has exacerbated individual weather events.

There is still much uncertainty as to how these events amplified by human-induced climate change will develop over time, but it is at least possible that they will do so in catastrophic fashion, at the very least setting back gains in human development over the last few decades, and at worst making large areas

DOI: 10.4324/9781003438007-14

of land effectively uninhabitable and human civilisation all but impossible as countries grapple with food insecurity, water stress, migration and possible conflicts that may arise.

Economists have struggled to express the possible impacts of climate change in economic terms, and monetary estimates of these impacts vary widely. However, the largest of these estimates suggest that current levels of emissions over the next few decades could reduce levels of global economic output by up to 35%. Social order in many countries, especially poorer countries, would collapse under such pressures.

In contrast, climate scientists have been clear about the risks posed to human life and livelihoods by unabated climate change, especially given the risks of 'tipping points' – changes in the Earth system that give rise to feedback effects that amplify global warming and effectively put it on a runaway trajectory that is unresponsive to any future human mitigation action.

The response to these risks by governments and the people they represent is, so far, orders of magnitude less than they accord to other risk areas of human life, such as pandemics, accidents in aviation and road transport, or defence planning. Whereas in all these other areas worst-case planning is a common way in which countries seek to avert disasters, in climate the predominant attitude seems to be to do little and hope for the best. The unfolding tragedies around floods, storms, heat waves and droughts that have already been experienced suggest such a reaction is foolhardy at best and could turn out to be utterly disastrous.

The global response

The global institutional response to climate change was to set up in 1992 the United Nations Framework Convention on Climate Change (UNFCCC), with the objective of preventing dangerous human interference in the climate. Twenty-seven COP (Conference of the Parties) meetings later, it is clear that the UNFCCC has failed in this objective. Climate change has already taken lives in many countries.

The most important achievement of the UNFCCC was the signing at COP21 (2015) of the Paris Agreement, which committed all countries to do what was required to keep the average increase in global temperatures over pre-industrial levels 'well below 2°C' and to aim for 1.5°C. This would require global GHG emissions to be 'net zero' (i.e. any remaining GHG emissions to be exactly balanced by the removal of emissions from the atmosphere and their secure storage in perpetuity), by or soon after mid-century.

To this end countries have agreed to make commitments to emissions reduction, called 'Nationally Determined Contributions' (NDCs), which collectively are intended to achieve the Paris Agreement's temperature targets. The current NDC commitments do not achieve these targets and the most recent COP, COP27 in Egypt in 2022, signally failed to ramp them up from the previous year.

COP26 in 2021 had seen, in addition to new NDCs from many countries, a number of additional 'pledges' in respect of forests, methane and coal use. The most optimistic estimate of the global average temperature increase, if all the NDC commitments and extra pledges were met, is 1.8°C, consistent with the actual stated Paris Agreement target, but not with its 1.5°C aspiration. However, no country, even those that have made commitments to 'net zero' by 2050, is currently on track to achieve this, so that on the basis of current commitments and actions to implement them it seems unlikely that the average global temperature increases will be kept to 2°C, let alone 1.5°C.

In November 2022, after the COP27 meeting, the Executive Director of the International Energy Agency (IEA) stressed that the 1.5°C target was still feasible and that giving up on it politically would be 'a gift to carbon boosters'.[1] While both statements may be correct, in the absence of a considerable ramping up of key emissions reduction measures in most high emitters over the next couple of years, the 1.5°C target will pass definitively out of reach, and continuing to espouse it will lack credibility.

There are many reasons why countries find it so difficult to make and keep commitments to emission reduction. Most emissions come from burning fossil fuels, which have traditionally been a main driver of economic development, and many countries consider that this continues to be the case. In addition, there are very large differences in the per person emissions of different countries, both now and cumulatively since the industrial revolution. Moreover, with few exceptions, the countries with the largest per person emissions are also the richest countries, which has led to an attitude that climate change is a rich country problem, in that they have done most to cause it and they have the most resources with which to address it. There is justice in this assessment, and the UNFCCC recognises that there is 'common but differentiated responsibility' to address climate change.

A number of old industrial countries are now reducing their emissions, but nothing like fast enough to keep global warming to the Paris targets. And many countries not in this category, including China, the biggest emitter of all, are still increasing their emissions. The industrial/developing country divide is widened still further by the fact that some poor countries are now experiencing extensive losses and damage from climate change, to which they have contributed almost nothing, and they are demanding that the rich countries make payments to offset this loss and damage. This issue dominated the COP27 meeting, largely to the exclusion of further commitments to reduce emissions. There is a real danger that future COPs will devote more and more energy to arguing about how to share out the loss and damage from current climate change rather than seeking to reduce future damages through emission reduction.

---

1 www.theguardian.com/environment/2022/nov/30/giving-up-on-15c-climate-target-would-be-gift-to-carbon-boosters-says-iea-head. Accessed May 6, 2023.

A final divisive issue is the sharing out of the global carbon budget – the quantity of GHGs that can be emitted globally, consistent with any particular temperature target. The remaining budget for a 1.5°C target is now around 10 years of global emissions at their current level, with emissions still rising.

Emission trajectories to 2050

There has been an explosion in recent years of projections of GHG emissions into the future, on the basis of different assumptions about a wide range of social, economic and technological developments, seeking to understand what needs to happen for the Paris Agreement temperature targets to be met (i.e. for the emissions to be within the associated carbon budgets).

The models that make these projections differentiate between the major GHG-emitting sectors (power generation, industry, transport, buildings, agriculture and waste treatment) and deploy a very large range of different technologies that both reduce emissions and that remove carbon dioxide from the atmosphere. The former include energy efficiency, different industrial processes, electric vehicles (EVs), heat pumps, and technologies that use hydrogen or capture and store carbon, while the latter include nature-based actions, like peat regeneration or afforestation, and technologies of machine-based carbon removal.

While there are many differences between these model projections, also a number of strong common conclusions emerge. Energy efficiency plays an important role in all of them. Nearly all of them require substantial use of carbon capture and storage (CCS) and carbon dioxide removal (CDR) technologies. Even so, many of the trajectories result in global temperature increases beyond 1.5°C in the second part of this century and it is only through CDR that the temperature is brought back to 1.5°C – an effect about which there is also considerable uncertainty, especially if any of the major 'tipping points' have been surpassed at that stage.

Other common characteristics in these 'emission pathways' to net zero by 2050 are: the widespread electrification of the economy – in industry, in transport through electric vehicles (EVs), and in buildings through heat pumps; the provision of this electricity through a huge increase in solar photovoltaic (PV) panels and wind turbines, both onshore and offshore; and a more variable deployment of nuclear power, hydrogen technologies and bioenergy. Some of these technologies are well developed, and the challenge consists in their large-scale and speedy deployment. Others of them still require considerable development to increase their performance, or reduce their costs, or both.

Finally, all scenarios show some emissions reduction from people and organisations changing their behaviour, becoming willing, for example, to forgo certain high energy-using activities that are resistant to decarbonisation (e.g. flying), to change their diets to eat less meat, or to shift from personal vehicles to walking, cycling and public transport in cities. However, these trends are not pronounced in most scenarios, and the heavy lifting of emission reduction is done by technology, in line with evidence that many people are not prepared to reduce their emissions by giving up on or even scaling back significantly lifestyle attributes that they value.

Energy efficiency

The more efficiently that energy can be used, the less zero-carbon energy supply will be required to deliver the goods and services that people want. There are many opportunities to increase energy efficiency in buildings, transport and industry. Some of these opportunities make economic sense at current or projected energy prices, but are not taken up for a wide variety of reasons: people and businesses are not aware of the available technologies (an information deficit); the necessary infrastructures are not in place (e.g. EV charging points, or cycle lanes in towns and cities); the relevant installers do not have the skills (e.g. heat pumps); people and businesses cannot borrow the necessary upfront costs at interest rates they can afford (e.g. for home retrofits); regulations (e.g. for new buildings) are too often not enforced and therefore not followed; and institutional misalignment prevents the necessary investments (e.g. the landlord/tenant problem).

All these issues can be and have been resolved in some places. But they require determined and long-term policy measures, which in many countries, including the UK, are conspicuously lacking. Until governments grasp and address these issues, they will struggle to deliver the available energy savings, and end up either building more energy supply capacity than they need, so increasing the cost of the clean energy transition, or the transition itself will stall.

Fossil fuels

One of the most significant findings from the different emissions trajectories explored in Chapter 2 was that from the IEA's Net Zero Emissions (NZE) scenario: no new fossil fuel infrastructure or extraction activities (i.e. new coal mines or oil and gas wells) were consistent with achieving the 1.5°C Paris target. Notwithstanding, the world's fossil fuel companies continue to commit billions of dollars to new oil and gas exploration and production, and many countries are aiding and abetting them in developing new resources – Canada still invests heavily to produce oil from its carbon-intensive oil sands,[2] Norway seems determined to produce new gas from the Arctic[3] and the UK continues to license new North Sea oil fields,[4] and has even decided to open a new coal

---

2 www.capp.ca/news-releases/capp-projects-investment-in-canadas-natural-gas-and-oil-sector-will-rise-to-32-8-billion-in-2022/. Accessed May 6, 2023.
3 www.reuters.com/business/energy/equinor-partners-invest-144-bln-arctic-gas-field-2022-11-22/. Accessed May 6, 2023.
4 www.theguardian.com/business/2022/oct/07/uk-offers-new-north-sea-oil-and-gas-licences-despite-climate-concerns. Accessed May 6, 2023.

mine.[5] It is clear that, whatever these companies may say and write in their annual reports about being committed to 'net zero' and the Paris targets, through their investment strategies they are betting against policymakers actually implementing the required policies to achieve them, and policymakers so far are proving them right in this judgement.

The sobering evidence from this book suggests that the richest and most powerful industry the world has ever seen is not going to acquiesce in a majority of its fossil fuel reserves staying underground as 'stranded assets'. It will rather do everything it can to produce them, even if this involves crashing the climate. As future generations struggle with the freak weather that climate instability will bring, it is hard to imagine that these companies, who have effectively chosen not to invest in climate solutions that were affordable and available, because they could make more money out of fossil fuels, will not be viewed as climate criminals.

Renewables

The recent cost reductions and performance of renewables, especially wind and solar power, offer one of the few bright prospects in the energy transition landscape. It seems likely that renewable power will drive coal out of power generation on cost grounds, rather than because of climate concerns – a recent report (Solomon et al., 2023) estimated that the cost of new renewable power in the US is already below the cost of 99% of generation from existing coal-fired plants, and is likely to fall further given the support available from the US Inflation Reduction Act (IRA). The book by the Stanford University academic Mark Jacobson (Jacobson, 2023) explains how a 100% renewable energy system is both technically and economically possible.

Such a conclusion is controversial and has been widely criticised for downplaying the problems associated with renewables. For example, one review noted criticism of Jacobson's work from a number of academics and concluded:

> Decarbonizing the electricity sector will likely require significant investments in new transmission and storage capacity, advanced nuclear technology, carbon capture and storage, building efficiency and flexibility, and the research, development, and demonstrations required to drive down technology costs and make it all work.[6]

The evidence from this book suggests that renewables, especially wind and solar, will indeed do the heavy lifting in the decarbonisation of the electricity

---

5　www.theguardian.com/environment/2022/dec/07/uk-first-new-coalmine-for-30-years-gets-go-ahead-in-cumbria. Accessed May 6, 2023.

6　https://blogs.scientificamerican.com/plugged-in/landmark-100-percent-renewable-energy-study-flawed-say-21-leading-experts/. Accessed May 6, 2023.

system. It also suggests that transforming the electricity grid in different countries to be able to absorb and utilise large quantities of variable renewables generated both in large arrays and by numerous small-scale installations, and providing adequate storage to cover periods of non-availability of renewables, are both technically demanding and, at current prices, expensive tasks. It is not necessary to insist on 100% renewable power systems. Different countries have different renewable resources in different quantities. Solar and wind in the great majority of them are plentiful enough for these renewables to play the largest role in future power generation, but in many of them CCS (and especially bioenergy with CCS or BECCS), nuclear power and fossil methane gas (with CCS) to produce blue hydrogen, may also have a smaller role to play.

Nuclear power

New nuclear power stations are currently much more expensive than renewables where these are available, and it is far from clear why countries would invest in them rather than the grid strengthening and energy storage that will enable them to take full advantage of their renewable resources. However, the nuclear industry has always been very successful in persuading governments that they need to subsidise new nuclear power, and this trend continues in the UK with government commitments to build a new reactor at Sizewell, despite the new reactor at Hinkley Point being many years over time and billions of pounds over its initial budget. At Sizewell the government has taken a financial stake in the new plant, and, unlike with Hinkley Point, energy consumers and/or taxpayers will start paying for the reactor during its construction, before it starts generating. It is unlikely that this will prove a cost-effective public investment given the much lower costs of renewables and greater energy efficiency, the latter of which could make the new reactor unnecessary.

Small modular reactors (SMRs) are the latest nuclear flavour of the month in the UK, but it is still too soon to know whether these will be cheaper per kWh generated than either big nuclear reactors, or renewables, though the latter seems unlikely. It is also uncertain whether they will be more acceptable to the public. While SMRs are smaller than the earlier reactors, at 400–500 MW they are still sizeable industrial installations. Provided they are confined to existing nuclear sites, they may avoid popular protest. UK Government support for this technology has at least one eye on export market possibilities for the currently under-performing Rolls Royce company[7] that is developing it.

7 www.energylivenews.com/2023/01/30/rolls-royce-new-boss-brands-companys-performance-unsustainable/. Accessed May 6, 2023.

Bioenergy

Given pressures on land for food production, carbon sequestration (including through forest conservation) and the need to leave natural habitats for bio-diversity, there are clear limits to the amount of bioenergy that can be produced from land, and its production from oceans (e.g. from seaweed) is still in its infancy. Some small BECCS plants (e.g. 500 MW), drawing mainly on local sources of biomass, could be ecologically sustainable, and their flexibility may help with grid balancing, but large plants such as Drax in the UK, drawing their biomass from far away, are unlikely to be widely replicated both because of the limited supply of forestry waste to fuel them, and the environmental opposition and charges of 'greenwashing' to which they give rise.[8]

Some biomass will be used to make liquid biofuels, although these too are controversial when they use food crops ('first generation' biofuels), and the manufacture of second-generation biofuels, from woody biomass and food and agriculture waste, does not yet take place at scale. Given the land constraints on bioenergy generally, it is likely that the majority of biofuels will be used in aviation, or other sectors which are hard or impossible to decarbonise through electrification.

Hydrogen

Hydrogen is attracting increasing attention and investment in many coun-tries as a potential fuel for power generation (from fuel cells), transport and heating. Currently produced from fossil methane gas without CCS, the fitting of CCS to such plants is required for it to become a low-carbon energy source (blue hydrogen), and its production from electrolysis using renewable energy is required for it to be zero-carbon (green hydrogen). The cost of CCS and electrolysers, and the energy losses incurred in producing, transporting and storing hydrogen, mean that it is expensive and will only be cost-effective for uses which cannot use renewable or other zero-carbon fuels directly. In the future, if there are large quantities of renewable electricity (e.g. from offshore wind) available at times when it is not needed, so that it would otherwise be wasted, producing hydrogen with this electricity may allow large-scale green hydrogen production. Premium uses for this hydrogen are likely to be energy storage for times when renewable electricity production is low and, as in winter, demand is high; the manufacture of synthetic fuels (synfuels, or e-fuels); and certain transport applications where there is a need for zero emissions of local air pollutants (SO_2, NO_x) (e.g. forklift trucks working indoors). Unless there is a revolution in how it can be produced, it is unlikely that it will be used on a mass scale for such purposes as home heating.

---

8 www.clientearth.org/projects/the-greenwashing-files/drax/. Accessed May 6, 2023.

Carbon capture, use and storage (CCUS)

While the use of CCS in enhanced oil recovery (EOR) is quite widespread in the oil industry, the use of CCS just to store CO_2 underground has developed very slowly. For it to play a significant role in reducing emissions before 2050 its scale-up will need to be very fast. It is not clear that the large investment that would be required for this would not be better used in building new renewable generation facilities. The main beneficiaries of CCS would be the fossil fuel companies, whose fuels could continue to be used with a fraction of the carbon emissions, and it is telling that they have so far refused to invest in large-scale CCS without large public subsidies. It is possible that, as shown in Chapter 7, CCS is on the cusp of major investment, but it remains to be seen what proportion of the projected facilities is actually built. The U (of CCUS) refers to the use of CO_2 that would otherwise be emitted. While such uses exist (e.g. greenhouses, synthetic fuels) it is very unlikely that these will be scaled up such that they make a significant contribution to decarbonisation overall.

Carbon dioxide/greenhouse gas removal

The large-scale removal of GHGs, especially CO_2, from the atmosphere, is an essential condition for reaching net zero in mid-century and the 1.5°C Paris temperature target thereafter, and even more so for reaching real zero and returning CO_2 concentrations back to safe levels.

Unfortunately, with deforestation ongoing in many tropical countries, the danger is that, far from land increasing its absorption of human CO_2 emissions, it will both reduce it and itself start emitting more CO_2 to the atmosphere. A key issue is whether the Amazon rainforest can be conserved at its current extent (with already 20% lost over the past 50 years[9]), such that its large-scale conversion to savannah can be avoided. This in turn will depend on whether the relatively new President Lula can stop the ravages of the forest that took place under his predecessor, President Bolsonaro, and cement conservation of the Amazon deep into the culture and institutions of Brazil,[10] so that another Bolsonaro becomes unthinkable.

The Amazon's tropical forests are not the only ones being lost. Deforestation is also ongoing in Western and Central Africa, with the Democratic Republic of the Congo (DRC) intending to sell off licences for oil and gas exploration on forested land,[11] and in South East Asia, especially Thailand, Vietnam, Malaysia and Indonesia, which has lost 420,000 km² of forest since 1990, and continue to lose it at 1.2% per year.[12]

9 www.cfr.org/amazon-deforestation/#/en. Accessed May 6, 2023.
10 www.politico.eu/article/luiz-inacio-lula-da-silva-amazon-team-save-rainforest/. Accessed May 6, 2023.
11 www.theguardian.com/environment/2022/nov/01/democratic-republic-of-congo-sale-oil-gas-drilling-permits-threatens-vast-peat-carbon-sink-rainforest-aoe. Accessed May 6, 2023.
12 https://earth.org/deforestation-in-southeast-asia/. Accessed May 6, 2023.

Such forest loss, and its associated CO_2 emissions, dwarfs the attempts being made elsewhere to remove CO_2 from the atmosphere and sequester it on land or in the oceans, as described in Chapter 7. Reversing the loss, and massively increasing the removal, remain among the most difficult tasks that need to be accomplished to stop climate change.

Carbon offsets

A huge voluntary carbon market is coming into existence. There are two main drivers: the companies which are buying carbon credits so that they can keep emitting CO_2 from their operations (or, in the case of fossil fuel companies, from the combustion of their products) while claiming to be *en route* for 'net zero'; and projects or intermediaries which are keen to take their money, while claiming that they are using it to avoid or remove emissions elsewhere, thereby 'offsetting' the emissions of the purchasers of the carbon credits.

The evidence is now overwhelming that the great majority of so-called 'carbon offsets' are nothing of the kind. They may salve the consciences or reputations of the purchasers, and some of the payments for them may find their way into good causes. But they definitely do not 'offset' in any meaningful sense the emissions of those who have paid for them. There is a real prospect in the development of the offset market that the sale of carbon offsets increases, but so do the emissions they are supposed to be offsetting. And the offsetting process itself can have other negative impacts, on land prices or indigenous peoples, as described in Chapter 7.

The high-level attempts to remedy this situation, also described in Chapter 7, are doubtless well-meaning, but my conclusion is that they are unlikely to be successful, especially in respect of offsets involving 'nature-based solutions'. The problems and complexities involved in 'monitoring, reporting and verification' (MRV) of this kind of carbon offset, and the incentives for cheating on them, are simply too great. If the money that goes into carbon offsetting is actually essential for afforestation, peatland regeneration, or avoiding deforestation, then some other means will need to be found for raising it. Economists would generally recommend a global carbon tax on emissions for such a purpose, but unfortunately this solution too, though much easier to implement than MRV of offsets, has so far been unacceptable to the world community and is not even discussed at COP meetings. No wonder emissions keep rising.

Geoengineering

The world is moving inexorably towards geoengineering, notwithstanding the huge risks and uncertainties that still exist in many of the methods that have been proposed. On current trends, it is only a matter of time before some country, tormented by impacts from climate change, decides to engage in, for example, injecting particles into the upper atmosphere to block out some of the sun, as described in Chapter 7. The only way in which this eventuality seems likely to be avoided is if governments, the people they represent, and businesses engage urgently and determinedly with the real-zero agenda, with a single-minded focus that has so far been conspicuously absent.

The demand sectors: transport

The shift to EVs now seems well under way and will only accelerate over the next ten years until EVs become the great majority of new car sales. It may be that batteries develop to such an extent that they are able to substitute for internal combustion engines more widely than is currently expected (e.g. heavy goods vehicles, aviation, shipping) but at present it seems more likely that these transport areas will be decarbonised later than cars, and through different technologies (e.g. hydrogen fuel cells, biofuels, synthetic fuels, including ammonia). A major potential obstacle to mass-market EVs may be a failure to invest adequately in the minerals that are required for the batteries, so that their costs remain high, and EVs remain too expensive for most consumers. This risk could be addressed by mining becoming more environmentally and socially acceptable to the countries in which the mines are located, by recycling rates for these minerals in batteries being dramatically increased, and by far fewer people in urban areas owning cars because improvements in public transport, and walking and cycling infrastructure, provide an attractive alternative.

The demand sectors: buildings

The priority for buildings in the future will be that they can be adequately cooled in summer and warmed in winter with zero-carbon energy. Cooling through air-conditioning is, of course, already electrified, so that the priority here, to reduce the zero-carbon energy infrastructure required, is to minimise the amount of energy for cooling that is required through building design and through very efficient air conditioners. Heating is potentially a more difficult issue to resolve, especially in those countries, like the UK, that have historically heated the great majority of their buildings using coal, oil and gas. In these contexts, buildings will need to be brought up to a certain minimum (high) level of energy efficiency so that they can be kept adequately warm through heat pumps, while ensuring that they do not overheat in the hotter summers that are to be expected, and are adequately ventilated to prevent mould and excess condensation, and remain healthy. Smart meters are already enabling people to smooth their consumption away from peak periods, and be paid for doing so, while, in commercial buildings, digital building information systems will enable building energy managers to make greater energy savings.

The demand sectors: industry

Many industrial processes can be electrified. Those that cannot are likely to use hydrogen or synthetic fuels to reach the high process temperatures that are sometimes required. Steel, cement and chemicals are particularly problematic. In steel, replacement of coking coal with direct hydrogen reduction seems likely to be well under way this decade in Europe, and may spread to the rest of the world if the economics make it viable. For chemicals, it is possible to synthesise ammonia and methanol using hydrogen and, for the latter, captured CO_2 as well, but scale-up of these processes to the current level of production of these chemicals presents a formidable challenge. For cement the options for decarbonisation are perhaps even more limited, with substitution by laminated timber possible on the scale permitted by land constraints for growing the timber, and a great need for the development of low-carbon cement chemistries. Refurbishing, rather than demolishing, existing buildings could also substantially reduce emissions from cement and other energy-intensive building materials.

Of course, the situation with all these sectors could be improved by making far more efficient use of both the energy and materials used in production. The book *Sustainable Materials* (Allwood et al., 2015) has lots of fascinating suggestions for achieving greater material efficiency resulting in lower material use. And of course lifestyles could become much less demanding of 'stuff'. Strategies include repairing, reusing and recycling more, producing less (e.g. single-use plastics), using products for longer (e.g. clothes), stopping making things (e.g. cars) bigger and bigger (a full-size SUV weighs more than twice as much as a small car and nearly 50% more than a large car), owning fewer things (e.g. just one and not second or more homes), and sharing more things (from cars to appliances). These latter suggestions, of course, strike at the heart of conspicuous consumption, which seems to have been characteristic of human societies since the start of recorded human history. So it might be more effective to concentrate on incentivising innovation to make the manufacture of these products zero-carbon rather than exhorting frugality.

Agriculture and land use

Food production contributes significantly to climate change, but is also under significant threat from it. The contribution to climate change comes largely from methane from livestock, and from intensive fertiliser use. The dominant system of agriculture is also the major destroyer of biodiversity and cause of water stress and pollution in many countries.

There is no way out of this cycle of environmental impacts without substantial reductions of meat consumption by those people (mainly but not only in richer countries) who eat more meat than is good for their health, and without a shift to an agriculture that works more with natural processes than against them. Such an agriculture will, of course, be different in different countries and regions, but everywhere it will need to build soil fertility and organic carbon, make better use of fertilisers and minimum use of pesticides, and enhance rather than destroy biodiversity. At a global level, while there are many promising experiments and projects, these are far further from mainstream acceptance and deployment than in other areas of GHG emission reduction.

Future demands for food and biodiversity are likely to need both the conservation of substantial areas of high and vulnerable biodiversity, with minimum agriculture, and large areas of agriculture where farming takes account of and, as far as possible, stimulates biodiversity, while at the same time producing high yields of food. Some farmers already manage the latter but they are still relatively rare. The methods that they have pioneered should be given the support by public policy that they need to become the dominant agriculture of the 21st century.

At the same time there has been an explosion of meat substitutes and companies that produce them, whose products have already found their way onto the shelves of mainstream supermarkets. It is wholly possible that the shift away from meat will be market-driven by climate and health concerns in rich countries, and then price-driven in poorer countries, once the former trend has undermined the cultural association of meat-eating with affluence and 'development'.

Technology and innovation

This is the area in which decarbonisation policy has so far been most successful, and the area that provides the most hope that climate change can indeed be stopped. The price reductions of the main renewables technologies (solar PV, wind) have been spectacular, and the means by which they have been brought about are well understood: the provision of market incentives for the large-scale introduction and deployment of a new technology. Of course, this does not guarantee cost-reduction on the scale seen for solar PV and wind, but it is the best single chance of achieving it for the other technologies that seem necessary for real zero, particularly EVs, heat pumps, electrolysers for green hydrogen and machines for direct air capture of carbon dioxide from the atmosphere. In addition, there needs to be innovation in institutions and finance, and capacity building to enable the technologies to be widely built, deployed and used, in developing as well as industrialised countries.

Finance and investment

Bloomberg New Energy Finance reported that investment in clean energy topped US$1 trillion for the first time in 2022,[13] nearly five times what it was ten years earlier, with around half the investment going into renewable power and most of the rest going into EVs, with storage and grid investments only getting around 0.5% (US$23 billion). However, while the overall total is definitely going

---

13 www.bloomberg.com/news/articles/2023-01-26/global-clean-energy-investments-match-fossil-fuel-for-first-time?leadSource=uverify%20wall. Accessed May 6, 2023.

in the right direction, this is still well short of what is required to reach the 1.5°C target, as confirmed by the IEA's 2022 *World Energy Investment*[14] report. The IEA report also shows that only about 20% of the clean energy investment is going into emerging market and developing economies (apart from China). At least part of the reason for the low investment share in developing economies is the high cost of capital in many of them. Priorities for the future are to greatly increase the quantity of investment going into emerging market and developing economies, including by finding ways to reduce the cost of capital in these countries, as well as ramping up investment further in old industrial economies, and to increase energy storage and grid investments, which will be essential for grid stability as the share of variable renewable energy increases.

Economic growth, employment and health co-benefits

There has been a steady rise in employment in low-carbon industries, especially in renewables, and there is every prospect that this can be increased very much further in the push for real zero. Alongside this, the realisation seems to be growing that, not only will decarbonisation not be costly in macro-economic terms, as was once assumed, but it seems likely to provide the foundation for the new industries and technology that, with widespread deployment, will underpin a new phase of economic growth. As this book was being completed, there was much speculation about decarbonisation fuelling a 'race to net zero', driven by the US$370 billion Inflation Reduction Act (IRA) in the United States,[15] to which the European Union was poised to respond with its own 'Green Deal Industrial Plan',[16] in a clear desire to capture its share of the global investment in low-carbon technologies that real zero will require. In the UK the Skidmore Review of January 2023 said it had "heard loud and clear that net zero is the economic opportunity of the 21st century ... it is clear we are in an international race for capital, skills, and the industries of the future" (Skidmore, 2023, p. 7).

The benefits of going for real zero stretch well beyond the economy and employment. The most important co-benefits will be related to people's health, as the air pollution from burning fossil fuels declines, as people in cold countries live in warmer, better insulated buildings, and as people adopt healthier diets and engage in more active travel.

Climate justice

The UNFCCC and its Paris Agreement are explicit that considerations of justice must be at the heart of mitigating climate change. This will need to

---

14 www.iea.org/reports/world-energy-investment-2022/overview-and-key-findings. Accessed May 6, 2023.
15 www.whitehouse.gov/cleanenergy/inflation-reduction-act-guidebook/. Accessed May 6, 2023.
16 https://ec.europa.eu/commission/presscorner/detail/en/IP_23_510. Accessed May 6, 2023.

apply across all the issues connected with reducing GHG emissions: the distribution of the remaining carbon budget to stay within the Paris temperature targets; the production of the fossil fuels that can be burned within those targets; the transition to a zero-carbon economy; and payments for the loss and damage that climate change is already causing, and will cause further in the future. Around each of these issues there is now a lively international debate, most recently with the agreement at COP27 to set up a special fund for loss and damage payments.

However, none of these issues is yet being addressed with the urgency that both climate justice and decarbonisation towards real zero require. The rich world does not yet seem to realise that its support for decarbonisation in emerging and developing countries is an absolute prerequisite for those countries to develop in a low-carbon way. Without it they will develop in a high-carbon way and then, even if by 2050 there is a 'real-zero' North America and a 'real-zero' Europe, GHG emissions from these other countries will take the world way past the Paris temperature targets. Real zero requires a climate 'Marshall Plan', whereby the old industrial countries enable the rest of the world to build decent lives for their populations through a sustained process of zero-carbon development.

Policy for real zero

There has been huge innovation in climate policy over the last two or three decades. There can be little doubt that policymakers now know how to reduce emissions in ways that are both effective and affordable. The problem is that they, and the publics they represent, are not ready to implement these policies to anything like the extent or in the strength that is now required for emissions reductions to meet the Paris temperature targets.

A paradigmatic case in point is carbon pricing. One of the very few points of agreement among economists is that some element of carbon pricing, and preferably carbon taxation because it is far less administratively complex than carbon trading, is essential for serious carbon emission reduction. But a global carbon tax is not even on the *agenda* of the climate COP meetings, let alone anywhere near implementation. And yet a global carbon tax, the revenues from which were retained by national governments, would provide a source of investment in their energy systems, which is so desperately needed, as well as acting as an important spur to low-carbon innovation throughout the economy. Making the carbon tax global would also ensure that carbon pricing did not lead to competitive disadvantage in those countries that moved first, fears of which have acted as an effective veto on carbon pricing, especially in respect of heavily-traded products from carbon-intensive industries. And so we are left with the EU's proposed Carbon Border Adjustment Mechanism (CBAM), which is entirely necessary under the circumstances, if the EU is to withdraw the free allowances for its emissions trading system (ETS) that it currently gives its energy-intensive industries, but which will be administratively cumbersome, internationally contentious and act as a constraint on trade. A global carbon tax would face none of these disadvantages.

Moreover, some revenues from a global carbon tax could be allocated to international climate finance, which countries have so far found it so difficult to provide. Of course plenty of other potential sources of this finance have been proposed that would align with ability to pay, for example a tax on international flights or a financial transactions tax (the 'Tobin tax', after the economist who first proposed it many years ago). But so far the politics has stymied all such proposals, so that the global community is still scrabbling around for the US$100 billion in climate finance which it promised back at the Copenhagen COP in 2009. To put that number in context, it is estimated that US$100 billion is less than half the cost of the 2022 football World Cup in Qatar.[17] Such priorities will ensure that net zero, let alone real zero, will remain a fantasy.

Beyond the particular policy instruments, many of which were covered in Chapter 12 or earlier in this book, all decarbonisation policy has to exhibit four cardinal characteristics, given the long-term nature of such policy and the need to reassure investors that their investments will yield the long-term returns without which they will not be made. The '4-C' characteristics were well articulated in the Skidmore review (Skidmore, 2023, Figure 1, p. 40):

Continuity: Across people, policy areas and time
Certainty: On funding, business environment and regulation
Consistency: Creating a level playing field
Clarity: Data to inform future decision-making

Of course, no government can guarantee all these characteristics all the time. Governments change, and policies necessarily change with them, but as a multi-decade transition, climate policy demands more of these characteristics than other policy areas, and many governments do not even seem to have tried very hard to deliver them. A particularly egregious example, to illustrate the point, was when, after the UK General Election in 2015, the then UK Chancellor George Osborne cancelled the Zero Carbon Homes policy for new houses, when the policy, having been legislated in 2006, was nine years into a ten-year preparation period. Many progressive builders had spent the intervening years preparing for the policy, only to have the rug pulled from under them at the last moment.[18] The policy reversal was condemned by substantial businesses, 246 of which (including E.ON, Whitbread, Saint-Gobain and the British Property Federation) urged the Chancellor in a letter "to reverse his decision to drop zero carbon building targets, because it had 'undermined industry confidence'

---

17 www.dw.com/en/qatar-world-cup-will-be-the-most-expensive-of-all-time/a-63681083. Accessed May 6, 2023.
18 www.theguardian.com/environment/2015/jul/10/uk-scraps-zero-carbon-home-target. Accessed May 6, 2023.

and could damage investment in technological innovation". Prophetically in the light of gas and electricity prices in 2022, the letter said: "The weakening of standards will mean our future homes, offices, schools and factories will be more costly to run, locking future residents and building users into higher energy bills".[19]

The decision by UK Prime Minister Rishi Sunak in September 2023 to row back on other UK climate targets shows that, even now, some politicians are not prepared to put the '4-C' conditions for strong and effective climate policy before perceptions of short-term political advantage. It can only be hoped that UK voters will prove him wrong in the General Election expected in 2024.

Conclusion

Humanity could, even now, stop causing climate change by 2050, and do much to re-stabilise the climate by 2100. Doing so would require it to stop using fossils fuels, halt deforestation and eliminate GHGs from agriculture by 2050, and then use the subsequent 50 years systematically to remove CO_2 from the atmosphere, so that by 2100 the atmospheric concentration of GHGs was back to 350 ppm. The technologies are available to do this. The money is available to deliver the technologies, provided that projects to do so have a financially viable risk/return ratio. Financial viability does not depend just on the technologies themselves, many of which can already compete with fossil fuels in many parts of the world, and many more of which will become cheaper as they are deployed so that they can also compete. Where the technologies remain uneconomic, this is likely to be because of institutional, governance, skills and capacity issues that result, for example, in excessive policy uncertainty, high costs of capital and uncertainty of delivery. These problems are, in principle, solvable. But to solve them requires a level of commitment to addressing climate change that policymakers and the people they represent currently do not have. If they can find it, they will need to give it expression through a new-found focus on multi-lateral cooperation and national '4-C' climate policy across all areas of government decision-making. The rich countries will need in addition to make good on the 'common but differentiated' language in the UNFCCC and demonstrate the enlightened self-interest in the provision of resources that allows the developing world to dream of, and over time to realise, a post-carbon development that allows their people to live healthy and fulfilled lives, as envisaged by the United Nations' Sustainable Development Goals (SDGs).[20] So far the signs that humanity collectively is up to this challenge are few. But the transformation required to address it effectively remains possible.

---

19 www.cibsejournal.com/news/government-scraps-zero-carbon-targets/. Accessed May 6, 2023.
20 https://sdgs.un.org/goals

References

Allwood, J., Cullen, J., Carruth, M., Cooper, D., McBrien, M., Milford, R., Moynihan, M. & Patel, A. (2015). *Sustainable Materials Without the Hot Air*. UIT Cambridge, UK.

Jacobson, M. (2023). *No Miracles Needed: How Today's Technology Can Save Our Climate and Clean Our Air*. Cambridge University Press, Cambridge/New York.

Skidmore, C. (2023). *Mission Zero: Independent Review of Net Zero*. Her Majesty's Government, London. https://assets.publishing.service.gov.uk/government/uploads/system/uploads/attachment_data/file/1128689/mission-zero-independent-review.pdf

Solomon, M., Gimon, E., O'Boyle, M., Paliwal, U., & Phadke, A. (2023). *Coal Cost Crossover 3.0: Local Renewables Plus Storage Create New Opportunities for Customer Savings and Community Reinvestment*. Energy Innovation Policy and Technology, San Francisco. https://energyinnovation.org/publication/coal-cost-crossover-3-0-local-renewables-plus-storage-create-new-opportunities-for-customer-savings-and-community-reinvestment/

Index

Note: Footnotes are indicated by the page number followed by "n" and the note number e.g., 109n26 refers to note 26 on page 109. Page locators in **bold** and *italics* represents tables and figures, respectively.

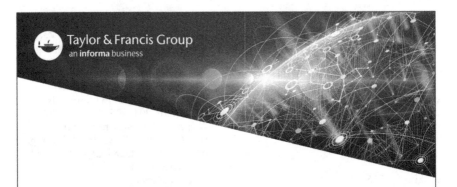

Printed in the United States
by Baker & Taylor Publisher Services